STUDY GUIDE
AND
SELECTED SOLUTIONS

Karen C. Timberlake
Los Angeles Valley College

CHEMISTRY

An Introduction to General, Organic, and Biological Chemistry

TENTH EDITION

PEARSON
Prentice Hall

Upper Saddle River, NJ 07458

Editor-in-Chief, Science: Nicole Folchetti
Senior Editor: Kent Porter Hamann
Assistant Editor: Jessica Neumann
Assistant Managing Editor, Science: Gina M. Cheselka
Project Manager, Science: Maureen Pancza
Supplement Cover Manager: Paul Gourhan
Supplement Cover Designer: Victoria Colotta
Operations Specialist: Amanda A. Smith
Director of Operations: Barbara Kittle
Cover Photos: *Diver photo*: Jeff Hunter/Photographer's Choice/Getty
Images, Inc.; *Coral photo*: Wayne Lawler/Ecosene/CORBIS

© 2009 Pearson Education, Inc.
Pearson Prentice Hall
Pearson Education, Inc.
Upper Saddle River, NJ 07458

Pearson Prentice Hall™ is a trademark of Pearson Education, Inc.

The author and publisher of this book have used their best efforts in preparing this book. These efforts include the development, research, and testing of the theories and programs to determine their effectiveness. The author and publisher make no warranty of any kind, expressed or implied, with regard to these programs or the documentation contained in this book. The author and publisher shall not be liable in any event for incidental or consequential damages in connection with, or arising out of, the furnishing, performance, or use of these programs.

Printed in the United States of America

10   9   8   7   6   5   4   3   2   1

ISBN-13: 978-0-13-601999-2

ISBN-10:   0-13-601999-4

Pearson Education Ltd., *London*
Pearson Education Australia Pty. Ltd., *Sydney*
Pearson Education Singapore, Pte. Ltd.
Pearson Education North Asia Ltd., *Hong Kong*
Pearson Education Canada, Inc., *Toronto*
Pearson Educación de Mexico, S.A. de C.V.
Pearson Education—Japan, *Tokyo*
Pearson Education Malaysia, Pte. Ltd.

Table of Contents

This Study Guide and Solutions Manual is written to accompany *Chemistry: An Introduction to General, Organic, and Biological Chemistry,* Tenth Edition. The purpose of this Study Guide is to provide you with additional learning resources that increase your understanding of the concepts. Each section in the Study Guide is correlated with a chapter in the text. Within each section, there are Learning Exercises that focus on problem solving, which promotes an understanding of the chemical principles of the Learning Goals. Following the Learning Exercises, a Check List of learning goals and a multiple choice Practice Test provide a review of the entire chapter content. Finally, the Answers and Solutions to Selected Problems give the worked out solutions to all the odd-numbered problems from each chapter in the text.

I hope that this study guide will help you to learn chemistry. If you wish to make comments or corrections, or ask questions, you can send me an e-mail message at khemist@aol.com.

Karen C. Timberlake
Los Angeles Valley College

"One must learn by doing the thing;
though you think you know it, you
have no certainty until you try."

—Sophocles

Here you are in a chemistry class with your textbook in front of you. Perhaps you have already been assigned some reading or some problems to do in the book. Looking through the chapter, you may see words, terms, and pictures that are new to you. This may very well be your first experience with a science class like chemistry. At this point you may have some questions about what you can do to learn chemistry. This Study Guide is written with those considerations in mind.

Learning chemistry is similar to learning a new skill such as tennis or skiing or driving. If I asked you how you learn to play tennis or ski or drive a car, you would probably tell me that you would need to practice every day. It is the same with learning chemistry. Learning the chemical ideas and learning to solve the problems depends on the time and effort you invest in it. If you practice every day, you will find that learning chemistry is an exciting experience and a way to understand the current issues of the environment, health, and medicine.

Manage Your Study Time

I often recommend a study system in which you read one section of the text and immediately practice the questions and problems that go with it. In this way, you concentrate on a small amount of information and actively use what you learned to answer questions. This helps you to organize and review the information without being overwhelmed by the entire chapter. It is important to understand each section, because the concepts build like steps. Information presented in each chapter proceeds from the basic to the more complex. Perhaps you will only study three or four sections of the chapter. As long as you also practice doing some problems at the same time, the information will stay with you.

Form a Study Group

I highly recommend that you form a study group in the first week of your chemistry class. Working with your peers will help you use the language of chemistry. Schedule a time to meet each week to study and prepare to discuss problems. You will be able to teach some things to the other students in the group, and sometimes they will help you understand a topic that puzzles you. You won't always understand a concept right away. Your group will help you see your way through it. Most of all, a study group creates a strong support system whereby students can help each other complete the class successfully.

Go to Office Hours

Finally, go to your tutor's and/or professor's office hours. Your professor wants you to understand and enjoy learning this material and should have office hours. Often a tutor is assigned to a class or there are tutors available at your college. Don't be intimidated. Going to see a tutor or your professor is one of the best ways to clarify what you need to learn in chemistry.

Using This Study Guide

Now you are ready to sit down and study chemistry. Let's go over some methods that can help you learn chemistry. This Study Guide is written specifically to help you understand and practice the chemical concepts that are presented in your class and in your text. Some of the exercises teach basic skills; others encourage you to extend your scientific curiosity. The following features are part of this Study Guide:

1. Study Goals
The Study Goals give you an overview of what the chapter is about and what you can expect to accomplish when you complete your study and learning of a chapter.

2. Think About It
Each chapter in the Study Guide has a group of questions that encourage you to think about some of the ideas and practical applications of the chemical concepts you are going to study. You may find that you already have knowledge of chemistry in some of the areas. That will be helpful to you. Other questions give you an overview of the chemistry ideas you will be learning.

3. Key Terms
Each chapter in the Study Guide introduces Key Terms. As you complete the description of the Key Terms, you will have an overview of the topics you will be studying in that chapter. Because many of the Key Terms may be new to you, this is an opportunity to determine their meaning.

4. Chapter Sections
Each section of the chapter begins with a list of key concepts to illustrate the important ideas in that section. The summary of concepts is written to guide you through each of the learning activities. When you are ready to begin your study, read the matching section in the textbook and review the sample exercises in the text.

5. Learning Exercises
The Learning Exercises give you an opportunity to practice problem solving related to the chemical principles in the chapter. The Study Notes highlight important concepts in each chapter. There is room for you to answer the questions or complete the exercise. The answers are found immediately following each exercise. (Sometimes they will be located at the top of the next page.) Check your answers right away. If they don't match the answer in the Study Guide, go back to the textbook and review the material again. It is important to make corrections before you go on. Chemistry involves a layering of skills such that each one must be understood before the next one can be learned. At various times, you will notice some essay questions that illustrate one of the concepts. I believe that writing out your ideas is a very important way of learning content. If you can put your problem-solving techniques into words, then you can understand the patterns of your thinking and you will find that you have to memorize less.

6. Check List
Use the Check List to review your understanding of the Study Goals. This gives you an overview of the major topics in the section. If something does not sound familiar, go back and review. One aspect of being a strong problem-solver is the ability to check your knowledge and understanding as you go along.

7. Practice Test
A Practice Test is found at the end of each chapter. When you have learned the material in a chapter, you can apply your understanding to the Practice Test. If the results of this test indicate that you know the material, you are ready to proceed to the next chapter. If, however, the results indicate further study is needed, you can repeat the Learning Exercises in the sections you still need to work on. Answers for all of the questions are included at the end of the Practice Test.

8. Answers
Answers and Solutions to Selected Text Problems give worked out solutions to the odd-numbered problems from each chapter in the text. Answers to odd-numbered problems from Combining Ideas are included at the end of Answers for Chapters 2, 5, 8, 12, 15, and 18.

This book is dedicated to my husband Bill
Thank you for your help, expertise, patience, and gourmet cooking
that make it possible for me to finish this book.

Prologue
Chemistry in Our Lives

Study Goals

- Define the term *chemistry* and identify substances as chemicals.
- Identify the activities that are part of the scientific method.
- Develop a study plan for learning chemistry.

Think About It

1. Why can we say that the salt and sugar we use on food are chemicals?

2. How can the scientific method help us make decisions?

3. What are some things you can do to help you study and learn chemistry?

Key Terms

Match each of the following key terms with the correct definition:

   1. scientific method **2.** experiment **3.** hypothesis **4.** theory **5.** chemistry

   a. _____ An explanation of nature validated by many experiments

   b. _____ The study of substances and how they interact

   c. _____ A possible explanation of a natural phenomenon

   d. _____ The process of making observations, writing a hypothesis, and testing with experiments

   e. _____ A procedure used to test a hypothesis

Answers   **a.** 4   **b.** 5   **c.** 3   **d.** 1   **e.** 2

P.1 Chemistry and Chemicals

- A chemical is any material used in or produced by a chemical process.
- A substance is a chemical containing one type of material that has the same composition and properties.
- In the sciences, physical quantities are described in units of the metric or International System (SI).

◆ Learning Exercise P.1

Indicate if each of the following is a chemical:

   a. _____ aluminum

   b. _____ heat

   c. _____ sodium fluoride in toothpaste

   d. _____ ammonium nitrate in fertilizer

   e. _____ time

Answers   **a.** yes   **b.** no   **c.** yes   **d.** yes   **e.** no

P.2 Scientific Method: Thinking Like a Scientist

- The scientific method is a process of making observations, writing a hypothesis, and testing the hypothesis with experiments.
- A theory develops when experiments that validate a hypothesis are repeated by many scientists with consistent results.

◆ Learning Exercise P.2

Identify each of the following as observation (O), hypothesis (H), or experiment (E).

a. _____ Sunlight is necessary for the growth of plants.

b. _____ Plants in the shade were shorter than plants in the sun.

c. _____ Plant leaves are covered with aluminum foil and their growth measured.

d. _____ Fertilizer added to plants accelerates their growth.

e. _____ Ozone slows plant growth by interfering with photosynthesis.

f. _____ Ozone causes brown spots on plant leaves.

Answers   **a.** H   **b.** O   **c.** E   **d.** E
              e. H   **f.** O

P.3 A Study Plan for Learning Chemistry

- Components of the text that promote learning include Looking Ahead, Learning Goals, Guides to Problem Solving (GPS), Study Checks, Health Notes, Green Chemistry Notes, Environmental Notes, Explore Your World, Questions and Problems, Concept Maps, Chapter Reviews, Key Terms, Understanding the Concepts, Additional Questions and Problems, Challenge Problems, and Answers to Selected Questions and Problems.
- An active learner continually interacts with chemical concepts while reading the text and attending lecture.
- Working with a study group clarifies ideas and illustrates problem solving.

◆ Learning Exercise P.3

Which of the following activities would be included in a successful study plan for learning chemistry?

a. _____ attending lecture once in a while

b. _____ working problems with friends from class

c. _____ attending review sessions

d. _____ planning a regular study time

e. _____ not doing the assigned problems

f. _____ going to the instructor's office hours

Answers   **a.** no   **b.** yes   **c.** yes   **d.** yes
              e. no   **f.** yes

Checklist for Prologue

You are ready to take the self-test for the Prologue. Be sure you have accomplished the following learning goals for this chapter. If you are not sure, review the section listed at the end of the goal. Then apply your new skills and understanding to the self-test.

After studying the Prologue, I can successfully:

_____ Describe a substance as a chemical (P.1).

_____ Identify the components of the scientific method (P.2).

_____ Design a study plan for successfully learning chemistry (P.3).

Practice Test for Prologue

1. Which of the following would be described as a chemical?
   A. sleeping   **B.** salt   **C.** singing   **D.** listening to a concert   **E.** energy

2. Which of the following is not a chemical?
   A. wool   **B.** sugar   **C.** feeling cold   **D.** salt   **E.** vanilla

For questions 3–7, identify each statement as observation (O), hypothesis (H), or experiment (E):

3. _____ Hot water can dissolve more sugar than cold water.

4. _____ Place 20 g of sugar each in a glass of cold water and a glass of hot water.

5. _____ Sugar is solid, white crystals.

6. _____ Water flows downhill.

7. _____ Drinking ten glasses of water a day will help me lose weight.

For questions 8–12, answer yes or no.

To learn chemistry, I will:

8. _____ work the problems in the chapter and check answers.

9. _____ attend some lectures, but not all.

10. _____ form a study group.

11. _____ set up a regular study time.

12. _____ wait until the night before the exam to start studying.

Answers to the Practice Test

1. B	**2.** C	**3.** O	**4.** E	**5.** O	**6.** O
7. H	**8.** yes	**9.** no	**10.** yes	**11.** yes	**12.** no

Answers and Solutions to Selected Text Problems

P.1   Many chemicals are listed on a vitamin bottle such as vitamin A, vitamin B_3, vitamin B_{12}, folic acid, etc.

P.3   No. All these ingredients are chemicals.

P.5   One advantage of a pesticide is that it gets rid of insects that bite or damage crops. One disadvantage is that a pesticide can destroy beneficial insects or be retained in a crop that is eventually eaten by animals or humans.

P.7   **a.** An observation (O) is a description or measurement of a natural phenomenon.

   b. A hypothesis (H) proposes a possible explanation for a natural phenomenon.

   c. An experiment (E) is a procedure that tests the validity of a hypothesis.

   d. An observation (O) is a description or measurement of a natural phenomenon.

e. An observation (O) is a description or measurement of a natural phenomenon.

f. A theory (T) confirms a hypothesis of a possible explanation for a natural phenomenon.

P.9   The following would help your friend learn chemistry.

a. Form a study group.   **c.** Visit the professor during office hours.

e. Become an active learner.

P.11   Yes. Sherlock's investigation includes observations (gathering data), formulating a hypothesis, testing the hypothesis, and modifying it until the hypothesis is validated.

P.13   **a.** Determination of a melting point is an observation.

b. Describing a reason for the extinction of dinosaurs is a hypothesis.

c. Measuring the speed of a race is an observation.

P.15   **a.** An observation (O) is a description or measurement of a natural phenomenon.

b. A hypothesis (H) proposes a possible explanation for a natural phenomenon.

c. An experiment (E) is a procedure that tests the validity of a hypothesis.

d. A hypothesis (H) proposes a possible explanation for a natural phenomenon.

Study Goals

- Learn the units and abbreviations for the metric (SI) system.
- Distinguish between measured numbers and exact numbers.
- Determine the number of significant figures in a measurement.
- Convert a standard number to scientific notation.
- Use prefixes to change base units to larger or smaller units.
- Write conversion factors from the units in an equality.
- In problem solving, convert the initial unit of a measurement to another unit.
- Round off a calculator answer to report an answer with the correct number of significant figures.
- Calculate the density of a substance; use density to convert between mass and volume.

Think About It

1. What kind of device would you use to measure each of the following: your height, your weight, and the quantity of water to make soup?

2. How do you determine the amount of money in your wallet?

3. When you make a measurement, why should you write down a number and a unit?

4. Why does oil float on water?

Key Terms

Match each of the following key terms with the correct definition.

a. metric system	**b.** exact number	**c.** significant figures
d. conversion factor	**e.** density	**f.** scientific notation

1. _____ All the numbers recorded in a measurement including the estimated digit.

2. _____ A fraction that gives the quantities of an equality in the numerator and denominator.

3. _____ A form of writing a number using a coefficient and a power of 10.

4. _____ The relationship of the mass of an object to its volume is usually expressed as g/mL.

5. _____ A number obtained by counting items or from a definition.

6. _____ A decimal system of measurement used throughout the world.

Answers   **1.** c   **2.** d   **3.** f   **4.** e   **5.** b   **6.** a

1.1 Units of Measurement

- In the sciences, physical quantities are described in units of the metric or International System (SI).
- Length or distance is measured in meters (m), volume in liters (L), mass in grams (g), and temperature in Celsius degrees (°C) or kelvins (K).

- A number written in scientific notation has two parts: a coefficient between 1 and 10 followed by a power of 10.
- For large numbers greater than 10, the decimal point is moved to the left to give a positive power of ten. For small numbers less than 1, the decimal point is moved to the right to give a negative power of ten.

◆ Learning Exercise 1.1

Indicate the type of measurement in each of the following:

1. length   **2.** mass   **3.** volume   **4.** temperature

   a. _____ 45 g   **c.** _____ 215 °C   **e.** _____ 825 K

   b. _____ 8.2 m   **d.** _____ 45 L   **f.** _____ 8.8 kg

Answers   **a.** 2   **b.** 1   **c.** 4   **d.** 3   **e.** 4   **f.** 2

1.2 Scientific Notation

Study Note

1. The number 2.5×10^3 means that 2.5 is multiplied by 10^3 (1000).

$$2.5 \times 1000 = 2500$$

The number 8.2×10^{-2} means that 8.2 is multiplied by 10^{-2} (0.01).

$$8.2 \times 0.01 = 0.082$$

2. For a number greater than 10, the decimal point is moved to the left to give a number 1 to 9 and a positive power of ten. For a number less than 1, the decimal point is moved to the right to give a number 1 to 9 and a negative power of ten.

◆ Learning Exercise 1.2

Write the following measurements in scientific notation:

   a. 240 000 cm _____   **e.** 0.002 m _____

   b. 825 m _____   **f.** 0.000 001 5 g _____

   c. 230 000 kg _____   **g.** 0.084 kg _____

   d. 53 000 yr _____   **h.** 0.000 15 s _____

Answers   **a.** 2.4×10^5 cm   **b.** 8.25×10^2 m   **c.** 2.3×10^5 kg   **d.** 5.3×10^4 yr

            e. 2×10^{-3} m   **f.** 1.5×10^{-6} g   **g.** 8.4×10^{-2} kg   **h.** 1.5×10^{-4} s

1.3 Measured Numbers and Significant Figures

- A measured number is obtained when a measuring device is used to determine an amount of some item.
- An exact number is obtained by counting items or from a definition that relates units in the same measuring system.
- There is uncertainty in every measured number, but not in exact numbers.
- Significant figures in a measured number are all the reported figures including the estimated digit.
- Zeros written in front of a nonzero number or zeros used as placeholders in a large number without a decimal point are not significant digits.

◆ **Learning Exercise 1.3A**

Are the numbers in each of the following statements measured (M) or exact (E)?

a. _____ There are 7 days in one week.   **d.** _____ The potatoes have a mass of 2.5 kg.

b. _____ A concert lasts for 73 minutes.   **e.** _____ A student has 26 CDs.

c. _____ There are 1000 g in 1 kg.   **f.** _____ The snake is 1.2 m long.

Answers   **a.** E (counted)   **b.** M (use a watch)   **c.** E (metric definition)
            d. M (use a balance)   **e.** E (counted)   **f.** M (use a metric ruler)

Study Note
Significant figures (SFs) are all the numbers reported in a measurement including the estimated digit. Zeros are significant unless they are placeholders appearing at the beginning of a decimal number or in a large number without a decimal point. 4.255 g (four SFs)   0.0042 (two SFs)   46 500 (three SFs)

◆ **Learning Exercise 1.3B**

State the number of significant figures in the following measured numbers:

a. 35.24 g   _____   **e.** 5.025 L   _____

b. 0.000 080 m _____   **f.** 0.006 kg   _____

c. 55 000 m   _____   **g.** 268 200 mm _____

d. 805 mL   _____   **h.** 25.0 °C   _____

Answers   **a.** 4   **b.** 2   **c.** 2   **d.** 3
            e. 4   **f.** 1   **g.** 4   **h.** 3

1.4 Significant Figures in Calculations

- In multiplication or division, the final answer must have the same number of significant figures as the measurement with the fewest significant figures.
- In addition or subtraction, the final answer must have the same number of decimal places as the measurement with the fewest decimal places.
- When evaluating a calculator answer, it is important to count the significant figures in the measurements, and round off the calculator answer properly.
- Answers in chemical calculations rarely use all the numbers that appear in the calculator. Exact numbers are not included in the determination of the number of significant figures (SFs).

Study Note
1. To round off a number less than 5, keep the digits you need and drop all the digits that follow. Round 42.8254 to three SFs ⟶ 42.8 (drop 254) **2.** If the first number dropped is 5 or greater, keep the proper number of digits and increase the last retained digit by 1. Round 8.4882 to two SFs ⟶ 8.5 **3.** In large numbers, maintain the value of the answer by adding nonsignificant zeros. Round 356 835 to three SFs ⟶ 357 000

◆ Learning Exercise 1.4A

Round off each of the following to give **two** significant figures:

a. 88.75 m _____

b. 0.002 923 g _____

c. 50.525 g _____

d. 1.6726 m _____

e. 0.001 055 8 kg _____

f. 82.08 L _____

Answers

a. 89 m

b. 0.0029 g

c. 51 g

d. 1.7 m

e. 0.0011 kg

f. 82 L

Study Note

1. An answer from multiplying and dividing has the same number of significant figures (SFs) as the measurement that has the smallest number of significant figures.

$$1.5 \times 32.546 = 48.819 \rightarrow 49 \quad \textit{Answer rounded to two SFs}$$
$$\text{two SFs} \quad \text{five SFs}$$

2. An answer from adding or subtracting has the same number of decimal places as the initial number with the fewest decimal places.

$$82.223 + 4.1 = 86.323 \rightarrow 86.3 \quad \textit{Answer rounded to one decimal place.}$$

◆ Learning Exercise 1.4B

Solve each problem and give the answer with the correct number of significant figures:

a. $1.3 \times 71.5 =$

b. $\dfrac{8.00}{4.00} =$

c. $\dfrac{0.082 \times 25.4}{0.116 \times 3.4} =$

d. $\dfrac{3.05 \times 1.86}{118.5} =$

e. $\dfrac{376}{0.0073} =$

f. $38.520 - 11.4 =$

g. $4.2 + 8.15 =$

h. $102.56 + 8.325 - 0.8825 =$

Answers

a. 93

b. 2.00

c. 5.3

d. 0.0479

e. 52 000

f. 27.1

g. 12.4

h. 110.00

1.5 Prefixes and Equalities

- In the metric system, larger and smaller units use prefixes to change the size of the unit by factors of 10. For example, a prefix such as *centi* or *milli* preceding the unit meter gives a smaller length than a meter. A prefix such as *kilo* added to gram gives a unit that measures a mass that is 1000 times greater than a gram.
- Some of the most common metric (SI) prefixes are shown below:

Prefix	Symbol	Numerical Value	Scientific Notation	Equality
Prefixes That Increase the Size of the Unit				
tera	T	1 000 000 000 000	10^{12}	$1\ Tg = 1 \times 10^{12}\ g$
giga	G	1 000 000 000	10^{9}	$1\ Gm = 1 \times 10^{9}\ m$
mega	M	1 000 000	10^{6}	$1\ Mg = 1 \times 10^{6}\ g$
kilo	k	1 000	10^{3}	$1\ km = 1 \times 10^{3}\ m$
Prefixes That Decrease the Size of the Unit				
deci	d	0.1	10^{-1}	$1\ dL = 1 \times 10^{-1}\ L$ $1\ L = 10\ dL$
centi	c	0.01	10^{-2}	$1\ cm = 1 \times 10^{-2}\ m$ $1\ m = 100\ cm$
milli	m	0.001	10^{-3}	$1\ ms = 1 \times 10^{-3}\ s$ $1\ s = 1 \times 10^{3}\ ms$
micro	μ	0.000 001	10^{-6}	$1\ \mu g = 1 \times 10^{-6}\ g$ $1\ g = 1 \times 10^{6}\ \mu g$
nano	n	0.000 000 001	10^{-9}	$1\ nm = 1 \times 10^{-9}\ m$ $1\ m = 1 \times 10^{9}\ nm$

- An equality contains two units that measure the *same* length, volume, or mass.
- Some common metric equalities are: 1 m = 100 cm; 1 L = 1000 mL; 1 kg = 1000 g.
- Some useful metric-American equalities are:
  2.54 cm = 1 inch; 1 kg = 2.20 lb; 946 mL = 1 quart

◆ Learning Exercise 1.5A

Match the items in column **A** with those from column **B**.

A	**B**
1. _____ kilo-	**a.** millimeter
2. _____ one thousand liters	**b.** 0.1 L
3. _____ deciliter	**c.** one-millionth of a liter
4. _____ milliliter	**d.** kiloliter
5. _____ centimeter	**e.** 0.01 m
6. _____ one-tenth centimeter	**f.** 1000 g
7. _____ microliter	**g.** one-thousandth of a liter
8. _____ kilogram	**h.** one thousand times

Answers   **1.** h   **2.** d   **3.** b   **4.** g
　　　　　　　 5. e   **6.** a   **7.** c   **8.** f

◆ Learning Exercise 1.5B

Place the following units in order from smallest to largest.

a. kilogram   milligram   gram   _____

b. centimeter   kilometer   millimeter   _____

c. dL   mL   L   _____

d. kg   mg μg   _____

Answers   **a.** milligram, gram, kilogram   **b.** millimeter, centimeter, kilometer
                   c. mL, dL, L   **d.** μg, mg, kg

◆ Learning Exercise 1.5C

Complete the following metric relationships:

a. 1 L = _____ mL   **f.** 1 cm = _____ mm

b. 1 L = _____ dL   **g.** 1 mg = _____ μg

c. 1 m = _____ cm   **h.** 1 dL = _____ L

d. 1 dL = _____ mL   **i.** 1 m = _____ mm

e. 1 kg = _____ g   **j.** 1 cm = _____ m

Answers   **a.** 1000   **b.** 10   **c.** 100   **d.** 100   **e.** 1000
                  f. 10   **g.** 1000   **h.** 0.1   **i.** 1000   **j.** 0.01

1.6 Writing Conversion Factors

- Conversion factors are used in a chemical calculation to change from one unit to another. Each factor represents an equality that is expressed in the form of a fraction.
- Two forms of a conversion factor can be written for any equality. For example, the metric-U.S. equality 2.54 cm = 1 inch can be written as follows:

$$\frac{2.54 \text{ cm}}{1 \text{ inch}} \text{ and } \frac{1 \text{ inch}}{2.54 \text{ cm}}$$

Study Note

Metric conversion factors are obtained from metric prefixes. For example, the metric equality of 1 m = 100 cm gives the factors

$$\frac{1 \text{ m}}{100 \text{ cm}} \text{ and } \frac{100 \text{ cm}}{1 \text{ m}}$$

◆ Learning Exercise 1.6

Write two conversion factors for each of the following pairs of units:

    a. millimeters and meters   **b.** kilograms and grams

    c. kilograms and pounds   **d.** inches and centimeters

e. centimeters and meters   **f.** milliliters and quarts

g. deciliters and liters   **h.** millimeters and centimeters

Answers   **a.** $\dfrac{1000 \text{ mm}}{1 \text{ m}}$ and $\dfrac{1 \text{ m}}{1000 \text{ mm}}$ **b.** $\dfrac{1000 \text{ g}}{1 \text{ kg}}$ and $\dfrac{1 \text{ kg}}{1000 \text{ g}}$ **c.** $\dfrac{2.20 \text{ lb}}{1 \text{ kg}}$ and $\dfrac{1 \text{ kg}}{2.20 \text{ lb}}$

d. $\dfrac{2.54 \text{ cm}}{1 \text{ in.}}$ and $\dfrac{1 \text{ in.}}{2.54 \text{ cm}}$ **e.** $\dfrac{100 \text{ cm}}{1 \text{ m}}$ and $\dfrac{1 \text{ m}}{100 \text{ cm}}$ **f.** $\dfrac{946 \text{ mL}}{1 \text{ qt}}$ and $\dfrac{1 \text{ qt}}{946 \text{ mL}}$

g. $\dfrac{10 \text{ dL}}{1 \text{ L}}$ and $\dfrac{1 \text{ L}}{10 \text{ dL}}$ **h.** $\dfrac{10 \text{ mm}}{1 \text{ cm}}$ and $\dfrac{1 \text{ cm}}{10 \text{ mm}}$

1.7 Problem Solving

● Conversion factors from metric and/or U.S. relationships, percent, and density are used to change a quantity expressed in one unit to a quantity expressed in another unit.

Study Note

To solve a problem, identify the given quantity, write a plan, select the appropriate conversion factors, arrange the conversion factors to cancel the starting unit, and provide the desired unit.

◆ **Learning Exercise 1.7A**

Use metric-metric conversion factors to solve the following problems:

a. 189 mL = _____ L

   Example: Plan: mL → Setup: $189 \text{ mL} \times \dfrac{1 \text{ L}}{1000 \text{ mL}} = 0.189 \text{ L}$

b. 2.7 cm = _____ mm

c. 0.0025 L = _____ mL

d. 76 mg = _____ g

e. How many meters tall is a person whose height is 175 cm?

f. There are 285 mL in a cup of tea. How many liters is that?

g. The recommended daily value for calcium is 200 mg. How many grams are recommended?

h. You walked 1.5 km on the treadmill at the gym. How many meters did you walk?

Answers   **a.** 0.189 L   **b.** 27 mm   **c.** 2.5 mL   **d.** 0.076 g
         e. 1.75 m   **f.** 0.285 L   **g.** 0.2 g   **h.** 1500 m

◆ **Learning Exercise 1.7B**

Use metric-U.S. conversion factors to solve the following problems:

a. 18 inches = _____ cm

e. 150 lb = _____ kg

b. 4.0 qt = _____ L

f. 840 g = _____ lb

c. 275 mL = _____ qt

g. 15 ft = _____ cm

d. 1300 mg = _____ lb

h. 8.50 oz = _____ g

Answers   **a.** 46 cm   **b.** 3.8 L   **c.** 0.291 qt   **d.** 0.0029 lb
              e. 68 kg   **f.** 1.9 lb   **g.** 460 cm   **h.** 241 g

---

Study Note

1. For setups that require a series of conversion factors, it is helpful to write out the plan first. Work from the starting unit to the final unit. Then use a conversion factor for each unit change.

$$\text{Starting unit} \rightarrow \text{unit(1)} \rightarrow \text{unit (2)} = \text{final unit}$$

2. To convert from one unit to another, select conversion factors that cancel the given unit and provide a unit or the final unit for the problem. Several factors may be needed to work the units toward the final unit.

$$\cancel{\text{Starting unit}} \times \frac{\cancel{\text{unit (1)}}}{\cancel{\text{Starting unit}}} \times \frac{\text{unit (2)}}{\cancel{\text{unit (1)}}} = \text{final unit}$$

◆ **Learning Exercise 1.7C**

Use conversion factors to solve the following problems:

a. A piece of plastic tubing measures 120 mm. What is the length of the tubing in inches?

b. A statue weighs 232 pounds. What is the mass of the statue in kilograms?

c. Your friend has a height of 6 feet 3 inches. What is your friend's height in meters?

d. In a triple-bypass surgery, a patient requires 3.00 pints of whole blood. How many mL of blood were given if 1 quart = 2 pints?

e. A doctor orders 0.450 g of a sulfa drug. On hand are 150-mg tablets. How many tablets are needed?

Answers   **a.** 4.7 in.   **b.** 105 kg   **c.** 1.9 m   **d.** 1420 mL   **e.** 3 tablets

1.8 Density

- The density of a substance is a ratio of its mass to its volume usually in units of g/mL or g/cm^3 (1 mL is equal to 1 cm^3). For example, the density of sugar is 1.59 g/mL and silver is 10.5 g/mL.

$$\text{Density} = \frac{\text{mass of substance}}{\text{volume of substance}}$$

- Specific gravity (sp gr) is a unitless relationship of the density of a substance divided by the density of water, 1.00 g/mL. We can calculate the specific gravity of sugar as

$$\frac{1.59 \text{ g/mL (density of sugar)}}{1.00 \text{ g/mL (density of water)}} = 1.59 \text{ (sp gr of sugar)}$$

Study Note

Density can be used as a factor to convert between the mass (g) and volume (mL) of a substance. The density of silver is 10.5 g/mL. What is the mass of 6.0 mL of silver?

$$6.0 \text{ mL silver} \times \frac{10.5 \text{ g silver}}{1 \text{ mL silver}} = 63 \text{ g silver}$$

Density factor

What is the volume of 25 g of olive oil (D = 0.92 g/mL)?

$$25 \text{ g olive oil} \times \frac{1 \text{ mL olive oil}}{0.92 \text{ g olive oil}} = 27 \text{ mL olive oil}$$

Density factor

◆ Learning Exercise 1.8

Calculate the density or specific gravity, or use density as a conversion factor to solve each of the following:

a. What is the density (g/mL) of glycerol if a 200. mL sample has a mass of 252 g?

b. A person with diabetes may produce 5 to 12 liters of urine per day. Calculate the specific gravity of a 100.0 mL sample that has a mass of 100.2 g.

c. A small solid has a mass of 5.5 oz. When placed in a graduated cylinder with a water level of 25.2 mL, the object causes the water level to rise to 43.8 mL. What is the density of the object in g/mL?

d. A sugar solution has a density of 1.20 g/mL. What is the mass in grams of 0.250 L of the solution?

e. A piece of pure gold weighs 0.26 of a pound. If gold has a density of 19.3 g/mL, what is the volume in mL of the piece of gold?

f. Diamond has a density of 3.52 g/mL. What is the specific gravity of diamond?

g. A salt solution has a specific gravity of 1.15 and a volume of 425 mL. What is the mass in grams of the solution?

h. A 50.0-g sample of a glucose solution has a density of 1.28 g/mL. What is the volume in liters of the sample?

Answers

a. 1.26 g/mL	**b.** 1.002	**c.** 8.4 g/mL	**d.** 300. g
e. 6.1 mL	**f.** 3.52	**g.** 489 g	**h.** 0.0391 L

Checklist for Chapter 1

You are ready to take the practice test for Chapter 1. Be sure that you have accomplished the following learning goals for this chapter. If you are not sure, review the section listed at the end of the goal. Then apply your new skills and understanding to the practice test.

After studying Chapter 1, I can successfully:

_____ Write the names and abbreviations for the metric (SI) units of measurement (1.1).

_____ Write large or small numbers using scientific notation (1.2).

_____ Identify a number as a measured number or an exact number (1.3).

_____ Count the number of significant figures in measured numbers (1.4).

_____ Report an answer with the correct number of significant figures (1.5).

_____ Write a metric equality from the numerical values of metric prefixes (1.6).

_____ Use conversion factors to change from one unit to another (1.7).

_____ Calculate the density of a substance; use the density to calculate the mass or volume (1.8).

Practice Test for Chapter 1

Select the letter preceding the word or phrase that best answers the question.

1. Which of the following is a metric measurement of volume?
  A. kilogram   **B.** kilowatt   **C.** kiloliter   **D.** kilometer   **E.** kiloquart

2. The measurement 24 000 g written in scientific notation is
  A. 24 g   **B.** 24×10^3 g   **C.** 2.4×10^3 g   **D.** 2.4×10^{-3} g   **E.** 2.4×10^4 g

3. The measurement 0.005 m written in scientific notation is
  A. 5 m   **B.** 5×10^{-3} m   **C.** 5×10^{-2} m   **D.** 0.5×10^{-4} m   **E.** 5×10^3 m

4. The measured number in the following is
  A. 1 book   **B.** 2 cars   **C.** 4 flowers   **D.** 5 rings   **E.** 45 g

5. The number of significant figures in 105.4 m is
  A. 1   **B.** 2   **C.** 3   **D.** 4   **E.** 5

6. The number of significant figures in 0.00082 g is
  A. 1   **B.** 2   **C.** 3   **D.** 4   **E.** 5

7. The calculator answer 5.78052 rounded to two significant figures is
  A. 5 **B.** 5.7 **C.** 5.8 **D.** 5.78 **E.** 6.0

8. The calculator answer 3486.512 rounded to three significant figures is
  A. 4000 **B.** 3500 **C.** 349 **D.** 3487 **E.** 3490

9. The reported answer for the problem 16.0 ÷ 8.0 is
  A. 2 **B.** 2.0 **C.** 2.00 **D.** 0.2 **E.** 5.0

10. The reported answer for the problem 58.5 + 9.158 is
  A. 67 **B.** 67.6 **C.** 67.7 **D.** 67.66 **E.** 67.658

11. The reported answer for the problem $\dfrac{2.5 \times 3.12}{4.6}$ is

  A. 0.54 **B.** 7.8 **C.** 0.85 **D.** 1.7 **E.** 1.69

12. Which of these prefixes has the largest value?
  A. centi **B.** deci **C.** milli **D.** kilo **E.** micro

13. What is the decimal equivalent of the prefix *centi?*
  A. 0.001 **B.** 0.01 **C.** 0.1 **D.** 10 **E.** 100

14. Which of the following is the smallest unit of measurement?
  A. gram **B.** milligram **C.** kilogram **D.** decigram **E.** centigram

15. Which volume is the largest?
  A. mL **B.** dL **C.** cm^3 **D.** L **E.** kL

16. Which of the following is a conversion factor?
  A. 12 in. **B.** 3 ft **C.** 20 m **D.** $\dfrac{1000\ g}{1\ kg}$ **E.** $2\ cm^3$

17. Which is the correct conversion factor that relates milliliters and liters?
  A. $\dfrac{1000\ mL}{1\ L}$ **B.** $\dfrac{100\ mL}{1\ L}$ **C.** $\dfrac{10\ mL}{1\ L}$ **D.** $\dfrac{0.01\ mL}{1\ L}$ **E.** $\dfrac{0.001\ mL}{1\ L}$

18. Which is the correct conversion factor that relates millimeters and centimeters?
  A. $\dfrac{1\ mm}{1\ cm}$ **B.** $\dfrac{10\ mm}{1\ cm}$ **C.** $\dfrac{100\ cm}{1\ mm}$ **D.** $\dfrac{100\ mm}{1\ cm}$ **E.** $\dfrac{10\ cm}{1\ mm}$

19. 294 mm is equal to
  A. 2940 m **B.** 29.4 m **C.** 2.94 m **D.** 0.294 m **E.** 0.0294 m

20. The handle on a tennis racket measures 4.5 inches. What is that size in centimeters?
  A. 11 cm **B.** 1.8 cm **C.** 0.56 cm **D.** 450 cm **E.** 15 cm

21. What is the volume of 65 mL in liters?
  A. 650 L **B.** 65 L **C.** 6.5 L **D.** 0.65 L **E.** 0.065 L

22. What is the mass in kg of a 22-lb turkey?
  A. 10. kg **B.** 48 kg **C.** 10 000 kg **D.** 0.048 kg **E.** 22 000 kg

23. The number of milliliters in 2 deciliters is
  A. 20 mL **B.** 200 mL **C.** 2000 mL **D.** 20 000 mL **E.** 500 000 mL

24. A person who is 5 feet 4 inches tall would be
  A. 64 m **B.** 25 m **C.** 14 m **D.** 1.6 m **E.** 1.3 m

25. How many oz are in 1 500 g? (1 lb = 16 oz)
    A. 94 oz   **B.** 53 oz   **C.** 24 000 oz   **D.** 33 oz   **E.** 3.3 oz

26. How many qt of orange juice are in 255 mL of juice?
    A. 0.255 qt   **B.** 270 qt   **C.** 236 qt   **D.** 0.270 qt   **E.** 0.400 qt

27. An order for a patient calls for 0.020 g of medication. On hand are 4-mg tablets. How many tablets are needed for the patient?
    A. 2 tablets   **B.** 4 tablets   **C.** 5 tablets   **D.** 8 tablets   **E.** 200 tablets

28. A doctor orders 1500 mg of a sulfa drug. Tablets in stock are 0.500 g. How many tablets are needed?
    A. 1 tablet   **B.** $1\frac{1}{2}$ tablets   **C.** $\frac{1}{3}$ tablet   **D.** $2\frac{1}{2}$ tablets   **E.** 3 tablets

29. What is the density of a bone with a mass of 192 g and a volume of 120 cm^3?
    A. 0.63 g/mL   **B.** 1.4 g/cm^3 **C.** 1.6 g/cm^3 **D.** 1.9 g/cm^3 **E.** 2.8 g/cm^3

30. How many milliliters of a salt solution with a density of 1.8 g/mL are needed to provide 400 g of salt solution?
    A. 220 mL   **B.** 22 mL   **C.** 720 mL   **D.** 400 mL   **E.** 4.5 mL

31. The density of a solution is 0.85 g/mL. Its specific gravity is
    A. 222 mL   **B.** 8.5   **C.** 0.85 mL   **D.** 1.2   **E.** 0.85

32. Three liquids have densities of 1.15 g/mL, 0.79 g/mL, and 0.95 g/mL. When the liquids, which do not mix, are poured into a graduated cylinder, the liquid at the top is the one with a density of
    A. 1.15 g/mL   **B.** 1.00 g/mL   **C.** 0.95 g/mL   **D.** 0.79 g/mL   **E.** 0.16 g/mL

33. A sample of oil has a mass of 65 g and a volume of 80.0 mL. What is its specific gravity?
    A. 1.5   **B.** 1.4   **C.** 1.2   **D.** 0.90   **E.** 0.81

34. What is the mass of a 10.0 mL sample of urine with a specific gravity of 1.04?
    A. 104 g   **B.** 10.4 g   **C.** 1.04 g   **D.** 1.40 g   **E.** 9.62 g

35. Ethyl alcohol has a density of 0.790 g/mL. What is the mass of 0.250 L of the alcohol?
    A. 198 g   **B.** 158 g   **C.** 3.95 g   **D.** 0.253 g   **E.** 0.160 g

Answers to the Practice Test

1. C	**2.** E	**3.** B	**4.** E	**5.** D
6. B	**7.** C	**8.** E	**9.** B	**10.** C
11. D	**12.** D	**13.** B	**14.** B	**15.** E
16. D	**17.** A	**18.** B	**19.** D	**20.** A
21. E	**22.** A	**23.** B	**24.** D	**25.** B
26. D	**27.** C	**28.** E	**29.** C	**30.** A
31. E	**32.** D	**33.** E	**34.** B	**35.** A

Answers and Solutions to Selected Text Problems

1.1   In the U.S.,
        a. weight is measured in pounds (lb),   **b.** height in feet and inches,
        c. gasoline in gallons, and   **d.** temperature in Fahrenheit (°F).

    In Mexico,
        a. mass is measured in kilograms,   **b.** height in meters,
        c. gasoline in liters, and   **d.** temperature in Celsius (°C).

1.3   **a.** meters (used to measure length)   **b.** grams (mass)   **c.** milliliters (volume)
d. seconds (time)   **e.** Celsius degrees (temperature)

1.5   **a.** Move the decimal point left four decimal places: 5.5×10^4 m
b. 4.8×10^2 g
c. Move the decimal point right six decimal places: 5×10^{-6} cm
d. 1.4×10^{-4} s
e. 7.2×10^{-3} L

1.7   **a.** 7.2×10^3 **b.** 3.2×10^{-2} **c.** 1×10^4 **d.** 6.8×10^{-2}

1.9   Measured numbers are obtained using some kind of measuring tool. Exact numbers are numbers obtained by counting or from a definition in the metric or the U.S. measuring system.
a. measured   **b.** exact   **c.** exact   **d.** measured

1.11   Measured numbers are determined using a measuring device; exact numbers are counted or are from definitions.
a. 6 oz of meat   **b.** none
c. 0.75 lb, 350 g   **d.** none (definitions are exact)

1.13   Zeros are only significant when they occur after a nonzero digit. They are not significant if they are placeholders at the end of large numbers without a decimal point.
a. not significant   **b.** significant   **c.** significant
d. significant   **e.** not significant

1.15   **a.** five   **b.** two   **c.** two   **d.** three
e. four   **f.** three

1.17   Calculators carry out mathematical computations and display an answer without regard to significant figures. Our task is to round the calculator's answer to the number of significant figures allowed by the precision of the original data.

1.19   **a.** 1.85   **b.** 184   **c.** 0.004 74 (4.74×10^{-3})
d. 8810 (8.81×10^3)   **e.** 1.83×10^5

1.21   **a.** 5.08×10^3 L   **b.** 3.74×10^4 g   **c.** 1.05×10^5 m   **d.** 2.51×10^{-4} m

1.23   **a.** $45.7 \times 0.034 = 1.6$ **b.** $0.002\ 78 \times 5 = 0.01$
c. $\dfrac{34.56}{1.25} = 27.6$ **d.** $\dfrac{(0.2465)(25)}{1.78} = 3.5$

1.25   **a.** 45.48 cm $+ 8.057$ cm $= 53.54$ cm   **b.** 23.45 g $+ 104.1$ g $+ 0.025$ g $= 127.6$ g
c. 145.675 mL $- 24.2$ mL $= 121.5$ mL   **d.** 1.08 L $- 0.585$ L $= 0.50$ L

1.27   The km/hr markings indicate how many kilometers (how much distance) will be traversed in one hour's time if the speed is held constant. The mph markings indicate the same distance traversed *but measured in miles* during the one hour of travel.

1.29   Because the prefix *kilo* means one thousand times, a *kilo*gram is equal to 1000 grams.

1.31   **a.** mg   **b.** dL   **c.** km   **d.** kg
e. μL   **f.** ng

1.33   **a.** 0.01   **b.** 1000   **c.** 0.001   **d.** 0.1
e. 1 000 000   **f.** 1×10^{-12}

1.35   **a.** 100 cm   **b.** 1000 m   **c.** 0.001 m   **d.** 1000 mL

1.37   **a.** kilogram   **b.** milliliter   **c.** km   **d.** kL   **e.** nanometer

1.39   Because a conversion factor is unchanged when inverted, $\dfrac{1\ m}{100\ cm}$ and $\dfrac{100\ cm}{1\ m}$

1.41 Learning the relationships between the metric prefixes will help you write the following equalities and their resulting conversion factors.

    a. 1 m = 100 cm $\dfrac{1\ m}{100\ cm}$ and $\dfrac{100\ cm}{1\ m}$

    b. 1 g = 1000 mg $\dfrac{1\ g}{1000\ mg}$ and $\dfrac{1000\ mg}{1\ g}$

    c. 1 L = 1000 mL $\dfrac{1\ L}{1000\ mL}$ and $\dfrac{1000\ mL}{1\ L}$

    d. 1 dL = 100 mL $\dfrac{1\ dL}{100\ mL}$ and $\dfrac{100\ mL}{1\ dL}$

    e. 1 week = 7 days $\dfrac{1\ week}{7\ days}$ and $\dfrac{7\ days}{1\ week}$

1.43 One conversion factor is written as a ratio of the two quantities in an equality; the other conversion factor is the inverse.

    a. $\dfrac{3.5\ m}{1\ s}$ and $\dfrac{1\ s}{3.5\ m}$ **b.** $\dfrac{3500\ mg\ potassium}{1\ day}$ and $\dfrac{1\ day}{3500\ mg\ potassium}$

    c. $\dfrac{46.0\ km}{1.0\ gal}$ and $\dfrac{1.0\ gal}{46.0\ km}$ **d.** $\dfrac{50\ mg\ Atenolol}{1\ tablet}$ and $\dfrac{1\ tablet}{50\ mg\ Atenolol}$

    e. $\dfrac{29\ \mu g}{1\ kg}$ and $\dfrac{1\ kg}{29\ \mu g}$

1.45 When using a conversion factor you are trying to cancel existing units and arrive at a new (desired) unit. The conversion factor must be properly oriented so that unit cancellation (numerator to denominator) can be accomplished.

1.47   **a.** $175\ \cancel{cm} \times \dfrac{1\ m}{100\ \cancel{cm}} = 1.75\ m$ **b.** $5500\ \cancel{mL} \times \dfrac{1\ L}{1000\ \cancel{mL}} = 5.5\ L$

    c. $0.0055\ \cancel{kg} \times \dfrac{1000\ g}{1\ \cancel{kg}} = 5.5\ g$

1.49   **a.** Plan: qt $\rightarrow$ mL

        $0.500\ \cancel{qt} \times \dfrac{946\ mL}{1\ \cancel{qt}} = 473\ mL$

    b. Plan: lb $\rightarrow$ kg

        $145\ \cancel{lb} \times \dfrac{1\ kg}{2.20\ \cancel{lb}} = 65.9\ kg$

    c. Plan: kg $\rightarrow$ lb body weight $\rightarrow$ lb of fat

        $74\ \cancel{kg} \times \dfrac{2.20\ \cancel{lb}}{1\ \cancel{kg}} \times \dfrac{15\ lb\ fat}{100\ \cancel{lb\ body\ weight}} = 24\ lb\ of\ fat$

1.51   **a.** Plan: L $\rightarrow$ qt $\rightarrow$ gal

        $250\ \cancel{L} \times \dfrac{1.06\ \cancel{qt}}{1\ \cancel{L}} \times \dfrac{1\ gal}{4\ \cancel{qt}} = 66\ gal$

    b. Plan: g $\rightarrow$ mg $\rightarrow$ tablets

        $0.024\ \cancel{g} \times \dfrac{1000\ \cancel{mg}}{1\ \cancel{g}} \times \dfrac{1\ tablet}{8\ \cancel{mg}} = 3.0\ tablets$ (add significant zero)

c. Plan: lb $\rightarrow$ g $\rightarrow$ kg $\rightarrow$ mg of ampicillin

$$34 \text{ lb body weight} \times \frac{454 \text{ g}}{1 \text{ lb}} \times \frac{1 \text{ kg}}{1000 \text{ g}} \times \frac{115 \text{ mg ampicillin}}{1 \text{ kg body weight}} = 1.8 \times 10^3 \text{ mg}$$

1.53 Density is the mass of a substance divided by its volume. The densities of solids and liquids are usually stated in g/ml or g/cm³.

$$\text{Density} = \frac{\text{Mass (grams)}}{\text{Volume (mL)}}$$

a. $\dfrac{24.0 \text{ g}}{20.0 \text{ mL}} = 1.20 \text{ g/mL}$

b. $\dfrac{1.65 \text{ lb}}{170 \text{ mL}} \times \dfrac{454 \text{ g}}{1 \text{ lb}} = 4.4 \text{ g/mL}$

c. volume of gem $= 34.5 \text{ mL total} - 20.0 \text{ mL water} = 14.5 \text{ mL}$

density of gem $= \dfrac{45.0 \text{ g}}{14.5 \text{ mL}} = 3.10 \text{ g/mL}$

1.55 **a.** $150 \text{ mL} \times \dfrac{1.4 \text{ g}}{1 \text{ mL}} = 2.1 \times 10^2 \text{ g}$

b. Plan: L $\rightarrow$ mL $\rightarrow$ g

$$0.500 \text{ L} \times \frac{1000 \text{ mL}}{1 \text{ L}} \times \frac{1.15 \text{ g}}{1 \text{ mL}} = 575 \text{ g}$$

c. Plan: mL $\rightarrow$ g $\rightarrow$ lb $\rightarrow$ oz

$$225 \text{ mL} \times \frac{7.8 \text{ g}}{1 \text{ mL}} \times \frac{1 \text{ lb}}{454 \text{ g}} \times \frac{16 \text{ oz}}{1 \text{ lb}} = 62 \text{ oz}$$

1.57 **a.** $\dfrac{1.030 \text{ g/mL}}{1.000 \text{ g/mL (H}_2\text{O)}} = 1.030$

b. $\dfrac{45.0 \text{ g}}{40.0 \text{ mL}} = \dfrac{1.13 \text{ g/mL}}{1.000 \text{ g/mL (H}_2\text{O)}} = 1.13$

c. 0.85 (sp gr of oil) $\times$ 1.000 g/mL (H₂O density) = 0.85 g/mL density of oil

1.59 The following pairs have the same number of significant figures.
   c. 0.000 75 s and 75 000 s both have two significant figures.
   d. 255.0 L and 6.240×10^{-2} both have four significant figures.

1.61 **a.** The number of legs is a counted number; it is exact.
   b. The height is measured with a ruler or tape measure; it is a measured number.
   c. The number of chairs is a counted number; it is exact.
   d. The area is measured with a ruler or tape measure; it is a measured number.

1.63 **a.** length = 6.96 cm; width = 4.75 cm
   b. length = 69.6 mm; width = 47.5 mm
   c. There are three significant figures in the length measurement.
   d. There are three significant figures in the width measurement.
   e. 33.1 cm²
   f. Since there are three significant figures in the width and length measurements, there are three significant figures in the area.

1.65 The volume of the object is 23.1 mL − 18.5 mL = 4.6 mL

The mass is 8.24 g and the density is: $\dfrac{8.24 \text{ g}}{4.6 \text{ mL}} = 1.8 \text{ g/mL}$

1.67 Round to three significant figures by dropping any additional digits. If the first number dropped is 5 to 9, raise the last retained digit by 1. In large numbers, replace the dropped digits by zeros to keep the place value. If the number of digits is less than three, add zeros.

a. 0.000 012 6 L   **b.** 3.53×10^2 kg   **c.** 125 000 m

d. 58.7 g   **e.** 3.00×10^{-3} s   **f.** 0.010 8 g

1.69 This problem requires several conversion factors. Let's take a look first at a possible plan. When you write out the plan, be sure you know a conversion factor you can use for each step. Plan: ft → in. → cm → m → min

$$7500 \text{ ft} \times \frac{12 \text{ in.}}{1 \text{ ft}} \times \frac{2.54 \text{ cm}}{1 \text{ in.}} \times \frac{1 \text{ m}}{100 \text{ cm}} \times \frac{1 \text{ min}}{55.0 \text{ m}} = 42 \text{ min}$$

1.71 $\dfrac{1.75 \text{ Euro}}{\text{kg}} \times \dfrac{1 \text{ kg}}{2.20 \text{ lb}} \times \dfrac{\$1.36}{1 \text{ Euro}} = \dfrac{\$1.08}{\text{lb}}$

1.73 $4.0 \text{ lb} \times \dfrac{454 \text{ g}}{1 \text{ lb}} \times \dfrac{1 \text{ onion}}{115 \text{ g}} = 16 \text{ onions}$

1.75 **a.** $8.0 \text{ oz} \times \dfrac{6 \text{ crackers}}{0.50 \text{ oz}} = 96 \text{ crackers}$

b. $10 \text{ crackers} \times \dfrac{1 \text{ serving}}{6 \text{ crackers}} \times \dfrac{4 \text{ g fat}}{1 \text{ serving}} \times \dfrac{1 \text{ lb}}{454 \text{ g}} \times \dfrac{16 \text{ oz}}{1 \text{ lb}} = 0.2 \text{ oz of fat}$

c. $50 \text{ boxes} \times \dfrac{8.0 \text{ oz}}{1 \text{ box}} \times \dfrac{1 \text{ serving}}{0.50 \text{ oz}} \times \dfrac{140 \text{ mg sodium}}{1 \text{ serving}} \times \dfrac{1 \text{ g}}{1000 \text{ mg}} = 110 \text{ g of sodium}$

1.77 **b.** $10 \text{ days} \times \dfrac{4 \text{ tablets}}{1 \text{ day}} \times \dfrac{250 \text{ mg amoxicillin}}{1 \text{ tablet}} \times \dfrac{1 \text{ g}}{1000 \text{ mg}} \times \dfrac{1 \text{ lb}}{454 \text{ g}} \times \dfrac{16 \text{ oz}}{1 \text{ lb}}$

$= 0.35 \text{ oz of amoxicillin}$

1.79 This problem has two units. Convert g to mg, and convert L in the denominator to dL.

$$\dfrac{1.85 \text{ g}}{1 \text{ L}} \times \dfrac{1000 \text{ mg}}{1 \text{ g}} \times \dfrac{1 \text{ L}}{10 \text{ dL}} = 185 \text{ mg/dL}$$

1.81 The difference between the initial volume of the water and its volume with the lead object will give us the volume of the lead object. 285 mL total − 215 mL water = 70 mL of lead

Using the density of lead, we can convert mL to the mass in grams of the lead object.

$$70 \text{ mL} \times \frac{11.3 \text{ g}}{1 \text{ mL}} = 790 \text{ g}$$

1.83 Plan: L gas → mL gas → g gas → g oil → mL oil → cm³ of oil

$$1.00 \text{ L gas} \times \frac{1000 \text{ mL gas}}{1 \text{ L gas}} \times \frac{0.66 \text{ g gas}}{1 \text{ mL gas}} \times \frac{1 \text{ g oil}}{1 \text{ g gas}} \times \frac{1 \text{ mL oil}}{0.92 \text{ g oil}} \times \frac{1 \text{ cm}^3}{1 \text{ mL}} = 720 \text{ cm}^3 \text{ of oil}$$

1.85 $3.0 \text{ L fat} \times \dfrac{1000 \text{ mL fat}}{1 \text{ L fat}} \times \dfrac{0.94 \text{ g fat}}{1 \text{ mL fat}} \times \dfrac{1 \text{ lb fat}}{454 \text{ g fat}} = 6.2 \text{ lb of fat}$

1.87 **a.** 1.012 (sp gr urine) $\times$ 1.000 g/mL (H_2O density) = 1.012 g/mL (urine density)

b. 1.022 (sp gr urine) $\times$ 1.000 g/mL (H_2O density) = 1.022 g/mL (urine density)

$$5.00 \; \text{mL urine} \times \frac{1.022 \; \text{g urine}}{1 \; \text{mL urine}} = 5.11 \; \text{g urine}$$

1.89 You would record the mass as 34.075 g. Since your balance will weigh to the nearest 0.001 g, the mass value would be reported to 0.001 g.

1.91 $$3.0 \; \text{h} \times \frac{55 \; \text{mi}}{1 \; \text{h}} \times \frac{1 \; \text{km}}{0.621 \; \text{mi}} \times \frac{1 \; \text{L}}{11 \; \text{km}} \times \frac{1.06 \; \text{qt}}{1 \; \text{L}} \times \frac{1 \; \text{gal}}{4 \; \text{qt}} = 6.4 \; \text{gal}$$

1.93 Volume: $$1.50 \; \text{g} \times \frac{1 \; \text{cm}^3}{2.33 \; \text{g}} = 0.644 \; \text{cm}^3$$

Radius: $$3.00 \; \text{in} \times \frac{1}{2} \times \frac{2.54 \; \text{cm}}{1 \; \text{in}} = 3.81 \; \text{cm}$$

$$h = \frac{V}{\pi r^2} = \frac{0.644 \; \text{cm}^3}{3.14(3.81 \; \text{cm})^2} = 0.0141 \; \text{cm} \times \frac{10 \; \text{mm}}{1 \; \text{cm}} = 0.141 \; \text{mm}$$

1.95 **a.** $$180 \; \text{lb} \times \frac{1 \; \text{kg}}{2.20 \; \text{lb}} \times \frac{192 \; \text{mg}}{1 \; \text{kg}} \times \frac{6 \; \text{fl oz}}{100 \; \text{mg}} \times \frac{1 \; \text{cup}}{12 \; \text{fl oz}} = 79 \; \text{cups}$$

b. $$180 \; \text{lb} \times \frac{1 \; \text{kg}}{2.20 \; \text{lb}} \times \frac{192 \; \text{mg}}{1 \; \text{kg}} \times \frac{1 \; \text{can}}{50 \; \text{mg}} = 314 \; \text{cans}$$

c. $$180 \; \text{lb} \times \frac{1 \; \text{kg}}{2.20 \; \text{lb}} \times \frac{192 \; \text{mg}}{1 \; \text{kg}} \times \frac{1 \; \text{tablet}}{100 \; \text{mg}} = 157 \; \text{tablets}$$

2

Energy and Matter

Study Goals

- Describe potential and kinetic energy.
- Determine the kilocalories for food samples.
- Calculate temperature values in degrees Celsius and kelvins.
- Calculate the calories lost or gained by a specific amount of a substance for a specific temperature change.
- Determine the energy lost or gained during a change of state at the melting or boiling point.
- Identify the states of matter and changes of state on a heating or cooling curve.

Think About It

1. What kinds of activities did you do today that used *kinetic* energy?

2. What are some of the forms of energy you use in your home?

3. Why is the energy in your breakfast cereal *potential* energy?

4. Why is the high specific heat of water important to our survival?

5. How does perspiring during a workout help to keep you cool?

6. During a rain or snowfall, temperature rises. Why?

7. Why is a steam burn much more damaging to skin than a hot water burn?

Key Terms

Match the following terms with the statements below:

    a. change of state   **b.** kinetic energy   **c.** potential energy
    d. calorie   **e.** Kelvin

1. _____ The amount of heat needed to raise the temperature of 1 g of water by 1 °C.

2. _____ Water boiling at 100 °C.

3. _____ The energy of motion.

4. _____ Stored energy.

5. _____ SI unit of temperature.

Answers   **1.** d   **2.** a   **3.** b   **4.** c   **5.** e

2.1 Energy

- Energy is the ability to do work.
- Potential energy is stored energy; kinetic energy is the energy of motion.
- Some forms of energy include heat, mechanical, radiant, solar, electrical, chemical, and nuclear.

◆ Learning Exercise 2.1A

Match the words in column A with the descriptions in column B.

A	**B**
1. _____ kinetic energy	**a.** Inactive or stored energy
2. _____ potential energy	**b.** The ability to do work
3. _____ chemical energy	**c.** The energy of motion
4. _____ energy	**d.** The energy available in the bonds of chemical compounds

Answers   **1.** c   **2.** a   **3.** d   **4.** b

◆ Learning Exercise 2.1B

State whether the following statements describe potential (P) or kinetic (K) energy:

1. _____ A potted plant sitting on a ledge	6. _____ A ski jumper at the top of the ski jump
2. _____ Your breakfast cereal	7. _____ A jogger running
3. _____ Logs sitting in a fireplace	8. _____ A sky diver waiting to jump
4. _____ A piece of candy	9. _____ Water flowing down a stream
5. _____ An arrow shot from a bow	10. _____ A bowling ball striking the pins

Answers   **1.** P   **2.** P   **3.** P   **4.** P   **5.** K
               6. P   **7.** K   **8.** P   **9.** K   **10.** K

◆ Learning Exercise 2.1C

Match the words in column A with the descriptions in column B.

A	**B**
1. _____ calorie	**a.** 1000 calories
2. _____ specific heat	**b.** The heat needed to raise 1 g of water by 1°C
3. _____ kilocalorie	**c.** A measure of the ability of a substance to absorb heat

Answers   **1.** b   **2.** c   **3.** a

2.2 Energy and Nutrition

- A nutritional Calorie is the same amount of energy as 1 kcal or 1000 calories.
- When a substance is burned in a calorimeter, the water that surrounds the reaction chamber absorbs the heat given off. The calories absorbed by the water are calculated and the caloric value (energy per gram) is determined for the substance.

◆ Learning Exercise 2.2A

State the caloric value in kcal/g associated with the following:

a. amino acid _____   f. fat _____

b. protein _____   g. starch _____

c. sugar _____   h. lipid _____

d. sucrose _____   i. glucose _____

e. oil _____   j. lard _____

Answers   a. 4 kcal/g   b. 4 kcal/g   c. 4 kcal/g   d. 4 kcal/g   e. 9 kcal/g
f. 9 kcal/g   g. 4 kcal/g   h. 9 kcal/g   i. 4 kcal/g   j. 9 kcal/g

---

Study Note

The caloric content of a food is the sum of calories from carbohydrate, fat, and protein. It is calculated by using their number of grams in a food and the caloric values of 4 kcal/g for carbohydrate and protein, and 9 kcal/g for fat. (Round to the tens place.)

◆ Learning Exercise 2.2B

Calculate the kcal for the following foods using the following data:

Food	Carbohydrate	Fat	Protein	kcal
a. Peas, green, cooked	19 g	1 g	9 g	_____
b. Potato chips, 10 chips	10 g	8 g	1 g	_____
c. Cream cheese, 8 oz	5 g	86 g	18 g	_____
d. Hamburger, lean, 3 oz	0	10 g	23 g	_____
e. Salmon, canned	0	5 g	17 g	_____
f. Snap beans, 1 cup	7 g	2 g	30 g	_____
g. Banana, 1	26 g	0	1 g	_____

Answers   a. 120 kcal   b. 120 kcal   c. 870 kcal   d. 180 kcal
e. 110 kcal   f. 170 kcal   g. 110 kcal

♦ **Learning Exercise 2.2C**

Using caloric values, give answers for each of the following problems (for kcal round to the tens place):

1. How many kcal are in a single serving of pudding that contains 4 g protein, 31 g of carbohydrate, and 5 g of fat?

2. A can of tuna has a caloric value of 230 kcal. If there are 2 g of fat and no carbohydrate, how many grams of protein are contained in the can of tuna?

3. A serving of breakfast cereal provides 220 kcal. In this serving, there are 8 g of protein and 6 g of fat. How many grams of carbohydrate are in the cereal?

4. Complete the following table listing ingredients for a peanut butter sandwich.

	Protein	Carbohydrate	Fat	Kcal
a. 2 slices of bread	5 g	30 g	0	_____
b. 2 Tbsp of peanut butter	8 g	6 g	_____	170 kcal
c. 2 tsp of jelly	0	_____	0	40 kcal
d. 1 tsp of margarine	0	0	8 g	_____
e.			Total kcal in sandwich	_____

Answers

1. kcal protein + kcal carbohydrate + kcal fat = total kcal
   $(4 \times 4) + (31 \times 4) + (5 \times 9) = 190$ kcal (rounded to tens place)

2. (230 kcal − kcal of fat) ÷ 4 = g of protein
   (230 kcal − 2 g fat × 9 kcal/g) ÷ 4 = 53 g of protein

3. (220 kcal − kcal of protein − kcal of fat) ÷ 4 = g of carbohydrate
   (220 kcal − 8 g of protein × 4 kcal/g − 6 g of fat × 9 kcal/g) ÷ 4 = 34 g of carbohydrate

4. **a.** bread (5 g × 4 kcal/g) + (30 g × 4 kcal/g) = 140 kcal
   b. peanut butter (190 kcal − 8 g × 4 kcal/g − 6 g × 4 kcal/g) ÷ 9 kcal/g = 15 g of fat
   c. jelly 40 kcal ÷ 4 kcal/g = 10 g of carbohydrate
   d. margarine 8 g of fat × 9 kcal/g = 70 kcal
   e. Total kcal = 140 kcal + 190 kcal + 40 kcal + 70 kcal = 440 kcal
   Or 13 g of protein × 4 kcal/g + 46 g of carbohydrate × 4 kcal/g + 23 g of fat × 9 kcal/g = 440 kcal

2.3 Temperature Conversions

- A nutritional Calorie is the same amount of energy as 1 kcal or 1000 calories.
- In the sciences, temperature is measured in Celsius units, °C, or kelvins, K. In the United States, the Fahrenheit scale, °F, is still in use.
- The equation °F = 1.8 °C + 32 is used to convert a Celsius temperature to a Fahrenheit temperature. When rearranged for °C, the equation is used to convert from °F to °C.

$$°C = \frac{(°F - 32)}{1.8}$$

- The temperature on the Celsius scale is related to the Kelvin scale: K = °C + 273.

◆ Learning Exercise 2.3

Calculate the temperatures in the following problems:

a. To prepare yogurt, milk is warmed to 68 °C. What Fahrenheit temperature is needed to prepare the yogurt?

b. On a cold day in Alaska, the temperature drops to −12 °C. What is that temperature on a Fahrenheit thermometer?

c. A patient has a temperature of 39 °C. What is that temperature in °F?

d. On a hot summer day, the temperature is 95 °F. What is the temperature on the Celsius scale?

e. A pizza is cooked at a temperature of 425 °F. What is the °C temperature?

f. A research experiment requires the use of liquid nitrogen to cool the reaction flask to −45 °C. What temperature will this be on the Kelvin scale?

Answers   **a.** 154 °F   **b.** 10. °F   **c.** 102 °F   **d.** 35 °C   **e.** 218 °C   **f.** 228 K

2.4 Specific Heat

- Specific heat is the amount of energy required to raise the temperature of 1 g of a substance by 1 °C.
- The specific heat for liquid water is 1.00 cal/g °C or 4.184 J/g °C.

Study Note

The heat lost or gained by a substance is calculated from the mass, temperature change, and specific heat of the substance.

    Heat (calories) = **mass** (g) × **temperature change** (ΔT) × **specific heat** (cal/g °C)

There are 4.184 joules in one calorie.

Number of calories × 4.184 joules/calorie = number of joules

◆ Learning Exercise 2.4

Calculate the calories (cal) gained or released during the following:

1. Heating 20. g of water from 22 °C to 77 °C

2. Heating 10. g of water from 12 °C to 97 °C

3. Cooling 4.00 kg of water from 80.0 °C to 35.0 °C

4. Cooling 125 g of water from 72.0 °C to 45.0 °C

Answers   **1.** 1100 cal   **2.** 850 cal   **3.** 180 000 cal   **4.** 3380 cal

2.5 States of Matter

- Matter is anything that has mass and occupies space.
- The three states of matter are solid, liquid, and gas.
- Physical properties such as shape, state, or color can change without affecting the identity of a substance.

◆ Learning Exercise 2.5

Indicate whether the following statements describe a gas (G), a liquid (L), or a solid (S).

1. _____ There are no attractions among the molecules.

2. _____ Particles are held close together in a definite pattern.

3. _____ The substance has a definite volume, but no definite shape.

4. _____ The particles are moving extremely fast.

5. _____ This substance has no definite shape and no definite volume.

6. _____ The particles are very far apart.

7. _____ This material has its own volume, but takes the shape of its container.

8. _____ The particles of this material bombard the sides of the container with great force.

9. _____ The particles in this substance are moving very, very slowly.

10. _____ This substance has a definite volume and a definite shape.

Answers   **1.** G   **2.** S   **3.** L   **4.** G   **5.** G
              6. G   **7.** L   **8.** G   **9.** S   **10.** S

2.6 Changes of State

- A substance undergoes a physical change when its shape, size, or state changes, but not the type of substance itself.
- A substance melts and freezes at its melting (freezing) point.
- As long as a substance is changing state during boiling, or condensation, the temperature remains constant.
- The *heat of fusion* is the heat energy required to change 1 g of solid to liquid. For water to freeze at 0 °C, the heat of fusion is 80. cal. This is also the amount of heat lost when 1 gram of water freezes at 0 °C.
- Sublimation is the change of state from a solid directly to a gas.
- When water boils at 100 °C, 540 cal (the heat of vaporization) is required to change 1 g of liquid to gas (steam); it is also the amount of heat released when 1 g of water vapor condenses at 100 °C.
- A heating or cooling curve illustrates the changes in temperature and states as heat is added to or removed from a substance.

◆ **Learning Exercise 2.6A**

Identify each of the following as

   1. melting   **2.** freezing   **3.** sublimation.

   a. _____ A liquid changes to a solid.

   b. _____ Ice forms on the surface of a lake in winter.

   c. _____ Dry ice in an ice cream cart changes to a gas.

   d. _____ Butter in a hot pan turns to liquid.

Answers   **a.** 2   **b.** 2   **c.** 3   **d.** 1

Study Note

The amount of heat needed or released during melting or freezing can be calculated using the heat of fusion:

$$Heat(cal) = mass(g) \times \text{heat of fusion}$$

◆ **Learning Exercise 2.6B**

Calculate the energy required or released when the following substances melt or freeze:

   a. How many calories are needed to melt 15 g ice at 0 °C?

   b. How much heat in kilocalories is released when 325 g of water freezes at 0 °C?

   c. How many grams of ice would melt when 4000 calories of heat were absorbed?

Answers   **a.** 1200 cal   **b.** 26 kcal   **c.** 50 g

◆ **Learning Exercise 2.6C**

Calculate the energy required or released for the following substances undergoing boiling or condensation:

   a. How many calories are needed to completely change 10. g of water to vapor at 100 °C?

   b. How many kilocalories are released when 515 grams of steam at 100 °C condense to form liquid water at 100 °C?

   c. How many grams of water can be converted to steam at 100 °C when 272 kcal of energy are absorbed?

Answers   **a.** 5400 cal   **b.** 278 kcal   **c.** 504 g

◆ **Learning Exercise 2.6D**

On each heating or cooling curve, indicate the portion that corresponds to a solid, liquid, gas, and the changes in state.

1. Draw a heating curve for water that begins at −20 °C and ends at 120 °C. Water has a melting point of 0 °C and a boiling point of 100 °C.

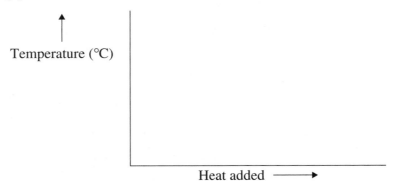

2. Draw a heating curve for bromine from −25 °C to 75 °C. Bromine has a melting point of −7 °C and a boiling point of 59 °C.

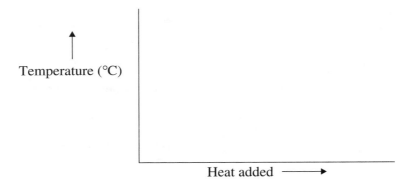

3. Draw a cooling curve for sodium from 1000 °C to 0 °C. Sodium has a freezing point of 98 °C and a boiling (condensation) point of 883 °C.

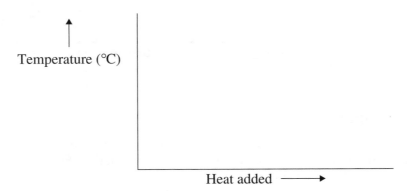

Answers:

1.

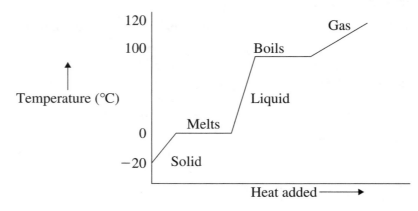

2.

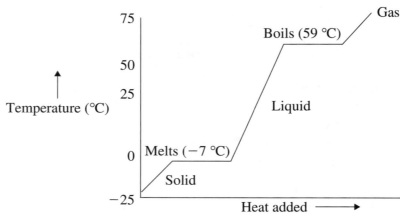

3.

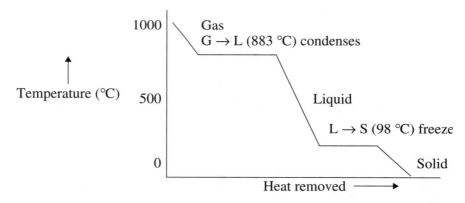

Checklist for Chapter 2

You are ready to take the practice test for Chapter 2. Be sure that you have accomplished the following learning goals for this chapter. If you are not sure, review the section listed at the end of the goal. Then apply your new skills and understanding to the practice test.

After studying Chapter 2, I can successfully:

_____ Describe some forms of energy (2.1).

_____ Calculate the energy of a food sample (2.2).

_____ Given a temperature, calculate a corresponding temperature on another scale (2.3).

_____ Given the mass of a sample, specific heat, and the temperature change, calculate the heat lost or gained (2.4).

_____ Identify the physical state of a substance as a solid, liquid, or gas (2.5).

_____ Calculate the heat change for the boiling and condensation of a specific amount of a substance (2.6).

_____ Draw heating and cooling curves using the melting and boiling points of a substance (2.6).

Practice Test for Chapter 2

1. Which of the following would be described as potential energy?
   A. A car going around a racetrack   B. A rabbit hopping   C. Oil in an oil well
   D. A moving merry-go-round   E. A bouncing ball

2. Which of the following would be described as kinetic energy?
   A. A car battery   B. A can of tennis balls   C. Gasoline in a car fuel tank
   D. A box of matches   E. A tennis ball crossing over the net

3. The number of calories needed to raise the temperature of 5.0 g water from 25 °C to 55 °C is
   A. 5 cal   B. 30 cal   C. 50 cal   D. 80 cal   E. 150 cal

4. The number of calories (kcal) released when 150 g of water cools from 58 °C to 22 °C is
   A. 1.1 kcal   B. 4.2 kcal   C. 5.4 kcal   D. 6.9 kcal   E. 8.7 kcal

For questions 5 through 8, consider a cup of milk with a caloric value of 165 kcal. In the cup of milk, there are 9 g of fat, 12 g of carbohydrate, and some protein.

5. The number of kcal provided by the carbohydrate is
   A. 4 kcal   B. 9 kcal   C. 36 kcal   D. 48 kcal   E. 81 kcal

6. The number of kcal provided by the fat is
   A. 4 kcal   B. 9 kcal   C. 36 kcal   D. 48 kcal   E. 81 kcal

7. The number of kcal provided by the protein is
   A. 4 kcal   B. 9 kcal   C. 36 kcal   D. 48 kcal   E. 81 kcal

8. Which of the following describes a liquid?
   A. A substance that has no definite shape and no definite volume.
   B. A substance with particles that are far apart.
   C. A substance with a definite shape and a definite volume.
   D. A substance containing particles that are moving very fast.
   E. A substance that has a definite volume, but takes the shape of its container.

Identify the statements in questions 9 through 12 as
   A. evaporation   B. heat of fusion
   C. heat of vaporization   D. boiling

9. _____ The energy required to convert a gram of solid to liquid.

10. _____ The heat needed to boil a liquid.

11. _____ The conversion of liquid molecules to gas at the surface of a liquid.

12. _____ The formation of a gas within the liquid as well as on the surface.

13. _____ Ice cools down a drink because
    A. the ice is colder than the drink and heat flows into the ice cubes.
    B. heat is absorbed from the drink to melt the ice cubes.
    C. the heat of fusion of the ice is higher than the heat of fusion for water.
    D. Both A and B
    E. None of the above

14. The number of kilocalories needed to convert 400 g of ice to liquid at 0 °C is
    A. 400 kcal   B. 320 kcal   C. 80 kcal   D. 40 kcal   E. 32 kcal

For questions 15 through 18, consider the heating curve below for p-toluidine. Answer the following questions when heat is added to p-toluidine at 20 °C where toluidine is below its melting point.

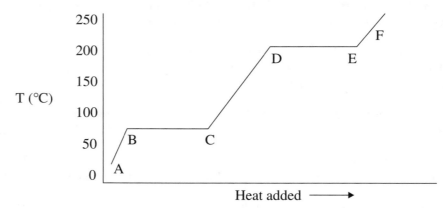

15. On the heating curve, segment BC indicates
    A. solid   B. melting   C. liquid   D. boiling   E. gas

16. On the heating curve, segment CD shows toluidine as
    A. solid   B. melting   C. liquid   D. boiling   E. gas

17. The boiling point of toluidine would be
    A. 20 °C   B. 45 °C   C. 100 °C   D. 200 °C   E. 250 °C

18. On the heating curve, segment EF shows toluidine as
    A. solid   B. melting   C. a liquid   D. boiling   E. a gas

19. 105 °F = _____ °C
    A. 73 °C   B. 41 °C   C. 58 °C   D. 90 °C   E. 189 °C

20. The melting point of gold is 1064 °C. The Fahrenheit temperature needed to melt gold would be
    A. 129 °C   B. 623 °F   C. 1031 °F   D. 1913 °F   E. 1947 °F

21. The average daytime temperature on the planet Mercury is 683 K. What is this temperature on the Celsius scale?
    A. 956 °C   B. 715 °C   C. 680 °C   D. 410 °C   E. 303 °C

Answers to the Practice Test

1. C	2. E	3. E	4. C	5. D
6. E	7. C	8. E	9. B	10. C
11. A	12. D	13. D	14. E	15. B
16. C	17. D	18. E	19. B	20. E
21. D				

Answers and Solutions to Selected Text Problems

2.1 At the top of the hill, all of the energy of the car is in the form of potential energy. As it descends the hill, potential energy is being converted into kinetic energy. When the car reaches the bottom, all of its energy is in the form of motion (kinetic energy).

2.3   **a.** potential   **b.** kinetic   **c.** potential   **d.** potential

2.5   **a.** $20 \text{ matches} \times \dfrac{1.1 \times 10^3 \text{ J}}{1 \text{ match}} \times \dfrac{1 \text{ kJ}}{1000 \text{ J}} = 22 \text{ kJ}$

     b. $415 \text{ J} \times \dfrac{1 \text{ cal}}{4.184 \text{ J}} = 99.2 \text{ cal}$ $20 \text{ matches} \times \dfrac{1.1 \times 10^3 \text{ J}}{1 \text{ match}} \times \dfrac{1 \text{ cal}}{4.184 \text{ J}} = 5300 \text{ cal}$

        $28 \text{ cal} \times \dfrac{4.184 \text{ J}}{1 \text{ cal}} = 120 \text{ J}$

     c. $20 \text{ matches} \times \dfrac{1.1 \times 10^3 \text{ J}}{1 \text{ match}} \times \dfrac{1 \text{ cal}}{4.184 \text{ J}} \times \dfrac{1 \text{ kcal}}{1000 \text{ cal}} = 5.3 \text{ kcal}$

        $4.5 \text{ J} \times \dfrac{1 \text{ cal}}{4.184 \text{ J}} \times \dfrac{1 \text{ kcal}}{1000 \text{ cal}} = 1100 \text{ cal}$

2.7   **a.** Because the orange juice contains both carbohydrate and protein, two calculations will be needed.

       $26 \text{ g carbohydrate} \times \dfrac{4 \text{ kcal}}{1 \text{ g carbohydrate}} \times \dfrac{1 \text{ Cal}}{1 \text{ kcal}} = 100 \text{ Cal}$

       $2 \text{ g protein} \times \dfrac{4 \text{ kcal}}{1 \text{ g protein}} \times \dfrac{1 \text{ Cal}}{1 \text{ kcal}} = 8 \text{ Cal}$

       Total: 100 Cal + 8 Cal = 110 Cal

     b. With only carbohydrate present, a single calculation is all that is required.

       $72 \text{ kcal} \times \dfrac{1 \text{ g carbohydrate}}{4 \text{ kcal}} = 18 \text{ g carbohydrate}$

     c. With only fat present, a single calculation is all that is required.

       $14 \text{ g fat} \times \dfrac{9 \text{ kcal}}{1 \text{ g fat}} \times \dfrac{1 \text{ Cal}}{1 \text{ kcal}} = 130 \text{ Cal}$

     d. Three calculations are needed:

       $30. \text{ g carbohydrate} \times \dfrac{4 \text{ kcal}}{1 \text{ g carbohydrate}} \times \dfrac{1 \text{ Cal}}{1 \text{ kcal}} = 120 \text{ Cal}$

       $15 \text{ g fat} \times \dfrac{9 \text{ kcal}}{1 \text{ g fat}} \times \dfrac{1 \text{ Cal}}{1 \text{ kcal}} = 140 \text{ Cal}$

       $5 \text{ g protein} \times \dfrac{4 \text{ kcal}}{1 \text{ g protein}} \times \dfrac{1 \text{ Cal}}{1 \text{ kcal}} = 20 \text{ Cal}$

       Total: 120 Cal + 140 Cal + 20 Cal = 280 Cal

2.9 Three calculations are needed:

     $9 \text{ g protein} \times \dfrac{4 \text{ kcal}}{1 \text{ g protein}} = 36 \text{ kcal rounds to } 40 \text{ kcal}$

     $12 \text{ g fat} \times \dfrac{9 \text{ kcal}}{1 \text{ g fat}} = 108 \text{ kcal rounds to } 110 \text{ kcal}$

     $16 \text{ g carbohydrate} \times \dfrac{4 \text{ kcal}}{1 \text{ g carbohydrate}} = 64 \text{ kcal rounds to } 60 \text{ kcal}$

Total: 40 kcal + 110 kcal + 60 kcal = 210 kcal

2.11 The Fahrenheit temperature scale is still used in the United States. A normal body temperature is 98.6 °F on this scale. To convert her temperature to the equivalent reading on the Celsius scale, the following calculation must be performed:

$$\frac{(99.8\ °F - 32)}{1.8} = 37.7\ °C\ (32\ \text{and}\ 1.8\ \text{are exact})$$

Because a normal body temperature is 37.0 on the Celsius scale, her temperature of 37.7 °C would be a mild fever.

2.13 **a.** $1.8(37.0\ °C) + 32 = 66.6 + 32 = 98.6\ °F$

b. $\dfrac{(65.3\ °F - 32)}{1.8} = \dfrac{33.3}{1.8} = 18.5\ °C\ (32\ \text{and}\ 1.8\ \text{are exact})$

c. $-27\ °C + 273 = 246\ K$

d. $62\ °C + 273 = 335\ K$

e. $\dfrac{(114\ °F - 32)}{1.8} = \dfrac{82}{1.8} = 46\ °C$

f. $\dfrac{(72\ °F - 32)}{1.8} = \dfrac{40.}{1.8} = 22\ °C;\ 22\ °C + 273 = 295\ K$

2.15 **a.** $\dfrac{(106\ °F - 32)}{1.8} = \dfrac{74}{1.8} = 41\ °C$

b. $\dfrac{(103\ °F - 32)}{1.8} = \dfrac{71}{1.8} = 39\ °C$

No, there is no need to phone the doctor. The child's temperature is less than 40.0 °C.

2.17 Copper has the lowest specific heat of the samples and will reach the highest temperature.

2.19 **a.** $\Delta T = 25\ °C - 15\ °C = 10\ °C \qquad 25\ \cancel{g}\ \text{fat} \times \dfrac{1.00\ \text{cal}}{\cancel{g}\,\cancel{°C}} \times 10\ \cancel{°C} = 250\ \text{cal}$

b. $75\ \cancel{g} \times \dfrac{4.184\ J}{\cancel{g}\,\cancel{°C}} \times 44\ \cancel{°C} = 14\,000\ J$

c. $150\ \cancel{g} \times \dfrac{1.00\ \text{cal}}{\cancel{g}\,\cancel{°C}} \times 62\ \cancel{°C} = 9.3\ \text{kcal}$

d. $175\ \cancel{g} \times \dfrac{0.385\ \cancel{J}}{\cancel{g}\,\cancel{°C}} \times 160.\ \cancel{°C} \times \dfrac{1\ kJ}{1000\ \cancel{J}} = 10.8\ kJ$

2.21 The heat required is given by the relationship: Heat $= m \times \Delta T \times SH$.

a. Heat $= m \times \Delta T \times SH = 25.0\ g \times (25.7\ °C - 12.5\ °C) \times 4.184\ J/g\ °C =$

$25.0\ \cancel{g} \times 13.2\ \cancel{°C} \times 4.184\ J/\cancel{g}\,\cancel{°C} = 1380\ J \qquad 1380\ \cancel{J} \times \dfrac{1\ \text{cal}}{4.184\ \cancel{J}} = 330.\ \text{cal}$

b. Heat $= m \times \Delta T \times SH = 38.0\ g \times (246\ °C - 122\ °C) \times 0.385\ J/g\ °C =$

$38.0\ \cancel{g} \times 124\ \cancel{°C} \times 0.385\ J/\cancel{g}\,\cancel{°C} = 1810\ J \qquad 1810\ \cancel{J} \times \dfrac{1\ \text{cal}}{4.184\ \cancel{J}} = 434\ \text{cal}$

c. Heat $= m \times \Delta T \times SH = 15.0\ g \times (65.0\ °C + 42.0\ °C) \times 2.46\ J/g\ °C =$

$15.0\ \cancel{g} \times 107\ \cancel{°C} \times 2.46\ J/\cancel{g}\,\cancel{°C} = 3780\ J \qquad 3780\ \cancel{J} \times \dfrac{1\ \text{cal}}{4.184\ \cancel{J}} = 904\ \text{cal}$

d. Heat $= m \times \Delta T \times SH = 125\ g \times (118\ °C - 55\ °C) \times 0.450\ J/g\ °C =$

$125\ \cancel{g} \times 63\ \cancel{°C} \times 0.450\ J/\cancel{g}\,\cancel{°C} = 3500\ J \qquad 3500\ \cancel{J} \times \dfrac{1\ \text{cal}}{4.184\ \cancel{J}} = 850\ \text{cal}$

2.23 **a.** $505 \text{ g} \times \dfrac{1.00 \text{ cal}}{\text{g} \,°\text{C}} \times 10.5 \,°\text{C} \times \dfrac{1 \text{ kcal}}{1000 \text{ cal}} = 5.30 \text{ kcal}$

b. $4980 \text{ g} \times \dfrac{1.00 \text{ cal}}{\text{g} \,°\text{C}} \times 42 \,°\text{C} \times \dfrac{1 \text{ kcal}}{1000 \text{ cal}} = 208 \text{ kcal}$

2.25 **a.** gas   **b.** gas   **c.** solid

2.27 **a.** melting   **b.** sublimation   **c.** freezing

2.29 **a.** $65 \text{ g ice} \times \dfrac{80. \text{ cal}}{1 \text{ g ice}} = 5200 \text{ cal (absorbed)}$

b. $17.0 \text{ g ice} \times \dfrac{80. \text{ cal}}{1 \text{ g ice}} = 1400 \text{ cal (absorbed)}$

c. $225 \text{ g water} \times \dfrac{80. \text{ cal}}{1 \text{ g water}} \times \dfrac{1 \text{ kcal}}{1000 \text{ cal}} = 18 \text{ kcal (released)}$

2.31 **a.** condensation   **b.** evaporation   **c.** boiling   **d.** condensation

2.33 **a.** The liquid water in perspiration absorbs heat and changes to vapor. The heat needed for the change is removed from the skin.
b. On a hot day, there are more liquid water molecules in the damp clothing that have sufficient energy to become water vapor. Thus, water evaporates from the clothes more readily on a hot day.

2.35 **a.** $10.0 \text{ g water} \times \dfrac{540 \text{ cal}}{1 \text{ g water}} = 540 \text{ cal (absorbed)}$

b. $50.0 \text{ g water} \times \dfrac{540 \text{ cal}}{1 \text{ g water}} \times \dfrac{1 \text{ kcal}}{1000 \text{ cal}} = 27 \text{ kcal (absorbed)}$

c. $8.0 \text{ kg steam} \times \dfrac{1000 \text{ g}}{1 \text{ kg}} \times \dfrac{540 \text{ cal}}{1 \text{ g steam}} \times \dfrac{1 \text{ kcal}}{1000 \text{ cal}} = 4300 \text{ kcal (released)}$

2.37

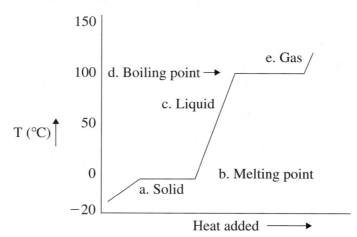

2.39 **a.** The smallest mark of 1 °C allows estimation of 0.1 °C to give 61.4 °C.
b. The smallest mark of 0.1 °C allows estimation of 0.01 °C to give 53.80 °C.
c. The smallest mark of 1 °C allows estimation of 0.1 °C to give 4.8 °C

2.41. $\dfrac{155 \,°\text{F} - 32°}{1.8} = \dfrac{123}{1.8} = 68 \,°\text{C}$

$68 \,°\text{C} + 273 = 341 \text{ K}$

2.43   45 g of protein (4 kcal/g) + 49 g of fat (9 kcal/g) + 120 g of carbohydrate (4 kcal/g)
   = 180 kcal + 440 kcal + 480 kcal = 1100 kcal in the meal

2.45   gold 250 J or 59 cal; aluminum 240 J or 58 cal; silver 250 J or 59 cal The heat needed for
   10.0-cm^3 samples of the metals are almost the same.

2.47   **a.** −60 °C   **b.** 60 °C
   c. A represents the solid state. B represents the change from solid to liquid or melting of the
      substance. C represents the liquid state as temperature increases. D represents the change from
      liquid to gas or boiling of the liquid.
   d. At −80 °C, solid; at −40 °C, liquid; at 25 °C, liquid; at 80 °C, gas

2.49   **a.** $\dfrac{(134 \text{ °F} - 32°)}{1.8} = 56.7 \text{ °C}$

   b. $\dfrac{(-70. \text{ °F} - 32°)}{1.8} = -56.7 \text{ °C}$

2.51 $\dfrac{-15 \text{ °F} - 32°}{1.8} = \dfrac{-47}{1.8} = -26 \text{ °C}$ $-26 \text{ °C} + 273 = 247 \text{ K}$

2.53   From Table 2.6, we see that liquid water has a high specific heat (1.00 cal/g °C), which means that a
   large amount of energy is required to cause a significant temperature change. Sand, on the other
   hand, has a low specific heat (0.19 cal/g °C), so even a small amount of energy will cause a significant temperature change in the sand.

2.55   Both water condensation (formation of rain) and deposition (formation of snow) from the gaseous
   moisture in the air are exothermic processes (heat is released). The heat released in either of these
   processes warms the surrounding air, and so the air temperature is in fact raised.

2.57   **a.** For 15% of the total Calories (kcal) supplied by protein, a conversion factor of 15 kcal from pro-
      tein/100 kcal total in the daily diet will be used in the calculation. Similar factors will be used for
      the carbohydrate and fat calculations.

$$1200 \text{ kcal} \times \frac{15 \text{ kcal}}{100 \text{ kcal}} \times \frac{1 \text{ g protein}}{4 \text{ kcal}} = 45 \text{ g of protein}$$

$$1200 \text{ kcal} \times \frac{45 \text{ kcal}}{100 \text{ kcal}} \times \frac{1 \text{ g carbohydrate}}{4 \text{ kcal}} = 140 \text{ g of carbohydrate}$$

$$1200 \text{ kcal} \times \frac{40. \text{ kcal}}{100 \text{ kcal}} \times \frac{1 \text{ g fat}}{4 \text{ kcal}} = 53 \text{ g of fat}$$

   b. The calculations for part b differ from part a only in the total kcal per day.

$$1900 \text{ kcal} \times \frac{15 \text{ kcal}}{100 \text{ kcal}} \times \frac{1 \text{ g protein}}{4 \text{ kcal}} = 71 \text{ g protein}$$

$$1900 \text{ kcal (total)} \times \frac{45 \text{ kcal}}{100 \text{ kcal}} \times \frac{1 \text{ g carbohydrate}}{4 \text{ kcal}} = 210 \text{ g carbohydrate}$$

$$1900 \text{ kcal} \times \frac{40. \text{ kcal}}{100 \text{ kcal}} \times \frac{1 \text{ g fat}}{9 \text{ kcal}} = 84 \text{ g of fat}$$

   c. The calculations for part c again differ only in the total kcal per day.

$$2600 \text{ kcal} \times \frac{15 \text{ kcal}}{100 \text{ kcal}} \times \frac{1 \text{ g protein}}{4 \text{ kcal}} = 98 \text{ g of protein}$$

$$2600 \text{ kcal} \times \frac{45 \text{ kcal}}{100 \text{ kcal}} \times \frac{1 \text{ g carbohydrate}}{4 \text{ kcal}} = 290 \text{ g of carbohydrate}$$

$$2600 \text{ kcal} \times \frac{40. \text{ kcal}}{100 \text{ kcal}} \times \frac{1 \text{ g fat}}{9 \text{ kcal}} = 120 \text{ g of fat}$$

2.59 Because each gram of body fat contains 15% water, a person actually loses 85 grams of fat per hundred grams of body fat. (We consider 1 lb of fat as exactly 1 lb.)

$$1 \text{ lb body fat} \times \frac{454 \text{ g}}{1 \text{ lb}} \times \frac{85 \text{ g fat}}{100 \text{ g body fat}} \times \frac{9 \text{ kcal}}{1 \text{ g fat}} = 3500 \text{ kcal}$$

2.61 A solid has a definite shape and definite volume. A liquid has a definite volume, but takes the shape of the container. A gas takes the shape and volume of the container.
   a. solid   **b.** gas   **c.** liquid   **d.** gas   **e.** solid

2.63 $725 \text{ g} \times \dfrac{1 \text{ cal}}{\text{g}\,^\circ\text{C}} \times 28\,^\circ\text{C} \times \dfrac{1 \text{ kcal}}{1000 \text{ cal}} = 20.\text{ kcal}$

2.65

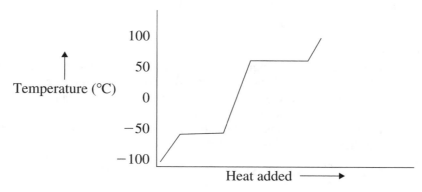

   a. solid   **b.** solid chloroform melts   **c.** liquid
   d. gas   **e.** −64 °C

2.67 $T_F = 1.8(60.0\,^\circ\text{C}) + 32^\circ = 140.\,^\circ\text{F}$ The liquids are at the same temperature.

2.69. $50.\text{ g} \times 12\,^\circ\text{C} \times \dfrac{1.00 \text{ cal}}{\text{g}\,^\circ\text{C}} = 600 \text{ cal } (6.0 \times 10^2 \text{ cal})$

specific heat $= \dfrac{600 \text{ cal}}{(25 \text{ g})(71\,^\circ\text{C})} = 0.34 \text{ cal/g }^\circ\text{C}$

2.71 condense steam : $75 \text{ g} \times \dfrac{540 \text{ cal}}{1 \text{ g}} = 40\,500 \text{ cal}$

temperature change: $75 \text{ g} \times 100\,^\circ\text{C} \times \dfrac{1.00 \text{ cal}}{\text{g}\,^\circ\text{C}} = 7\,500 \text{ cal}$

freeze water: $75 \text{ g} \times \dfrac{80.\text{ cal}}{1 \text{ g}} = 6\,000 \text{ cal}$

$40\,500 \text{ cal} + 7\,500 \text{ cal} + 6\,000 \text{ cal} = 54\,000 \text{ cal}$

$54\,000 \text{ cal} \times \dfrac{1 \text{ kcal}}{1000 \text{ cal}} = 54 \text{ kcal}$

2.73 **a.** $150 \text{ kg} \times \dfrac{1000 \text{ g}}{1 \text{ kg}} \times \dfrac{4.184 \text{ J}}{1 \text{ g}\,^\circ\text{C}} \times 78\,^\circ\text{C} \times \dfrac{1 \text{ lb}}{2.4 \times 10^7 \text{ J}} \times \dfrac{1 \text{ kg}}{2.20 \text{ lb}} = 0.93 \text{ kg of oil}$

   b. $150 \text{ kg} \times \dfrac{1000 \text{ g}}{1 \text{ kg}} \times \dfrac{540 \text{ cal}}{1 \text{ g}} \times \dfrac{4.184 \text{ J}}{1 \text{ cal}} \times \dfrac{1 \text{ lb}}{2.4 \times 10^7 \text{ J}} \times \dfrac{1 \text{ kg}}{2.20 \text{ lb}} = 6.4 \text{ kg of oil}$

Answers to Combining Ideas from Chapters 1 and 2

CI.1   **a.** 4 significant figures

       b. $20.17 \ \cancel{lb} \times \dfrac{1 \ kg}{2.20 \ \cancel{lb}} = 9.17 \ kg$

       c. $9.17 \ \cancel{kg} \times \dfrac{1000 \ \cancel{g}}{1 \ \cancel{kg}} \times \dfrac{1 \ cm^3}{19.3 \ \cancel{g}} = 475 \ cm^3$

       d. $1064 \ °C \ (1.8) + 32° = 1947 \ °F$
          $1064 \ °C + 273 = 1337 \ K$

       e. $145 \ °C = -18 \ °C = 127 \ °C$ temperature change

          $9170 \ \cancel{g} \times \dfrac{0.129 \ \cancel{J}}{\cancel{g} \ \cancel{°C}} \times 127 \ \cancel{°C} \times \dfrac{1 \ \cancel{kJ}}{1000 \ \cancel{J}} \times \dfrac{1 \ kcal}{4.184 \ \cancel{kJ}} = 35.9 \ kcal$

CI.3   **a.** B

       b. The volumes of both A and B remain the same.

       c. A is liquid water represented by diagram 2. In liquid water, the water particles are in a random arrangement, but close together. B is solid water represented by diagram **1.** In solid water, the water particles are fixed in a definite arrangement.

       d. freezing; freezing point; 0 °C and melting; melting point; 0 °C

       e. The solid water particles break apart from their fixed arrangement to have a more random arrangement, but they are still close together.

          $19.8 \ \cancel{g} \times 45 \ \cancel{°C} \times \dfrac{4.184 \ J}{\cancel{g} \ \cancel{°C}} = 3700 \ J$

       f. $19.8 \ \cancel{g} \times \dfrac{80. \ \cancel{cal}}{\cancel{g}} \times \dfrac{4.184 \ J}{\cancel{cal}} = 6600 \ J$

          total: $3700 \ J + 6600 \ J = 10 \ 300 \ J$

CI.5   **a.** $0.25 \ \cancel{lb} \times \dfrac{454 \ \cancel{g}}{\cancel{lb}} \times \dfrac{1 \ cm^3}{7.86 \ \cancel{g}} = 14 \ cm^3$

       b. $30 \ \cancel{nails} \times \dfrac{14 \ cm^3}{75 \ \cancel{nails}} = 5.6 \ cm^3 = 5.6 \ mL$

          17.6 mL water + 5.6 mL = 23.2 mL new water level

       c. $125 \ °C - 16 \ °C = 109 \ °C$ temperature change

          $0.25 \ \cancel{lb} \times \dfrac{454 \ \cancel{g}}{\cancel{lb}} \times 109 \ \cancel{°C} \times \dfrac{0.450 \ J}{\cancel{g} \ \cancel{°C}} = 5600 \ J$

3
Atoms and Elements

Study Goals

- Classify an example of matter as a pure substance or mixture.
- Write the name of an element from its symbol or its period and group number.
- Classify an element as a metal or nonmetal.
- Describe the three important particles in the atom, their location, charges, and relative masses.
- Describe Rutherford's gold-foil experiment and how it led to the current model of the atom.
- Use atomic number and mass number of an atom to determine the number of protons, neutrons, and electrons in the atom.
- Understand the relationship of isotopes to the atomic mass of an element on the periodic table.
- Use the percents abundance and masses of isotopes to calculate the atomic mass of an element.
- Write the electron level arrangements for elements 1–18 in the periodic table.
- Describe orbitals in the energy levels of atoms.
- Explain the relationship between electron arrangement, group number, and periodic law.
- Use the electron arrangement to explain periodic trends of the elements.

Think About It

1. Name some of the elements you have seen today.

2. How are the symbols of the elements related to their names?

3. What are some elements that are part of your vitamins?

Key Terms

Match each of the following key terms with the correct definition.

     a. element   **b.** atom   **c.** atomic number   **d.** mass number   **e.** isotope

1. _____ The number of protons and neutrons in the nucleus of an atom

2. _____ The smallest particle of an element

3. _____ A primary substance that cannot be broken down into simpler substances

4. _____ An atom of an element that has a different number of neutrons than another atom of the same element

5. _____ The number of protons in an atom

Answers   **1.** d   **2.** b   **3.** a   **4.** e   **5.** c

3.1 Classification of Matter

- A pure substance, element or compound, has a definite composition.
- Elements are the simplest type of matter; compounds consist of a combination of two or more elements.
- Mixtures contain two or more substances that are physically, not chemically, combined.
- Mixtures are classified as homogeneous or heterogeneous.

◆ Learning Exercise 3.1A

Identify each of the following as an element (E) or compound (C):

1. _____ carbon
2. _____ carbon dioxide
3. _____ potassium iodide

4. _____ silver
5. _____ aluminum
6. _____ table salt (sodium chloride)

Answers 1. E 2. C 3. C 4. E 5. E 6. C

◆ Learning Exercise 3.1B

Identify each of the following as a pure substance (P) or mixture (M):

1. _____ bananas and milk
2. _____ sulfur
3. _____ gold

4. _____ a bag of raisins and nuts
5. _____ water
6. _____ sand and water

Answers 1. M 2. P 3. P 4. M 5. P 6. M

◆ Learning Exercise 3.1C

Identify each of the following mixtures as homogeneous (HO) or heterogeneous (HE):

1. _____ chocolate milk
2. _____ sand and water
3. _____ apple juice without pulp

4. _____ a bag of raisins and nuts
5. _____ air
6. _____ vinegar

Answers 1. HO 2. HE 3. HO 4. HE 5. HO 6. HO

3.2 Elements and Symbols

- Elements are the primary substances of matter.
- Chemical symbols are one- or two-letter abbreviations for the names of the elements.
- The physical properties of an element are those characteristics such as color, density, and melting and boiling points that we can observe or measure without changing the identity of the element.

Study Note

Now is the time to learn the names of the elements and their symbols. Practice saying and writing the names of the elements in the periodic table with atomic numbers 1–54 and Cs, Ba, Hg, Au, and Pb. Cover the symbols in the list of elements and practice writing the symbols for the elemental names.

◆ **Learning Exercise 3.2A**

Write the symbols for each of the following elements:

1. carbon _____	**6.** nitrogen _____	**11.** calcium _____
2. iron _____	**7.** iodine _____	**12.** gold _____
3. sodium _____	**8.** sulfur _____	**13.** copper _____
4. phosphorus _____	**9.** potassium _____	**14.** neon _____
5. oxygen _____	**10.** lead _____	**15.** chlorine _____

Answers

1. C	**2.** Fe	**3.** Na	**4.** P	**5.** O
6. N	**7.** I	**8.** S	**9.** K	**10.** Pb
11. Ca	**12.** Au	**13.** Cu	**14.** Ne	**15.** Cl

◆ **Learning Exercise 3.2B**

Write the names of the elements represented by each of the following symbols:

1. Mg _____	**7.** Ag _____
2. K _____	**8.** Br _____
3. Au _____	**9.** Zn _____
4. F _____	**10.** Al _____
5. Cu _____	**11.** Ba _____
6. Be _____	**12.** Li _____

Answers

1. magnesium	**2.** potassium	**3.** gold
4. fluorine	**5.** copper	**6.** beryllium
7. silver	**8.** bromine	**9.** zinc
10. aluminum	**11.** barium	**12.** lithium

◆ **Learning Exercise 3.2C**

Gold (Au, *aurum,* atomic number 79) has been highly valued from ancient times. Gold, which has a density of 19.3 g/mL, melts at 1064 °C and boils at 2857 °C. It is found free in nature usually along with quartz or pyrite deposits. Gold is a beautiful metal with a yellow color. It is a soft metal that is also the most malleable metal, which accounts for its use in jewelry and gold leaf. Gold is also a good conductor of heat and electricity. List all the physical properties that describe gold.

Answer

density 19.3 g/mL	melting point 1064 °C
boiling point 2857 °C	metal
yellow color	soft
highly malleable	good conductor of heat and electricity

3.3 The Periodic Table

- The periodic table is an arrangement of the elements by increasing atomic number.
- Each vertical column contains a *group* of elements, which have similar properties.
- A horizontal row of elements is called a *period.*
- On the periodic table, the *metals* are located on the left of the heavy zigzag line, the nonmetals are to the right, and metalloids are next to the zigzag line.

◆ Learning Exercise 3.3A

Study Note
1. The periodic table consists of horizontal rows called *periods* and vertical columns called *groups*. 2. Elements in Group 1A (1) are *alkali metals*. Elements in Group 2A (2) are *alkaline earth metals*, and Group 7A (17) contains *halogens*. Elements in Group 8A (18) are *noble gases*.

Indicate whether the following elements are in a group (G), period (P), or neither (N):

a. Li, C, and O _____

b. Br, Cl, and F _____

c. Al, Si, and Cl _____

d. C, N, and O _____

e. Mg, Ca, and Ba _____

f. C, S, and Br _____

g. Li, Na, and K _____

h. K, Ca, and Br _____

Answers **a.** P **b.** G **c.** P **d.** P
e. G **f.** N **g.** G **h.** P

◆ Learning Exercise 3.3B

Complete the list of elements, group numbers, and period numbers in the following table:

Element and Symbol	Group Number	Period Number
	2A (2)	3
Silicon, Si		
	5A (15)	2
Aluminum, Al		
	4A (14)	5
	1A (1)	6

Answers

Element and Symbol	Group Number	Period Number
Magnesium, Mg	2A (2)	3
Silicon, Si	4A (14)	3
Nitrogen, N	5A (15)	2
Aluminum, Al	3A (13)	3
Tin, Sn	4A (14)	5
Cesium, Cs	1A (1)	6

◆ **Learning Exercise 3.3C**

Identify each of the following elements as a metal (M), nonmetal (NM), or metalloid (ML):

1. Cl _____   **2.** N _____   **3.** Fe _____   **4.** K _____   **5.** Sb _____   **6.** C _____
7. Ca _____   **8.** Ge _____   **9.** Ag _____   **10.** Mg _____   **11.** Au _____   **12.** I _____

Answers   **1.** NM   **2.** NM   **3.** M   **4.** M   **5.** ML   **11.** M
            6. NM   **7.** M   **8.** ML   **9.** M   **10.** M   **12.** NM

◆ **Learning Exercise 3.3D**

Match the names of the chemical groups with the elements K, Cl, He, Fe, Mg, Ne, Li, Cu, and Br.

1. Halogens   _____

2. Noble gases   _____

3. Alkali metals   _____

4. Alkaline earth metals   _____

5. Transition metals   _____

Answers   **1.** Cl, Br   **2.** He, Ne   **3.** K, Li   **4.** Mg   **5.** Fe, Cu

3.4 The Atom

- An atom is the smallest particle that retains the characteristics of an element.
- Atoms are composed of three subatomic particles. Protons have a positive charge ($+$), electrons carry a negative charge ($-$), and neutrons are electrically neutral.
- The protons and neutrons each with a mass of about 1 amu are found in the tiny, dense nucleus. The electrons are located outside the nucleus.

◆ **Learning Exercise 3.4A**

Indicate whether each of the following statements is consistent with atomic theory (true or false):

1. _____ All matter is composed of atoms.

2. _____ All atoms of an element are identical.

3. _____ Atoms combine to form compounds.

4. _____ Most of the mass of the atom is in the nucleus.

Answers   **1.** true   **2.** false   **3.** true   **4.** true

◆ **Learning Exercise 3.4B**

Match the following terms with the correct statements:

    a. proton   **b.** neutron   **c.** electron   **d.** nucleus

1. _____ Found in the nucleus of an atom   **4.** _____ Has a mass of 1 amu

2. _____ Has a -1 charge   **5.** _____ The small, dense center of the atom

3. _____ Found outside the nucleus   **6.** _____ Is neutral

Answers   **1.** a and b   **2.** c   **3.** c   **4.** a and b   **5.** d   **6.** b

3.5 Atomic Number and Mass Number

- The *atomic number* is the number of protons in every atom of an element. In neutral atoms, the number of electrons is equal to the number of protons.
- The *mass number* is the total number of protons and neutrons in an atom.

◆ **Learning Exercise 3.5A**

Give the number of protons in each of the following neutral atoms:

    a. An atom of carbon   _____

    b. An atom of the element with atomic number 15   _____

    c. An atom with a mass number of 40 and atomic number 19   _____

    d. An atom with 9 neutrons and a mass number of 19   _____

    e. A neutral atom that has 18 electrons   _____

Answers   **a.** 6   **b.** 15   **c.** 19   **d.** 10   **e.** 18

◆ **Learning Exercise 3.5B**

Study Note

1. The *atomic number* is the number of protons in every atom of an element. In neutral atoms, the number of electrons equals the number of protons.
2. The *mass number* is the total number of neutrons and protons in the nucleus of an atom.
3. The number of neutrons is *mass number – atomic number.*

Give the number of neutrons in each of the following atoms:

    a. A mass number of 42 and atomic number 20   _____

    b. A mass number of 10 and 5 protons   _____

    c. $^{30}_{14}Si$ _____

    d. A mass number of 9 and atomic number 4   _____

    e. A mass number of 22 and 10 protons   _____

    f. A zinc atom with a mass number of 66   _____

Answers   **a.** 22   **b.** 5   **c.** 16   **d.** 5   **e.** 12   **f.** 36

Study Note

In the atomic symbol for a particular atom, the mass number appears in the upper left corner and the atomic number in the lower left corner.

Mass number → $^{32}_{16}S$ $^{26}_{13}Al$
Atomic number →

◆ Learning Exercise 3.5C

Complete the following table for neutral atoms.

Symbol	Atomic Number	Mass Number	Number of Protons	Number of Neutrons	Number of Electrons
	12			12	
			20	22	
		55		29	
	35			45	
		35	17		
$^{120}_{50}$Sn					

Answers

Symbol	Atomic Number	Mass Number	Number of Protons	Number of Neutrons	Number of Electrons
$^{24}_{12}$Mg	12	24	12	12	12
$^{42}_{20}$Ca	20	42	20	22	20
$^{55}_{26}$Fe	26	55	26	29	26
$^{80}_{35}$Br	35	80	35	45	35
$^{35}_{17}$Cl	17	35	17	18	17
$^{120}_{50}$Sn	50	120	50	70	50

3.6 Isotopes and Atomic Mass

- Atoms that have the same number of protons but different numbers of neutrons are called *isotopes*.
- The atomic mass of an element is the average mass of all the isotopes in a naturally occurring sample of that element.

◆ Learning Exercise 3.6A

Identify the sets of atoms that are isotopes.

   A. $^{20}_{10}$X   **B.** $^{20}_{11}$X   **C.** $^{21}_{11}$X   **D.** $^{19}_{10}$X   **E.** $^{19}_{9}$X

Answer   Atoms A and D are isotopes (At. No. 10); atoms B and C are isotopes (At. No. 11).

◆ **Learning Exercise 3.6B**

ESSAY: Copper has two naturally occurring isotopes, ^{63}Cu and ^{65}Cu. If that is the case, why is the atomic mass of copper listed as 63.55 on the periodic table?

Answer   Copper in nature consists of two isotopes with different atomic masses. The atomic mass is the average of the individual masses of the two isotopes and their percent abundance in the sample. The atomic mass does not represent the mass of any individual atom.

◆ **Learning Exercise 3.6C**

Rubidium is the element with atomic number 37.

  a. What is the symbol for rubidium? _____

  b. What is the Group number for rubidium? _____

  c. What is the name of the family that includes rubidium? _____

  d. Natural rubidium has two isotopes, $^{85}_{37}Rb$ and $^{87}_{37}Rb$. How many protons, neutrons, and electrons are in each isotope?

  e. In naturally occurring rubidium, 72.17% is $^{85}_{37}Rb$ with a mass of 84.91 amu; 27.83% is $^{87}_{37}Rb$ with a mass of 86.91 amu. What is the atomic mass of rubidium?

Answers

  a. Rb

  b. Rubidium is in Group 1A (1)

  c. Rubidium is in the family called the alkali metals.

  d. $^{85}_{37}Rb$ has 37 protons, 48 neutrons, and 37 electrons. $^{87}_{37}Rb$ has 37 protons, 50 neutrons, and 37 electrons.

  e. 84.91 amu $(72.17/100)$ + 86.91 amu $(27.83/100)$
      = 61.28 amu + 24.19 amu = 85.47 amu atomic mass Rb

3.7 Electron Energy Levels

• In an atom, the electrons of similar energy are grouped in specific energy levels. The first level nearest the nucleus can hold 2 electrons, the second level can hold 8 electrons, the third level will take up to 18 electrons.
• The electron arrangement is the number of electrons in each energy level beginning with the lowest energy level.
• In an atom, an electron occupies a region in space around the nucleus called an orbital.
• Energy level 1 consists of one *s* orbital; energy level 2 has two types of orbitals (one *s* and three *p* orbitals); energy level 3 has three types of orbitals (one *s*, three *p*, and five *d* orbitals).

◆ **Learning Exercise 3.7A**

Write the electron level arrangements for the following elements:

Element	Electron Level 1 2 3 4	Element	Electron Level 1 2 3 4
a. beryllium	_____	**e.** phosphorus	_____
b. carbon	_____	**f.** nitrogen	_____
c. potassium	_____	**g.** chlorine	_____
d. sodium	_____	**h.** silicon	_____

Answers   **a.** 2, 2   **b.** 2, 4   **c.** 2, 8, 8, 1   **d.** 2, 8, 1
             e. 2, 8, 5   **f.** 2, 5   **g.** 2, 8, 7   **h.** 2, 8, 4

◆ Learning Exercise 3.7B

State the maximum number of electrons for each of the following:

a. 3*s* orbital _____

b. 2*p* orbital _____

c. energy level 2 _____

d. three 4*p* orbitals _____

e. five 4*d* orbitals _____

f. energy level 3 _____

Answers **a.** 2 **b.** 2 **c.** 8 **d.** 6 **e.** 10 **f.** 18

3.8 Periodic Trends

- The changes in physical and chemical properties of the elements across each period is repeated in each successive period.
- Representative elements in a group have similar behavior.
- The group number of an element gives the number of valence electrons.
- The electron dot symbol shows each valence electron as a dot placed around the atomic symbol.
- The atomic radius of representative elements generally increases going down a group and decreases going across a period.
- The ionization energy generally decreases going down a group and increases going across a period.

◆ Learning Exercise 3.8A

State the number of valence electrons, the group number of each element, and the electron dot symbol for the following elements:

Element	Valence Electrons	Group Number	Electron Dot Symbol
a. sulfur			
b. oxygen			
c. magnesium			
d. hydrogen			
e. fluorine			
f. aluminum			

Answers

a. sulfur $6e^-$ Group 6A (16) $\cdot\ddot{S}:$

b. oxygen $6e^-$ Group 6A (16) $\cdot\ddot{O}:$

c. magnesium $2e^-$ Group 2A (2) $Mg\cdot$

d. hydrogen $1e^-$ Group 1A (1) $H\cdot$

e. fluorine $7e^-$ Group 7A (17) $\cdot\ddot{F}:$

f. aluminum $3e^-$ Group 3A (13) $\cdot\dot{Al}\cdot$

◆ **Learning Exercise 3.8B**

Indicate the element that has the larger atomic radius.

a. _____ Mg or Ca   e. _____ Li or Cs

b. _____ Si or Cl   f. _____ Li or N

c. _____ Sr or Rb   g. _____ N or P

d. _____ Br or Cl   h. _____ As or Ca

Answers   **a.** Ca   **b.** Si   **c.** Rb   **d.** Br
                e. Cs   **f.** Li   **g.** P   **h.** Ca

◆ **Learning Exercise 3.8C**

Indicate the element that has the lower ionization energy.

a. _____ Mg or Na   e. _____ Li or O

b. _____ P or Cl   f. _____ Sb or N

c. _____ K or Rb   g. _____ K or B

d. _____ Br or F   h. _____ S or Na

Answers   **a.** Na   **b.** P   **c.** Rb   **d.** Br
                e. Li   **f.** Sb   **g.** K   **h.** Na

Checklist for Chapter 3

You are ready to take the practice test for Chapter 3. Be sure that you have accomplished the following learning goals for this chapter. If you are not sure, review the section listed at the end of the goal. Then apply your new skills and understanding to the practice test.

After studying Chapter 3, I can successfully:

_____ Classify matter as a pure substance or a mixture (3.1).

_____ Write the correct symbol or name for an element (3.2).

_____ Use the periodic table to identify the group and period of an element, and describe it as a metal, non-metal, or metalloid (3.3).

_____ State the electrical charge, mass, and location of the protons, neutrons, and electrons in an atom (3.4).

_____ Given the atomic number and mass number of an atom, state the number of protons, neutrons, and electrons (3.5).

_____ Identify an isotope and describe the atomic mass of an element (3.6).

_____ Write the electron energy level arrangements for elements with atomic number 1–18 (3.7).

_____ Use the periodic table to predict periodic trends in atomic size and ionization energy (3.8).

Practice Test for Chapter 3

Instructions: Write or select the correct answer for each of the following questions.

Write the correct symbol for each of the elements listed:

1. potassium _____   4. carbon _____

2. phosphorus _____   5. sodium _____

3. calcium _____

Write the correct name for each of the symbols listed below:

6. Fe _____   **9.** Pb _____

7. Cu _____   **10.** Ag _____

8. Cl _____

11. The elements C, N, and O are part of a
    A. period   **B.** group   **C.** neither

12. The elements Li, Na, and K are part of a
    A. period   **B.** group   **C.** neither

13. What is the classification of an atom with 15 protons and 17 neutrons?
    A. metal   **B.** nonmetal   **C.** transition element
    D. noble gas   **E.** halogen

14. What is the group number of the element with atomic number 3?
    A. 1   **B.** 2   **C.** 3   **D.** 7   **E.** 8

For questions 15 through 18, consider an atom with 12 protons and 13 neutrons.

15. This atom has an atomic number of
    A. 12   **B.** 13   **C.** 23   **D.** 24   **E.** 25

16. This atom has a mass number of
    A. 12   **B.** 13   **C.** 23   **D.** 24   **E.** 25

17. This is an atom of
    A. carbon   **B.** sodium   **C.** magnesium   **D.** aluminum   **E.** manganese

18. The number of electrons in this atom is
    A. 12   **B.** 13   **C.** 23   **D.** 24   **E.** 25

For questions 19 through 22, consider an atom of calcium with a mass number of 42.

19. This atom of calcium has an atomic number of
    A. 20   **B.** 22   **C.** 40   **D.** 41   **E.** 42

20. The number of protons in this atom of calcium is
    A. 20   **B.** 22   **C.** 40   **D.** 41   **E.** 42

21. The number of neutrons in this atom of calcium is
    A. 20   **B.** 22   **C.** 40   **D.** 41   **E.** 42

22. The number of electrons in this atom of calcium is
    A. 20   **B.** 22   **C.** 40   **D.** 41   **E.** 42

23. Platinum, ^{195}Pt, has
    A. $78p^+$, $78e^-$, 78n   **B.** $195p^+$, $195e^-$, 195n   **C.** $78p^+$, $78e^-$, 195n
    D. $78p^+$, $78e^-$, 117n   **E.** $78p^+$, $117e^-$, 117n

For questions 24 and 25, use the following list of atoms.
$$^{14}_{7}V \quad ^{16}_{8}W \quad ^{19}_{9}X \quad ^{16}_{7}Y \quad ^{18}_{8}Z$$

24. Which atoms(s) are isotopes of an atom with 8 protons and 9 neutrons?
    A. W   **B.** W, Z   **C.** X, Y   **D.** X   **E.** Y

25. Which atom(s) are isotopes of an atom with 7 protons and 8 neutrons?
    A. V   **B.** W   **C.** V, Y   **D.** W, Z   **E.** none

26. Which element would you expect to have properties most like oxygen?
    A. nitrogen   **B.** carbon   **C.** chlorine   **D.** argon   **E.** sulfur

27. Which of the following is an isotope of nitrogen?
    A. $^{14}_{8}N$ **B.** $^{7}_{3}N$ **C.** $^{10}_{5}N$ **D.** $^{4}_{2}N$ **E.** $^{15}_{7}N$

28. Except for helium, the number of electrons in the outer levels of the noble gases is
    A. 3   **B.** 5   **C.** 7   **D.** 8   **E.** 12

29. The electron level arrangement for an oxygen atom is
    A. 2, 4   **B.** 2, 8   **C.** 2, 6   **D.** 2, 4, 2   **E.** 2, 6, 2

30. The electron level arrangement for aluminum is
    A. 2, 11   **B.** 2, 8, 5   **C.** 2, 8, 3   **D.** 2, 10, 1   **E.** 2, 2, 6, 3

Answers for the Practice Test

1. K	**2.** P	**3.** Ca	**4.** C	**5.** Na
6. iron	**7.** copper	**8.** chlorine	**9.** lead	**10.** silver
11. A	**12.** B	**13.** B	**14.** A	**15.** A
16. E	**17.** C	**18.** A	**19.** A	**20.** A
21. B	**22.** A	**23.** D	**24.** B	**25.** C
26. E	**27.** E	**28.** D	**29.** C	**30.** C

Answers and Solutions to Selected Text Problems

3.1   *Elements* are the simplest type of pure substance. *Compounds* contain two or more elements in the same ratio.

    a. Compound; combination of elements in a fixed ratio
    b. Element; one type of substance found in the list of elements
    c. Compound; combination of elements in a fixed ratio
    d. Element; one type of substance found in the list of elements

3.3   A *homogeneous mixture* has a uniform composition; a *heterogeneous mixture* does not have a uniform composition throughout the mixture.

    a. Homogeneous; uniform composition
    b. Heterogeneous; nonuniform composition
    c. Heterogeneous; nonuniform composition
    d. Homogeneous; uniform composition

3.5   A *pure substance* has a definite composition. A *mixture* has a variable composition.

    a. Mixture; heterogeneous with nonuniform composition
    b. Pure substance; element found in the list of elements
    c. Pure substance; compound with elements in a fixed ratio
    d. Mixture; heterogeneous with nonuniform composition

3.7   **a.** Cu   **b.** Si   **c.** K   **d.** N
         e. Fe   **f.** Ba   **g.** Pb   **h.** Sr

3.9   **a.** carbon   **b.** chlorine   **c.** iodine   **d.** mercury
         e. fluorine   **f.** argon   **g.** zinc   **h.** nickel

3.11 **a.** sodium and chlorine
  b. calcium, sulfur, and oxygen
  c. carbon, hydrogen, chlorine, nitrogen, and oxygen
  d. calcium, carbon, and oxygen

3.13 **a.** C, N, and O are in Period 2.
  b. He is the element at the top of Group 8A (18).
  c. The alkali metals are the elements in Group 1A (1).
  d. Period 2 is the horizontal row of elements that ends with neon (Ne).

3.15 **a.** alkaline earth metal   **b.** transition element   **c.** noble gas
  d. alkali metal   **e.** halogen

3.17 **a.** C   **b.** He   **c.** Na   **d.** Ca   **e.** Al

3.19 On the periodic table, *metals* are located to the left of the heavy zigzag line, *nonmetals* are elements to the right, and metalloids (B, Si, Ge, As, Sb, Te, Po, and At) are located along the line.

  a. metal   **b.** nonmetal   **c.** metal   **d.** nonmetal
  e. nonmetal   **f.** nonmetal   **g.** metalloid   **h.** metal

3.21 **a.** electron   **b.** proton   **c.** electron   **d.** neutron

3.23 The two most massive subatomic particles, protons and neutrons, are located in a very small region of the atom, which is called the nucleus.

3.25 **a.** True   **b.** True
  c. True   **d.** False; a proton is attracted to an electron

3.27 Because the hair strands repel one another, there must be like electrical charges on each strand.

3.29 **a.** atomic number   **b.** both   **c.** mass number   **d.** atomic number

3.31 **a.** lithium, Li   **b.** fluorine, F   **c.** calcium, Ca   **d.** zinc, Zn
  e. neon, Ne   **f.** silicon, Si   **g.** iodine, I   **h.** oxygen, O

3.33 **a.** 12   **b.** 30   **c.** 53   **d.** 19

3.35

Name of Element	Symbol	Atomic Number	Mass Number	Number of Protons	Number of Neutrons	Number of Electrons
Aluminum	Al	13	27	13	14	13
Magnesium	Mg	12	24	12	12	12
Potassium	K	19	39	19	20	19
Sulfur	S	16	31	16	15	16
Iron	Fe	26	56	26	30	26

3.37 **a.** Since the atomic number of aluminum is 13, every Al atom has 13 protons. An atom of aluminum with a mass number 27 and an atomic number 13 has 14 neutrons. $27 - 13 = 14$ n Therefore, 13 protons, 14 neutrons, 13 electrons

  b. 24 protons, 28 neutrons, 24 electrons
  c. 16 protons, 18 neutrons, 16 electrons
  d. 26 protons, 30 neutrons, 26 electrons

3.39 **a.** $^{31}_{15}P$ **b.** $^{80}_{35}Br$ **c.** $^{27}_{13}Al$ **d.** $^{35}_{17}Cl$ **e.** $^{202}_{80}Hg$

3.41 **a.** $^{32}_{16}S$ $^{33}_{16}S$ $^{34}_{16}S$ $^{36}_{16}S$
    b. They all have the same atomic number (the same number of protons and electrons).
    c. They have different numbers of neutrons, which is reflected in their mass numbers.
    d. The atomic mass of sulfur on the periodic table is the average atomic mass of all the naturally occurring isotopic masses.
    e. With atomic mass of 32.07, it is most likely that the ^{32}S is the most prevalent isotope of sulfur.

3.43 The electrons surrounding a nucleus have specific energies. Electrons with similar energies will be found grouped together within a specific energy level.

3.45 **a.** 8   **b.** 5   **c.** 8   **d.** 0   **e.** 8

3.47 **a.** 2, 4   **b.** 2, 8, 8   **c.** 2, 8, 6   **d.** 2, 8, 4
    e. 2, 8, 3   **f.** 2, 5

3.49 **a.** Li   **b.** Mg   **c.** H   **d.** Cl   **e.** O

3.51 absorb

3.53 **a.** boron: 2, 3 aluminum: 2, 8, 3
    b. Three
    c. Group 3A (13)

3.55 **a.** $2e^-$, Group 2A (2)   **b.** $7e^-$, Group 7A (17)   **c.** $6e^-$, Group 6A (16)
    d. $5e^-$, Group 5A (15)   **e.** $2e^-$, Group 2A (2)   **f.** $7e^-$, Group 7A (17)

3.57 Mg, Ca, and Sr have similar properties because they are members of the same group (family) in the periodic table. Chemical properties are related to the number of electrons in the outermost occupied energy level, and these three elements all have two electrons in their outermost level.

3.59 The number of dots is equal to the number of valence electrons as indicated by the group number.

    a. Sulfur has 6 valence electrons

    b. Nitrogen has 5 valence electrons

    c. Calcium has 2 valence electrons

    d. Sodium has 1 valence electron

    e. Potassium has 1 valence electron

3.61 **a.** M·   **b.** ·M·

3.63 Alkali metals are members of Group 1A (1), and each has one valence electron.

3.65 **a.** The atomic radius of representative elements decreases from Group 1A (1) to 8A (18): Mg, Al, Si
    b. The atomic radius of representative elements increases going down a group: I, Br, Cl
    c. The atomic radius of representative elements decreases from Group 1A (1) to 8A (18): Sr, Sb, I

3.67 The atomic radius of representative elements decreases going across a period from Group 1A (1) to 8A (18) and increases going down a group.

    a. In Period 3, Na, which is on the left, is larger than O.
    b. In Group 1A (1), Rb, which is further down the group, is larger than Na.
    c. In Period 3, Na, which is on the left, is larger than Mg.

3.69 **a.** The ionization energy decreases going down a group: Br, Cl, F
    b. Going across a period from left to right, the ionization energy generally increases: Na, Al, Cl
    c. The ionization energy decreases going down a group: Cs, K, Na

3.71 **a.** Br, which is above I in Group 7A (17), has a higher ionization energy than I.
    b. Ionization energy decreases down Group 2A (2), which gives Mg a higher ionization energy than Sr.
    c. Ionization energy increases from Group 4A (14) to Group 5A (15), which gives P a higher ionization energy than Si.

3.73 **a.** It is a compound with two different types of atoms combined in the same ratio.
b. It is a mixture because it consists of two different types of atoms.
c. It is an element because it has only one type of atom.

3.75 Mixture **a** is homogeneous because it has the same composition everywhere.

3.77 **a.** false   **b.** true   **c.** true   **d.** false

3.79 **a.** Atomic mass is the sum of the protons (1) and neutrons (2).
b. Atomic number is the number of protons (1).
c. The positive charge of the nucleus is due to the protons (1).
d. The negative charge is due to the electrons (3).
e. The mass number − atomic number gives the number of neutrons (2).

3.81 **a.** $^{16}_{8}X$ $^{17}_{8}X$ $^{18}_{8}X$ All have 8 protons.
b. $^{16}_{8}X$ $^{17}_{8}X$ $^{18}_{8}X$ All are isotopes of oxygen.
c. $^{16}_{8}X$ $^{16}_{9}X$ Mass number of 16
  $^{18}_{10}X$ $^{18}_{8}X$ Mass number of 18
d. $^{16}_{8}X$ $^{18}_{10}X$ Both have 8 neutrons.

3.83 **a.** K and Mg on the left of the zigzag line are metals.
b. Si on the zigzag line is a metalloid.
c. K in Group 1A (1) is an alkali metal.
d. Ar has the smallest size of this group.
e. S in Group 6A (16) has electron arrangement 2, 8, 6.

3.85 **a.** element   **b.** compound   **c.** mixture   **d.** element   **e.** mixture

3.87 **a.** Mg, magnesium **b.** Br, bromine   **c.** Al, aluminum   **d.** O, oxygen

3.89 **a.** halogen   **b.** noble gas   **c.** alkali metal   **d.** alkaline earth

3.91 **a.** False. A proton has a positive charge.
b. False. The neutron has about the same mass as a proton.
c. True.
d. False. The nucleus is the tiny, dense central core of an atom.
e. True.

3.93 **a.** protons   **b.** protons   **c.** alkali metals

3.95 **a.** lithium, Li   **b.** fluorine, F   **c.** calcium, Ca   **d.** arsenic, As
e. tin, Sn   **f.** cesium, Cs   **g.** gold, Au   **h.** oxygen, O

3.97 **a.** 13 protons, 14 neutrons, 13 electrons
b. 24 protons, 28 neutrons, 24 electrons
c. 16 protons, 18 neutrons, 16 electrons
d. 26 protons, 30 neutrons, 26 electrons
e. 54 protons, 82 neutrons, 54 electrons

3.99

Name	Nuclear Symbol	Number of Protons	Number of Neutrons	Number of Electrons
Sulfur	$^{34}_{16}S$	16	18	16
Zinc	$^{70}_{30}Zn$	30	40	30
Magnesium	$^{26}_{12}Mg$	12	14	12
Radon	$^{220}_{86}Rn$	86	134	86

3.101 a. $^{9}_{4}$Be   **b.** $^{26}_{12}$Mg   **c.** $^{46}_{20}$Ca   **d.** $^{70}_{30}$Zn   **e.** $^{63}_{29}$Cu

3.103 a. 82 protons, 126 neutrons, 82 electrons
    b. $^{214}_{82}$Pb   **c.** $^{214}_{83}$Bi, bismuth

3.105 a. N 2,5   Group 5A (15)
    b. Na 2,8,1   Group 1A (1)
    c. S 2,8,6   Group 6A (16)
    d. B 2,3   Group 3A (13)

3.107 Ca has a greater net nuclear charge than K. The least tightly bound electron in Ca is further from the nucleus than in Mg and needs less energy to be removed.

3.109 a. Na is on the far left of the heavy zigzag line. Na is a metal.
    b. Na at the beginning of Period 3 has the largest atomic radius.
    c. F at the top of Group 7A (17) and to the far right in Period 2 has the highest ionization energy.
    d. Na has the lowest ionization energy and loses an electron most easily.
    e. Cl is found in Period 3 in Group 7A (17).

3.111 a. X is Cl, chlorine. This isotope has 17 protons and 20 neutrons.
    b. X is Fe, iron. This isotope has 26 protons and 30 neutrons.
    c. X is Sn, tin. This isotope has 50 protons and 66 neutrons.
    d. X is Sn, tin. This isotope has 50 protons and 74 neutrons.
    e. X is Cd, cadmium. This isotope has 48 protons and 68 neutrons.
    f. **c** and **d** are both isotopes of Sn, tin.

3.113 a. In a group, the atomic radius increases going down a group, which gives O the smallest atomic radius in Group 6A (16).
    b. In a period, the atomic radius generally decreases going from left to right, which gives Ar the smallest atomic radius in Period 3.
    c. In a group, the ionization energy decreases going down a group, which gives N the highest ionization energy in Group 5A (15).
    d. In a period, the ionization energy generally increases going left to right, which gives Na the lowest ionization energy in Period 3.

3.115 $2.00 \text{ cm}^3 \times \dfrac{11.3 \text{ g}}{1 \text{ cm}^3} \times \dfrac{1 \text{ atom}}{3.4 \times 10^{-22} \text{ g}} = 6.6 \times 10^{22}$ atoms

3.117 68.926 amu (60.10/100) + 70.925 amu (39.90/100)
    = 41.42 amu + 28.30 amu
    = 69.72 amu atomic mass for Ga

4

Compounds and Their Bonds

Study Goals

- Write an electron dot formula for an atom of a representative element.
- Use the octet rule to determine the ionic charge of ions for representative elements.
- Use charge balance to write an ionic formula.
- Draw the electron dot formula for covalent compounds.
- Write the correct names for ionic and covalent compounds.
- Write ionic formulas and names of compounds with polyatomic ions.
- Use electronegativity to determine the polarity of a bond.
- Use VSEPR theory to determine the shape and bond angles of a molecule.
- Identify a covalent compound as polar or nonpolar.
- Describe the types of attractive forces that hold particles together.

Think About It

1. How does a compound differ from an element?

2. What are some compounds listed on the labels of your vitamins, toothpaste, and foods?

3. What makes salt an ionic compound?

4. Why are some covalent compounds polar and others nonpolar?

Key Terms

Match each the following key terms with the correct definition:

    a. molecule   **b.** ion   **c.** ionic bond   **d.** covalent bond   **e.** octet

1. _____ A sharing of valence electrons by two atoms

2. _____ An arrangement of 8 electrons in the outer energy level

3. _____ The attraction between positively and negatively charged particles

4. _____ The smallest unit of two or more atoms held together by covalent bonds

5. _____ An atom or group of atoms with a positive or negative charge

Answers   **1.** d   **2.** e   **3.** c   **4.** a   **5.** b

4.1 Octet Rule and Ions

- The stability of the noble gases is associated with 8 electrons, an octet, in their valence energy levels. Helium is stable with 2 electrons in its valence energy level.
  He   2   Ar   2, 8, 8   Ne   2, 8
  Atoms of elements other than the noble gases achieve stability by losing, gaining, or sharing their valence electrons with other atoms in the formation of compounds.
- A metal of the representative elements in Groups 1, 2, and 3 achieves a noble gas electron arrangement by losing its valence electrons to form a positively charged cation 1+, 2+, or 3+.
- When a nonmetal forms ions, electrons add to the valence energy level to give an octet, and form a negatively charged anion with a charge of 3−, 2−, or 1−.

Study Note

When an atom loses or gains electrons, it acquires the electron arrangement of its nearest noble gas. For example, sodium loses 1 electron, which gives the Na^+ ion an arrangement like neon. Oxygen gains 2 electrons to give the oxide ion O^{2-} an arrangement like neon.

◆ Learning Exercise 4.1A

The following elements lose electrons when they form ions. Indicate the group number, the number of electrons lost, and the ion (symbol and charge) for each of the following:

Element	Group Number	Electrons Lost	Ion Formed
Magnesium			
Sodium			
Calcium			
Potassium			
Aluminum			

Answers

Element	Group Number	Electrons Lost	Ion Formed
Magnesium	2A (2)	2	Mg^{2+}
Sodium	1A (1)	1	Na^+
Calcium	2A (2)	2	Ca^{2+}
Potassium	1A (1)	1	K^+
Aluminum	3A (13)	3	Al^{3+}

Study Note

The valence electrons are the electrons in the outermost energy level of an atom. For representative elements, you can determine the number of valence electrons by looking at the group number.

◆ Learning Exercise 4.1B

The following elements gain electrons when they form ions. Indicate the group number, the number of electrons gained, and the ion (symbol and charge) for each of the following:

Element	Group Number	Electrons Gained	Ion Formed
Chlorine			
Oxygen			
Nitrogen			
Fluorine			
Sulfur			

Answers

Element	Group Number	Electrons Gained	Ion Formed
Chlorine	7A (17)	1	Cl^-
Oxygen	6A (16)	2	O^{2-}
Nitrogen	5A (15)	3	N^{3-}
Fluorine	7A (17)	1	F^-
Sulfur	6A (16)	2	S^{2-}

4.2 Ionic Compounds

- In the formulas of ionic compounds, the total positive charge is equal to the total negative charge. For example, the compound magnesium chloride, $MgCl_2$, contains Mg^{2+} and $2\ Cl^-$. The sum of the charges is zero: $(2+) + 2(1-) = 0$.
- When two or more ions are needed for charge balance, that number is indicated by subscripts in the formula.

◆ Learning Exercise 4.2A

For this exercise, you may want to cut pieces of paper that represent typical positive and negative ions as shown below. To determine an ionic formula, place the pieces together with the positive ion(s) on the left. Add more positive or negative ions (squares or rectangles) to complete a geometric shape. Write the number of positive and negative ions as the subscripts for the formula.

Chapter 4

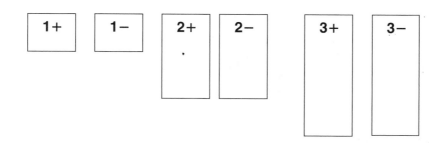

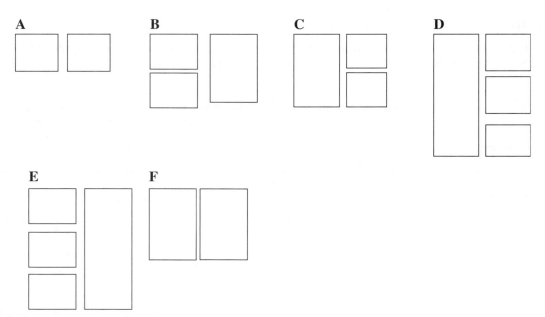

Give the letter (A, B, C, etc.) that matches the arrangement of ions in the following compounds:

Compound	Combination		Compound	Combination
1. $MgCl_2$	_____		5. K_3N	_____
2. Na_2S	_____		6. $AlBr_3$	_____
3. LiCl	_____		7. MgS	_____
4. CaO	_____		8. $BaCl_2$	_____

Answers
1. C	**2.** B	**3.** A	**4.** F
5. E	**6.** D	**7.** F	**8.** C

Study Note

You can check that the formula you write is electrically neutral by multiplying each of the ionic charges by their subscripts. When added together, their sum should equal zero. For example, the formula Na_2O gives $2(1+) + 1(2-) = (2+) + (2-) = 0$

◆ **Learning Exercise 4.2B**

Write the correct ionic formula for the compound formed from the following pairs of ions:

1. Na^+ and Cl^- _____

2. K^+ and S^{2-} _____

3. Al^{3+} and O^{2-} _____

4. Ba^{2+} and Cl^- _____

5. Ca^{2+} and S^{2-} _____

6. Al^{3+} and Cl^- _____

7. Li^+ and N^{3-} _____

8. Ba^{2+} and P^{3-} _____

Answers
1. NaCl	**2.** K_2S	**3.** Al_2O_3	**4.** $BaCl_2$
5. CaS	**6.** $AlCl_3$	**7.** Li_3N	**8.** Ba_3P_2

4.3 Naming and Writing Ionic Formulas

- In naming ionic compounds, the positive ion is named first, followed by the name of the negative ion. The name of a representative metal ion (Group 1A (1), 2A (2), or 3A (13)) is the same as its elemental name. The name of a nonmetal ion is obtained by replacing the end of its element name with *ide*.
- Most transition metals form cations with two or more ionic charges. Then the ionic charge must be written as a Roman numeral after the name of the metal. For example, the cations of iron, Fe^{2+} and Fe^{3+}, are named iron(II) and iron(III). The ions of copper are Cu^+, copper(I), and Cu^{2+}, copper(II).
- The only transition elements with fixed charges are zinc, Zn^{2+}, silver, Ag^+, and cadmium, Cd^{2+}.

◆ Learning Exercise 4.3A

Many of the transition metals form two or more ions with positive charge. Complete the table:

Name of ion	Symbol of ion		Name of ion	Symbol of ion
1. iron(III)	_____	**2.**	_____	Cu^{2+}
3. zinc	_____	**4.**	_____	Fe^{2+}
5. copper(I)	_____	**6.**	_____	Ag^+

Answers

1. Fe^{3+}		**2.** copper(II)		**3.** Zn^{2+}	
4. iron(II)		**5.** Cu^+		**6.** silver	

◆ Learning Exercise 4.3B

Write the ions and the correct ionic formula for the following ionic compounds:

Compound	Positive ion	Negative ion	Formula of compound
Aluminum sulfide			
Copper(II) chloride			
Magnesium oxide			
Iron(II) bromide			
Silver oxide			

Answers

Compound	Positive ion	Negative ion	Formula of compound
Aluminum sulfide	Al^{3+}	S^{2-}	Al_2S_3
Copper(II) chloride	Cu^{2+}	Cl^-	$CuCl_2$
Magnesium oxide	Mg^{2+}	O^{2-}	MgO
Iron(II) bromide	Fe^{2+}	Br^-	$FeBr_2$
Silver oxide	Ag^+	O^{2-}	Ag_2O

◆ Learning Exercise 4.3C

Write the name of each of the following ions:

1. Cl^- _____   **7.** S^{2-} _____

2. Fe^{2+} _____   **8.** Al^{3+} _____

3. Cu^+ _____   **9.** Fe^{3+} _____

4. Ag^+ _____   **10.** Ba^{2+} _____

5. O^{2-} _____   **11.** Cu^{2+} _____

6. Ca^{2+} _____   **12.** N^{3-} _____

Answers **1.** chloride **2.** iron(II) **3.** copper(I) **4.** silver **5.** oxide **6.** calcium
7. sulfide **8.** aluminum **9.** iron(III) **10.** barium **11.** copper(II) **12.** nitride

---

Study Note

The ionic charge of a metal that forms more than one positive ion is determined from the total negative charge in the formula. For example, in $FeCl_3$, the subscript 3 indicates 3 Cl^-, which gives a total of 3−. Therefore, the iron ion has an ionic charge of 3+ or Fe^{3+}, which is named iron(III).

◆ Learning Exercise 4.3D

Write the ions and a correct name for each of the following ionic compounds:

Formula	Ions		Name
1. $BaCl_2$	_____	_____	_____
2. $FeBr_3$	_____	_____	_____
3. Na_3P	_____	_____	_____
4. Al_2O_3	_____	_____	_____
5. CuO	_____	_____	_____
6. Mg_3N_2	_____	_____	_____

Answers **1.** Ba^{2+}, Cl^-, barium chloride   **2.** Fe^{3+}, Br^-, iron(III) bromide
3. Na^+, P^{3-}, sodium phosphide   **4.** Al^{3+}, O^{2-}, aluminum oxide
5. Cu^{2+}, O^{2-}, copper(II) oxide   **6.** Mg^{2+}, N^{3-}, magnesium nitride

4.4 Polyatomic Ions

- A polyatomic ion is a group of nonmetal atoms that carries an electrical charge, usually negative, 1−, 2−, or 3−.
- Polyatomic ions cannot exist alone, but are combined with an ion of the opposite charge.
- Ionic compounds containing polyatomic ions with three elements end with *-ate* or *-ite*.

Study Note

Learn the most common polyatomic ions: nitrate NO_3^-, carbonate CO_3^{2-}, sulfate SO_4^{2-}, and phosphate PO_4^{3-}. From each of these polyatomic ions, you can derive the related polyatomic ions. For example, the nitrite ion, NO_2^-, has one oxygen atom less than the nitrate, but the same charge.

◆ **Learning Exercise 4.4A**

Write the polyatomic ion (symbol and charge) for each of the following:

1. sulfate ion _____

2. hydroxide ion _____

3. carbonate ion _____

4. sulfite ion _____

5. ammonium ion _____

6. phosphate ion _____

7. nitrate ion _____

8. nitrite ion _____

Answers: 1. SO_4^{2-} 2. OH^- 3. CO_3^{2-} 4. SO_3^{2-}
5. NH_4^+ 6. PO_4^{3-} 7. NO_3^- 8. NO_2^-

◆ **Learning Exercise 4.4B**

Write the formula of each ion or polyatomic ion, and the correct formula for the following compounds:

Compound	Positive ion	Negative ion	Formula
Sodium phosphate			
Iron(II) hydroxide			
Ammonium carbonate			
Silver bicarbonate			
Iron(III) sulfate			
Iron(II) nitrate			
Potassium sulfite			
Barium phosphate			

Answers

Compound	Positive ion	Negative ion	Formula
Sodium phosphate	Na^+	PO_4^{3-}	Na_3PO_4
Iron(II) hydroxide	Fe^{2+}	OH^-	$Fe(OH)_2$
Ammonium carbonate	NH_4^+	CO_3^{2-}	$(NH_4)_2CO_3$
Silver bicarbonate	Ag^+	HCO_3^-	$AgHCO_3$
Iron(III) sulfate	Fe^{3+}	SO_4^{2-}	$Fe_2(SO_4)_3$
Iron(II) nitrate	Fe^{2+}	NO_3^-	$Fe(NO_3)_2$
Potassium sulfite	K^+	SO_3^{2-}	K_2SO_3
Barium phosphate	Ba^{2+}	PO_4^{3-}	$Ba_3(PO_4)_2$

4.5 Covalent Compounds

- In a covalent bond, atoms of nonmetals share electrons to achieve an octet. For example, oxygen with six valence electrons shares electrons with two hydrogen atoms to form the covalent compound water (H_2O).

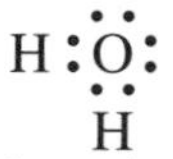

$$H:\overset{\displaystyle ..}{\underset{\displaystyle ..}{O}}:$$
$$H$$

- In a double bond, two pairs of electrons are shared between the same two atoms. In a triple bond, three pairs of electrons are shared.
- Covalent compounds are composed of nonmetals bonded together to give discrete units called molecules.
- The formula of a covalent compound is written using the symbol of the nonmetals in the name followed by subscripts given by the prefixes.

◆ **Learning Exercise 4.5A**

Write the electron dot formulas for the following covalent compounds (the central atom is underlined):

1. H_2 **3.** HCl   **5.** $H_2\underline{S}$

2. $\underline{N}Cl_3$ **4.** Cl_2 **6.** $\underline{C}Cl_4$

Answers **1.** H:H **2.**  :Cl:N:Cl: / :Cl: **3.**  H:Cl: **4.**  :Cl:Cl: **5.** 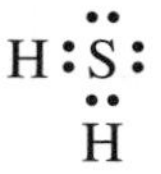 H:S: / H **6.** 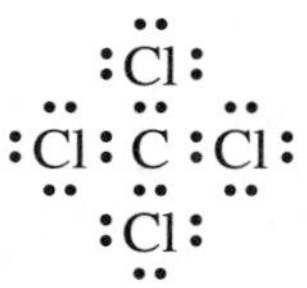 :Cl: / :Cl:C:Cl: / :Cl:

Study Note

Two nonmetals can form two or more different covalent compounds. In their names, prefixes are used to indicate the subscript in the formula. Some typical prefixes are mono (1), di (2), tri (3), tetra (4), and penta (5). The ending of the second nonmetal is changed to -*ide*.

◆ **Learning Exercise 4.5B**

Use the appropriate prefixes in naming the following covalent compounds:

1. CS_2 _____   **4.** SO_2 _____

2. CCl_4 _____   **5.** N_2O_4 _____

3. CO _____   **6.** PCl_3 _____

Answers   **1.** carbon disulfide   **2.** carbon tetrachloride   **3.** carbon monoxide
        4. sulfur dioxide   **5.** dinitrogen tetroxide   **6.** phosphorus trichloride

◆ Learning Exercise 4.5C

Write the formula of each of the following covalent compounds:

1. dinitrogen oxide _____
4. carbon dioxide _____

2. silicon tetrabromide _____
5. sulfur hexafluoride _____

3. nitrogen trichloride _____
6. oxygen difluoride _____

Answers   **1.** N_2O **2.** $SiBr_4$ **3.** NCl_3 **4.** CO_2 **5.** SF_6 **6.** OF_2

Summary of Writing Formulas and Names

- In both ionic and covalent compounds containing *two* different elements, the name of the element written first is named as the element. The ending of the name of the second element is replaced by "ide." For example, $BaCl_2$ is named *barium chloride*. If the metal is variable and forms two or more positive ions, a Roman numeral is added to its name to indicate the ionic charge in the compound. For example, $FeCl_3$ is named *iron(III) chloride*.
- In naming covalent compounds, a prefix before the name of an element indicates the numerical value of a subscript. For example, N_2O_3 is named *dinitrogen trioxide*.
- In ionic compounds with three or more elements, a group of atoms is named as a polyatomic ion. The names of negative polyatomic ions end in "ate" or "ite," except for hydroxide and cyanide. For example, Na_2SO_4 is named *sodium sulfate*. No prefixes are used.
- When a polyatomic ion occurs two or more times in a formula, its formula is placed inside parentheses, and the number of ions is shown as a subscript after the parentheses $Ca(NO_3)_2$.

◆ Learning Exercise 4.5D

Indicate the type of compound (ionic or covalent) formed from each pair of elements. If it is ionic, write the ions; if covalent, write the electron-dot formula of the molecule. Then give a formula and name for each.

Components	Ionic or covalent?	Ions or electron-dot formula	Formula	Name
Mg and Cl				
N and Cl				
K and SO_4				
Li and O				
C and Cl				
Na and PO_4				
H and S				
Ca and HCO_3				

Answers

Mg and Cl	ionic	Mg^{2+}, Cl^-	$MgCl_2$	Magnesium chloride
N and Cl	covalent	 $:\overset{..}{\underset{..}{Cl}}:N:\overset{..}{\underset{..}{Cl}}:$ $:\overset{..}{\underset{..}{Cl}}:$	NCl_3	Nitrogen trichloride
K and SO_4	ionic	K^+, SO_4^{2-}	K_2SO_4	Potassium sulfate
Li and O	ionic	Li^+, O^{2-}	Li_2O	Lithium oxide
C and Cl	covalent	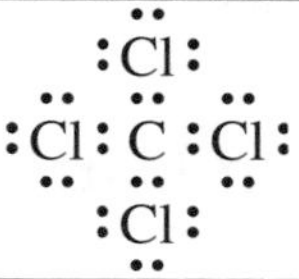 $:\overset{..}{\underset{..}{Cl}}:$ $:\overset{..}{\underset{..}{Cl}}:C:\overset{..}{\underset{..}{Cl}}:$ $:\overset{..}{\underset{..}{Cl}}:$	CCl_4	Carbon tetrachloride
Na and PO_4	ionic	Na^+, PO_4^{3-}	Na_3PO_4	Sodium phosphate
H and S	covalent	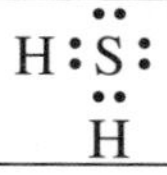 $H:\overset{..}{\underset{..}{S}}:$ H	H_2S	Dihydrogen sulfide
Ca and HCO_3	ionic	Ca^{2+}, HCO_3^-	$Ca(HCO_3)_2$	Calcium hydrogen carbonate (bicarbonate)

◆ Learning Exercise 4.5E

Write the formula of each ion or polyatomic ion, and the correct formula for each of the following compounds:

Compound	Positive ion	Negative ion	Formula
Sodium phosphate			
Iron(II) hydroxide			
Ammonium carbonate			
Silver bicarbonate			
Iron(III) sulfate			
Copper(II) nitrate			
Potassium sulfite			
Barium phosphate			

Answers

Compound	Positive ion	Negative ion	Formula
Sodium phosphate	Na^+	PO_4^{3-}	Na_3PO_4
Iron(II) hydroxide	Fe^{2+}	OH^-	$Fe(OH)_2$
Ammonium carbonate	NH_4^+	CO_3^{2-}	$(NH_4)_2CO_3$
Silver bicarbonate	Ag^+	HCO_3^-	$AgHCO_3$
Iron(III) sulfate	Fe^{3+}	SO_4^{2-}	$Fe_2(SO_4)_3$

Copper(II) nitrate	Cu^{2+}	NO_3^-	$Cu(NO_3)_2$
Potassium sulfite	K^+	SO_3^{2-}	K_2SO_3
Barium phosphate	Ba^{2+}	PO_4^{3-}	$Ba_3(PO_4)_2$

4.6 Electronegativity and Bond Polarity

- Electronegativity indicates the ability of an atom to attract electrons. In general, metals have low electronegativity values and nonmetals have high values.
- When atoms sharing electrons have the same or similar electronegativity values, electrons are shared equally and the bond is *nonpolar covalent.*
- Electrons are shared unequally in *polar covalent* bonds because they are attracted to the more electronegative atom.
- An electronegativity difference of 0 to 0.4 indicates a nonpolar covalent bond, while a difference of 0.5 to 1.7 indicates a polar covalent bond.
- An electronegativity difference of 1.8 or greater indicates a bond that is ionic.

◆ Learning Exercise 4.6

Using the table of electronegativity values in the text, determine the following:

1. The electronegativity difference for each pair
2. The type of bonding as (I) ionic, (PC) polar covalent, or (NP) nonpolar covalent

Elements	Electronegativity difference	Bonding	Elements	Electronegativity difference	Bonding
a. H and O	_____	_____	**e.** H and Cl	_____	_____
b. N and S	_____	_____	**f.** Cl and Cl	_____	_____
c. Al and O	_____	_____	**g.** S and F	_____	_____
d. Li and F	_____	_____	**h.** H and C	_____	_____

Answers	**a.** 1.4, PC	**b.** 0.5, PC	**c.** 2.0, I	**d.** 3.0, I
	e. 0.9, PC	**f.** 0.0, NP	**g.** 1.5, PC	**h.** 0.4, NP

4.7 Shapes and Polarity of Molecules

- Valence-shell electron-pair repulsion (VSEPR) theory predicts the geometry of a molecule by placing the electron pairs around a central atom as far apart as possible.
- A tetrahedral molecule has a central atom bonded to four atoms and no lone pairs. In a pyramidal molecule, a central atom is bonded to three atoms and one lone pair. In a bent molecule, a central atom is bonded to two atoms and two lone pairs.
- A polar bond with its charge separation is called a dipole.
- Molecules are nonpolar when the dipoles are in a symmetrical arrangement.
- In polar molecules, the dipoles do not cancel each other.

Study Note

Guidelines for predicting the shape of molecules:
1. Write the electron-dot formula.
2. Use VSEPR theory to predict the arrangement of the electron groups around the central atom.
3. Identify the shape of the molecule from the number of bonded atoms and electron pairs around the central atom.

◆ Learning Exercise 4.7A

Match the shape of a molecule with the following descriptions of the electron pairs around the central atoms and the number of bonded atoms.

      a. tetrahedral   **b.** pyramidal   **c.** bent

1. four electron groups with three bonded atoms _____

2. four electron groups with four bonded atoms _____

3. four electron groups with two bonded atoms _____

Answers   **1.** b   **2.** a   **3.** c

◆ Learning Exercise 4.7B

For each of the following, write the electron-dot formula, state the number of electron groups and bonded atoms, and predict the shape of the molecule.

Molecule or ion	Electron-dot formula	Number of electron groups	Number of bonded atoms	Shape
CCl_4				
NCl_3				
H_2S				

Answers

Molecule or ion	Electron-dot formula	Number of electron groups	Number of bonded atoms	Shape
CCl_4	:Cl: :Cl: C :Cl: :Cl:	4	4	Tetrahedral
NCl_3	:Cl: N :Cl: :Cl:	4	3	Pyramidal
H_2S	H:S: H	4	2	Bent

◆ Learning Exercise 4.7C

Write the symbols δ^+ and δ^- over the atoms in polar bonds.

1. H—O   **2.** N—N   **3.** C—Cl

4. O—F   **5.** N—F   **6.** Cl—Cl

Answers

1. $\overset{\delta^+\ \ \delta^-}{\text{H—O}}$ **2.** $\overset{\text{nonpolar}}{\text{N—N}}$ **3.** $\overset{\delta^+\ \ \delta^-}{\text{C—Cl}}$

4. $\overset{\delta^+\ \ \delta^-}{\text{O—F}}$ **5.** $\overset{\delta^+\ \ \delta^-}{\text{N—F}}$ **6.** $\overset{\text{nonpolar}}{\text{Cl—Cl}}$

◆ Learning Exercise 4.7D

Determine whether each of the following is polar or nonpolar:

1. CF_4 **2.** HCl

3. NH_3 **4.** OF_2

Answers

1. In the tetrahedral CF_4 molecule, all C—F dipoles cancel; CF_4 is a nonpolar molecule.
2. In HCl, H has a partial positive charge and Cl has a partial negative charge; HCl is a polar molecule.
3. In the pyramidal NH_3 molecule, the N—H dipoles do not cancel; NH_3 is a polar molecule.
4. In the bent OF_2 molecule, the O—F dipoles do not cancel; OF_2 is a polar molecule.

4.8 Attractive Forces in Compounds

- The interactions between particles in compounds determine their melting and boiling points.
- Ionic solids have high melting points due to strong ionic attractions between positive and negative ions.
- In polar substances, dipole–dipole attractions occur between the positive end of one molecule and the negative end of another.
- Hydrogen bonding, a type of dipole–dipole interaction, occurs between partially positive hydrogen atoms and strongly electronegative atoms of fluorine, oxygen, or nitrogen.
- Dispersion forces occur when temporary dipoles form within the nonpolar molecules, causing momentary attractions to other nonpolar molecules.

◆ Learning Exercise 4.8A

Indicate the major type of interactive force that occurs in each of the following substances:

a. ionic   **b.** dipole–dipole interaction   **c.** hydrogen bond   **d.** dispersion forces

1. _____ KCl   **4.** _____ Cl_2 **7.** _____ C_4H_{10}
2. _____ NCl_3 **5.** _____ HF   **8.** _____ Na_2O
3. _____ SBr_2 **6.** _____ H_2O

Answers   **1.** A   **2.** B   **3.** B   **4.** D
                  5. C   **6.** C   **7.** D   **8.** A

◆ **Learning Exercise 4.8B**

Identify the substance that would have the higher boiling point in each pair:

1. NaCl or HCl _____   4. CH_4 or CF_4 _____

2. Br_2 or HBr _____   5. $MgCl_2$ or OCl_2 _____

3. H_2O or H_2S _____   6. NH_3 or PH_3 _____

Answers
1. NaCl; An ionic compound has a higher boiling point than a compound (HCl) with dipole–dipole attractions.
2. HBr; A compound with dipole–dipole attractions has a higher boiling point than a compound (Br_2) with dispersion forces.
3. H_2O; A compound with hydrogen bonds has a higher boiling point than a compound (H_2S) with dipole–dipole attractions.
4. CF_4; A compound with dipole–dipole attractions has a higher boiling point than a compound (CH_4) with dispersion forces.
5. $MgCl_2$; An ionic compound has a higher boiling point than a compound (OCl_2) with dipole–dipole attractions.
6. NH_3; A compound with hydrogen bonds has a higher boiling point than a compound (PH_3) with dipole–dipole attractions.

◆ **Checklist for Chapter 4**

You are ready to take the self-test for Chapter 4. Be sure that you have accomplished the following learning goals for this chapter. If you are not sure, review the section listed at the end of the goal. Then apply your new skills and understanding to the practice test.
After studying Chapter 4, I can successfully:

_____ Illustrate the octet rule for the formation of ions (4.1).

_____ Write the formulas of compounds containing the ions of metals and nonmetals of representative elements (4.2).

_____ Use charge balance to write an ionic formula (4.2).

_____ Write the name of an ionic compound (4.3).

_____ Write the formula of a compound containing a polyatomic ion (4.4).

_____ Write the electron-dot formula of a covalent compound (4.5).

_____ Write the name and formula of a covalent compound (4.5).

_____ Classify a bond as nonpolar covalent, polar covalent, or ionic (4.6).

_____ Use VSEPR theory to determine the shape of a molecule (4.7).

_____ Classify a molecule as a polar or nonpolar molecule (4.7).

_____ Identify the attractive forces between particles (4.8).

Practice Test for Chapter 4

For questions 1 through 4, consider an atom of phosphorus.

1. It is in Group
   A. 2A (2)   **B.** 3A (13)   **C.** 5A (15)   **D.** 7A (17)   **E.** 8A (18)

2. How many valence electrons does it have?
    A. 2   **B.** 3   **C.** 5   **D.** 8   **E.** 15

3. To achieve an octet, the phosphorus atom will
    A. lose 1 electron   **B.** lose 2 electrons   **C.** lose 5 electrons
    D. gain 2 electrons   **E.** gain 3 electrons

4. As an ion, it has an ionic charge (valence) of
    A. 1+   **B.** 2+   **C.** 5+   **D.** 2−   **E.** 3−

5. To achieve an octet, a calcium atom
    A. loses 1 electron   **B.** loses 2 electrons   **C.** loses 3 electrons
    D. gains 1 electron   **E.** gains 2 electrons

6. To achieve an octet, a chlorine atom
    A. loses 1 electron   **B.** loses 2 electrons   **C.** loses 3 electrons
    D. gains 1 electron   **E.** gains 2 electrons

7. Another name for a positive ion is
    A. anion   **B.** cation   **C.** proton   **D.** positron   **E.** sodium

8. The correct ionic charge (valence) for a calcium ion is
    A. 1+   **B.** 2+   **C.** 1−   **D.** 2−   **E.** 3−

9. The silver ion has a charge of
    A. 1+   **B.** 2+   **C.** 1−   **D.** 2−   **E.** 3−

10. The correct ionic charge (valence) for a phosphate ion is
    A. 1+   **B.** 2+   **C.** 1−   **D.** 2−   **E.** 3−

11. The correct ionic charge (valence) for a fluoride is
    A. 1+   **B.** 2+   **C.** 1−   **D.** 2−   **E.** 3−

12. The correct ionic charge (valence) for a sulfate ion is
    A. 1+   **B.** 2+   **C.** 1−   **D.** 2−   **E.** 3−

13. When the elements magnesium and sulfur react,
    A. an ionic compound forms.
    B. a nonpolar covalent compound forms.
    C. no reaction occurs.
    D. the two repel each other and won't combine.
    E. none of the above.

14. An ionic bond typically occurs between
    A. two different nonmetals.
    B. two of the same type of nonmetals.
    C. two noble gases.
    D. two different metals.
    E. a metal and a nonmetal.

15. A nonpolar covalent bond typically occurs between
    A. two different nonmetals.
    B. two of the same type of nonmetals.
    C. two noble gases.
    D. two different metals.
    E. a metal and a nonmetal.

16. A polar covalent bond typically occurs between
    A. two different nonmetals.
    B. two of the same type of nonmetals.
    C. two noble gases.
    D. two different metals.
    E. a metal and a nonmetal.

17. The formula for a compound between carbon and chlorine is
    A. Cl   **B.** CCl_2 **C.** C_4Cl **D.** CCl_4 **E.** C_4Cl_2

18. The formula for a compound between sodium and sulfur is
    A. SoS   **B.** NaS   **C.** Na_2S **D.** NaS_2 **E.** Na_2SO_4

19. The formula for a compound between aluminum and oxygen is
    A. AlO   **B.** Al_2O **C.** AlO_3 **D.** Al_2O_3 **E.** Al_3O_2

20. The formula for a compound between barium and sulfur is
    A. BaS   **B.** BaS_2 **C.** BaS_2 **D.** Ba_2S_2 **E.** $BaSO_4$

21. The correct formula for iron(III) chloride is
    A. FeCl   **B.** $FeCl_2$ **C.** Fe_2Cl **D.** Fe_3Cl **E.** $FeCl_3$

22. The correct formula for ammonium sulfate is
    A. AmS   **B.** $AmSO_4$ **C.** $(NH_4)_2S$ **D.** NH_4SO_4 **E.** $(NH_4)_2SO_4$

23. The correct formula for copper(II) chloride is
    A. CoCl   **B.** CuCl   **C.** $CoCl_2$ **D.** $CuCl_2$ **E.** Cu_2Cl

24. The correct formula for lithium phosphate is
    A. $LiPO_4$ **B.** Li_2PO_4 **C.** Li_3PO_4 **D.** $Li_2(PO_4)_3$ **E.** $Li_3(PO_4)_2$

25. The correct formula for silver oxide is
    A. AgO   **B.** Ag_2O **C.** AgO_2 **D.** Ag_3O_2 **E.** Ag_3O

26. The correct formula for magnesium carbonate is
    A. $MgCO_3$ **B.** Mg_2CO_3 **C.** $Mg(CO_3)_2$ **D.** MgCO   **E.** $Mg_2(CO_3)_3$

27. The correct formula for copper(I) sulfate is
    A. $CuSO_3$ **B.** $CuSO_4$ **C.** Cu_2SO_3 **D.** $Cu(SO_4)_2$ **E.** Cu_2SO_4

28. The name of $AlPO_4$ is
    A. aluminum phosphide   **B.** alum phosphate   **C.** aluminum phosphate
    D. aluminum phosphorus oxide   **E.** aluminum phosphite

29. The name of CuS is
    A. copper sulfide   **B.** copper(I) sulfate   **C.** copper(I) sulfide
    D. cuprous sulfide   **E.** copper(II) sulfide

30. The name of $FeCl_2$ is
    A. ferric chloride   **B.** iron(II) chlorine   **C.** iron(II) chloride
    E. iron chlorine   **E.** iron(III) chloride

31. The name of $ZnCO_3$ is
    A. zinc(III) carbonate   **B.** zinc(II) carbonate   **C.** zinc bicarbonate
    D. zinc carbon trioxide   **E.** zinc carbonate

32. The name of Al_2O_3 is
    A. aluminum oxide   **B.** aluminum(II) oxide   **C.** aluminum trioxide
    D. dialuminum trioxide   **E.** aluminum oxygenate

33. The name of NCl_3 is
    A. nitrogen chloride   **B.** nitrogen trichloride   **C.** trinitrogen chloride
    D. nitrogen chlorine three   **E.** nitrogen chloride(III)

34. The name of CO is
    A. carbon monoxide   **B.** carbonic oxide   **C.** carbon oxide
    D. carbonious oxide   **E.** carboxide

For questions 35 through 40, indicate the type of bonding expected between the following elements:
    A. ionic   **B.** nonpolar covalent   **C.** polar covalent   **D.** none

35. _____ silicon and oxygen   **38.** _____ chlorine and chlorine

36. _____ barium and chlorine   **39.** _____ sulfur and oxygen

37. _____ aluminum and chlorine   **40.** _____ neon and oxygen

Determine the shape of each of the following molecules as

    A. tetrahedral   **B.** pyramidal   **C.** bent

    41. PCl_3 **42.** CBr_4 **43.** H_2S

For the compounds in questions 44 to 48, give the prevalent type of bonding using

    A. ionic bond   **B.** dipole–dipole   **C.** hydrogen bond   **D.** dispersion forces

    44. NCl_3 **45.** HF   **46.** LiCl

    47. H_2S **48.** N_2

Answers to the Practice Test

1. C	**2.** C	**3.** E	**4.** E	**5.** B
6. D	**7.** B	**8.** B	**9.** A	**10.** E
11. C	**12.** D	**13.** A	**14.** E	**15.** B
16. A	**17.** D	**18.** C	**19.** D	**20.** A
21. E	**22.** E	**23.** D	**24.** C	**25.** B
26. A	**27.** E	**28.** C	**29.** E	**30.** C
31. E	**32.** A	**33.** B	**34.** A	**35.** C
36. A	**37.** A	**38.** B	**39.** C	**40.** D
41. B	**42.** A	**43.** C	**44.** B	**45.** C
46. A	**47.** B	**48.** D		

Answers and Solutions to Selected Text Problems

4.1   Atoms with 1, 2, or 3 valence electrons lose those electrons to form ions.

    a. one   **b.** two   **c.** three   **d.** one   **e.** two

4.3   **a.** Li^+ **b.** F^- **c.** Mg^{2+} **d.** Fe^{3+} **e.** Zn^{2+}

4.5   **a.** Cl^- **b.** K^+ **c.** O^{2-} **d.** Al^{3+}

4.7   **a.** K with an electron in energy level 4 is larger than K^+.

    b. Cl^- with an added valence electron is larger than Cl.

    c. Ca with two more electrons in energy level 4 is larger than Ca^{2+}.

    d. K^+ with an octet in energy level 3 is larger than Li^+ with electrons in energy level 1.

4.9   **a.** (Li and Cl) and **c.** (K and O) form ionic compounds

4.11   **a.** Na_2O **b.** $AlBr_3$ **c.** BaO   **d.** $MgCl_2$ **e.** Al_2S_3

4.13   **a.** Na^+, S^{2-}, Na_2S **b.** K^+, N^{3-}, K_3N **c.** Al^{3+}, I^-, AlI_3 **d.** Li^+, O^{2-}, Li_2O

4.15   **a.** aluminum oxide   **b.** calcium chloride   **c.** sodium oxide

    d. magnesium nitride   **e.** potassium iodide

4.17   The Roman numeral is used to specify the positive charge on the transition metal in the compound. It is necessary for most transition metal compounds because many transition metals can exist as more than one cation; transition metals have variable ionic charges.

4.19   **a.** iron(II)   **b.** copper(II)   **c.** zinc

    d. lead(IV)   **e.** chromium(III)

4.21 a. tin(II) chloride   b. potassium oxide   c. copper(I) sulfide
    d. copper(II) sulfide   e. chromium(III) bromide   f. zinc chloride

4.23 a. Au^{3+} b. Fe^{3+} c. Pb^{4+} d. Sn^{2+}

4.25 a. $MgCl_2$ b. Na_2S c. Cu_2O d. Zn_3P_2
    e. AuN f. $CrCl_2$

4.27 a. HCO_3^- b. NH_4^+ c. PO_4^{3-} d. HSO_4^-

4.29 a. sulfate   b. carbonate   c. phosphate   d. nitrate

4.31

	OH^-	NO_2^-	CO_3^{2-}	HSO_4^-	PO_4^{3-}
Li^+	$LiOH$	$LiNO_2$	Li_2CO_3	$LiHSO_4$	Li_3PO_4
Cu^{2+}	$Cu(OH)_2$	$Cu(NO_2)_2$	$CuCO_3$	$Cu(HSO_4)_2$	$Cu_3(PO_4)_2$
Ba^{2+}	$Ba(OH)_2$	$Ba(NO_2)_2$	$BaCO_3$	$Ba(HSO_4)_2$	$Ba_3(PO_4)_2$

4.33 a. CO_3^{2-}, sodium carbonate   b. NH_4^+, ammonium chloride
    c. PO_4^{3-}, lithium phosphate   d. NO_2^-, copper(II) nitrite
    e. SO_3^{2-}, iron(II) sulfite

4.35 a. $Ba(OH)_2$ b. Na_2SO_4 c. $Fe(NO_3)_2$ d. $Zn_3(PO_4)_2$ e. $Fe_2(CO_3)_3$

4.37 a. :Br̈:Br̈:   b. H:H   c. H:F̈:   d. :F̈:Ö: :F̈:

4.39 a. phosphorus tribromide   b. carbon tetrabromide   c. silicon dioxide
    d. hydrogen fluoride

4.41 a. dinitrogen trioxide   b. nitrogen trichloride   c. silicon tetrabromide
    d. phosphorus pentachloride

4.43 a. CCl_4 b. CO c. PCl_3 d. N_2O_4

4.45 a. OF_2 b. BF_3 c. N_2O_3 d. SF_6

4.47 a. This is an ionic compound with Al^{3+} ion and the sulfate SO_4^{2-} polyatomic ion. The correct name is aluminum sulfate.
    b. This is an ionic compound with Ca^{2+} ion and the carbonate CO_3^{2-} polyatomic ion. The correct name is calcium carbonate.
    c. This is a covalent compound because it contains two nonmetals. Using prefixes, it is named dinitrogen oxide.
    d. This is an ionic compound with sodium ion Na^+ and the PO_4^{3-} polyatomic ion. The correct name is sodium phosphate.
    e. This ionic compound contains two polyatomic ions ammonium NH_4^+ and sulfate SO_4^{2-}. It is named ammonium sulfate.
    f. This is an ionic compound containing the variable metal ion Fe^{3+} and oxide ion O^{2-}. It is named using the Roman numeral as iron(III) oxide.

4.49 The electronegativity increases going across a period.

4.51 a. $K < Na < Li$ b. $Na < P < Cl$ c. $Ca < Se < O$

4.53 a. Si—Br electronegativity difference 1.0, polar covalent
    b. Li—F electronegativity difference 3.0, ionic
    c. Br—F electronegativity difference 1.2, polar covalent
    d. Br—Br electronegativity difference 0, nonpolar covalent
    e. N—P electronegativity difference 0.9, polar covalent
    f. C—O electronegativity difference 1.0, polar covalent

4.55 **a.** $\overset{\delta+ \quad \delta-}{N-F}$ **b.** $\overset{\delta+ \quad \delta-}{Si-Br}$ **c.** $\overset{\delta+ \quad \delta-}{C-O}$ **d.** $\overset{\delta+ \quad \delta-}{P-Br}$ **e.** $\overset{\delta+ \quad \delta-}{B-Cl}$

4.57 Tetrahedral. Four atoms bonded to the central atom form a tetrahedron.

4.59 The four electron groups in PCl_3 have a tetrahedral arrangement, but three bonded atoms around a central atom give a pyramidal shape.

4.61 In the electron-dot formula of PH_3, the central atom P has three bonded atoms and one lone pair, which give PH_3 a pyramidal shape. In the molecule NH_3, the central atom N is bonded to three atoms and one lone pair. The structure of the electron groups is tetrahedral, which gives NH_3 a pyramidal shape.

4.63 **a.** The central oxygen atom has four electron pairs with two bonded to fluorine atoms. Its shape is bent.
    b. The central atom C is bonded to four chlorine atoms; CCl_4 has a tetrahedral shape.

4.65 Cl_2 is a nonpolar molecule because there is a nonpolar covalent bond between Cl atoms, which have identical electronegativity values. In HCl, the bond is a polar covalent bond, which is a dipole and makes HCl a polar molecule.

4.67 **a.** polar   **b.** dipoles do not cancel; polar   **c.** four dipoles cancel; nonpolar

4.69 **a.** dipole–dipole attraction   **b.** ionic   **c.** hydrogen bond
    d. dispersion forces

4.71 **a.** hydrogen bond   **b.** dispersion forces   **c.** dipole–dipole attraction
    d. dipole–dipole attraction

4.73 **a.** By losing one valence electron from the third energy level, sodium achieves an octet in the second energy level.
    b. The sodium ion Na^+ has the same electron arrangement as Ne (2, 8).
    c. Group 1A (1) and 2A (2) elements acquire octets by losing electrons to form compounds. Group 8A (18) elements are stable with octets (or two electrons for helium).

4.75 **a.** P^{3-} ion   **b.** O atom   **c.** Zn^{2+} ion   **d.** Fe^{3+} ion

4.77 **a.** 2–pyramidal, polar   **b.** 1–bent, polar   **c.** 3–tetrahedral, nonpolar

4.79 **1.** H (E)   **2.** Li (C)   **3.** Li^+ (A)   **4.** H^+ (B)   **5.** N^{3-} (D)

4.81 **a.** 2, 8   **b.** 2, 8   **c.** 2, 8, 8   **d.** 2, 8   **e.** 2

4.83 **a.** An element that loses two electrons is in Group 2A (2)
    b. The atom with two valence electrons has two dots •X•
    c. Magnesium (Mg) is the Group 2A (2) element in Period 3.
    d. X^{2+} and N^{3-} form a compound with the formula X_3N_2.

4.85 **a.** Sn^{4+}
    b. 50 protons and 46 electrons
    c. Sn^{4+} and O^{2-} give a compound with the formula SnO_2
    d. Sn^{4+} and PO_4^{3-} give a compound with the formula $Sn_3(PO_4)_4$

4.87 **a.** Ions: Au^{3+} and $Cl^- \rightarrow AuCl_3$ **b.** Ions: Pb^{4+} and $O^{2-} \rightarrow PbO_2$
    c. Ions: Ag^+ and $Cl^- \rightarrow AgCl$ **d.** Ions: Ca^{2+} and $N^{3-} \rightarrow Ca_3N_2$
    e. Ions: Cu^+ and $P^{3-} \rightarrow Cu_3P$ **f.** Ions: Cr^{2+} and $Cl^- \rightarrow CrCl_2$

4.89 **a.** 1 N and 3 Cl $\rightarrow$ nitrogen trichloride   **b.** 1 S and 2 Cl $\rightarrow$ sulfur dichloride
    c. 2 N and 1 O $\rightarrow$ dinitrogen monoxide   **d.** 2 F $\rightarrow$ fluorine (named as the element)
    e. 1 P and 5 Cl $\rightarrow$ phosphorus pentachloride   **f.** 2 P and 5 O $\rightarrow$ diphosphorus pentoxide

4.91 **a.** 1 C and 1 O $\rightarrow$ CO   **b.** di (2) and penta (5) $\rightarrow P_2O_5$
    c. di (2) and 1 S $\rightarrow H_2S$ **d.** 1 S and di (2) Cl $\rightarrow SCl_2$

4.93   **a.** ionic, iron(III) chloride
    b. ionic, sodium sulfate
    c. covalent, 2 N and 1 O $\rightarrow$ dinitrogen oxide
    d. covalent, fluorine (named as the element)
    e. covalent, 1 P and 5 Cl $\rightarrow$ phosphorus pentachloride
    f. covalent, 1 C and 4 F $\rightarrow$ carbon tetrafluoride

4.95   **a.** Tin(II) is Sn^{2+}; carbonate is CO_3^{2-}. With charges balanced the formula is $SnCO_3$.
    b. Lithium is Li^+; phosphide is P^{3-}. Using three Li^+ for charge balance, the formula is Li_3P.
    c. Silicon has 4 valence electrons to share with four chlorine atoms to give $SiCl_4$.
    d. Iron(III) is Fe^{3+}; sulfide is S^{2-}. Charge is balanced with two Fe^{3+} and three S^{2-} to write the formula Fe_2S_3.
    e. Bromine Br_2 is a diatomic element of two bromine atoms each with seven valence electrons.
    f. Calcium is Ca^{2+}; bromide is Br^-. With charges balanced the formula is $CaBr_2$.

4.97   Determine the difference in electronegativity values:
    a. C—O (1.0)   C—N is less (0.5)   **b.** N—F (1.0)   N—Br is less (0.2)
    c. S—Cl (0.5)   Br—Cl is less (0.2)   **d.** Br—I (0.3)   Br—Cl is less (0.2)
    e. N—F (1.0)   N—O is less (0.5)

4.99   **a.** polar covalent (Cl 3.0 − Si 1.8 = 1.2)   **b.** nonpolar covalent (C 2.5 − C 2.5 = 0.0)
    c. ionic (Cl 3.0 − Na 0.9 = 2.1)   **d.** nonpolar covalent (C 2.5 − H 2.1 = 0.4)
    e. nonpolar covalent (F 4.0 − F 4.0 = 0.0)

4.101   **a.** Polar. A molecule with three bonded atoms with a lone pair is polar.
    b. Polar.
    c. Nonpolar. There are four equal Si-F bonds (four dipoles) in opposing directions in a tetrahedron. Thus, dipoles cancel and the molecule is nonpolar.

4.103   **a.** pyramidal, dipoles do not cancel, polar   **b.** bent, dipoles do not cancel, polar

4.105   **a.** bent, dipoles do not cancel, polar   **b.** pyramidal, dipoles do not cancel, polar

4.107   **a.** NH_3 (3) hydrogen bonds   **b.** HI (2) dipole–dipole interactions
    c. Br_2 (4) dispersion forces   **d.** Cs_2O (1) ionic bonds

4.109

Atom or Ion	Number of Protons	Number of Electrons	Electrons Lost/Gained
K^+	$19p^+$	$18e^-$	$1e^-$ lost
Mg^{2+}	$12p^+$	$10e^-$	$2e^-$ lost
O^{2-}	$8p^+$	$10e^-$	$2e^-$ gained
Al^{3+}	$13p^+$	$10e^-$	$3e^-$ lost

4.111   **a.** X = Group 1A(1); Y = Group 6A(16)   **b.** ionic   **c.** X^+ and Y^{2-} **d.** X_2Y
    e. X_2S **f.** YCl_2 **g.** covalent

4.113   **a.** Group 3A(13)   **b.** Group 6A(16)   **c.** Group 2A(2)

4.115   **a.** ionic; lithium oxide
    b. covalent; dinitrogen oxide
    c. covalent; carbon tetrafluoride
    d. ionic; chromium(II) nitrate
    e. ionic; magnesium bicarbonate or magnesium hydrogen carbonate
    f. covalent; nitrogen trifluoride
    g. ionic; calcium chloride
    h. ionic; potassium phosphate
    i. ionic; gold(III) sulfite
    j. covalent; iodine

Chemical Quantities and Reactions

Study Goals

- Determine the molar mass of a compound from its formula.
- Use the molar mass to convert between the grams of a substance and the number of moles.
- Classify a change in matter as a chemical change or a physical change.
- Show that a balanced equation has an equal number of atoms of each element on the reactant side and the product side.
- Write a balanced equation for a chemical reaction when given the formulas of the reactants and products.
- Classify an equation as a combination, decomposition, replacement, and/or combustion reaction.
- Describe the features of oxidation and reduction in an oxidation–reduction reaction.
- Using a given number of moles and a mole–mole conversion factor, determine the corresponding number of moles for a reactant or a product.
- Using a given mass of a substance in a reaction and the appropriate mole factor and molar masses, calculate the mass of a reactant or a product.

Think About It

1. What causes a slice of apple or an avocado to turn brown?

2. How is a recipe like a chemical equation?

3. Why is the digestion of food a series of chemical reactions?

Key Terms

Match the following terms with the statements below:

    a. chemical change   **b.** chemical equation   **c.** combination reaction   **d.** mole
    e. molar mass   **f.** physical change

1. _____ The amount of a substance that contains 6.02×10^{23} particles of that substance.

2. _____ A change that alters the composition of a substance producing a new substance with new properties.

3. _____ The mass in grams of an element or compound that is equal numerically to its atomic mass or sum of atomic masses.

4. _____ The type of reaction in which reactants combine to form a single product.

5. _____ A shorthand method of writing a chemical reaction with the formulas of the reactants written on the left side of an arrow and the formulas of the products on the right side.

Answers   **1.** d   **2.** a   **3.** e   **4.** c   **5.** b

5.1 The Mole

- A mole of any element contains Avogadro's number (6.02×10^{23}) of atoms; a mole of any compound contains 6.02×10^{23} molecules or formula units.
- The subscripts in a formula indicate the number of moles of each element in one mole of the compound.

◆ Learning Exercise 5.1A

Calculate each of the following:

a. number of P atoms in 1.50 moles of P

b. number of H_2S molecules in 0.0750 mole of H_2S

c. moles of Ag in 5.4×10^{24} atoms of Ag

d. moles of C_3H_8 in 8.25×10^{24} molecules of C_3H_8

Answers   **a.** 9.03×10^{23} P atoms   **b.** 4.52×10^{22} H_2S molecules
    c. 9.0 mole of Ag   **d.** 13.7 moles of C_3H_8

◆ Learning Exercise 5.1B

Consider the formula for vitamin C (ascorbic acid), $C_6H_8O_6$.

a. How many moles of carbon are in 2.0 moles of vitamin C?

b. How many moles of hydrogen are in 5.0 moles of vitamin C?

c. How many moles of oxygen are in 1.5 moles of vitamin C?

Answers   **a.** 12 moles of carbon (C)   **b.** 40. moles of hydrogen (H)
    c. 9.0 moles of oxygen (O)

◆ Learning Exercise 5.1C

For the compound ibuprofen ($C_{13}H_{18}O_2$) used in Advil™ AND Motrin™, determine the moles of each of the following:

a. Moles of carbon (C) atoms in 2.20 moles of ibuprofen.

b. Moles of hydrogen (H) in 0.5 mole of ibuprofen.

c. Moles of oxygen (O) in 0.75 mole of ibuprofen.

d. Moles of ibuprofen that contain 36 moles of hydrogen (H).

Answers   **a.** 28.6 moles of C   **b.** 9 moles of H
    c. 1.5 moles of O   **d.** 2.0 moles of ibuprofen

5.2 Molar Mass

- The molar mass (g/mole) of an element is numerically equal to its atomic mass in grams.
- The molar mass (g/mole) of a compound is the mass in grams equal numerically to the sum of the mass for each element in the formula. $MgCl_2$ has a molar mass that is the sum of the mass of 1 mole of Mg (24.3 g) and 2 moles of Cl (2 × 35.5 g) = 95.3 g/mole.
- The molar mass is useful as a conversion factor to change a given quantity in moles to grams.

$$\text{Moles of substance} \times \frac{\text{number of grams}}{1 \text{ moles of substance}} = \text{grams}$$

Study Note

The molar mass of an element or compound is determined as follows:
1. Determine the moles of each element (from subscripts) in the compound.
2. Calculate the total mass contributed by each element.
3. Total the masses of all the elements.

Example: What is the molar mass of silver nitrate, $AgNO_3$?

$$1 \text{ mole Ag} \times 107.9 \text{ g/mole} = 107.9 \text{ g}$$
$$1 \text{ mole N} \times 14.0 \text{ g/mole} = 14.0 \text{ g}$$
$$3 \text{ moles O} \times 16.0 \text{ g/mole} = \underline{48.0 \text{ g}}$$
$$\text{molar mass } AgNO_3 = 169.9 \text{ g}$$

◆ **Learning Exercise 5.2A**

Determine the molar mass for each of the following:

   a. K_2O 　　　　　　　　　　**b.** $AlCl_3$

   c. $C_{13}H_{18}O_2$ ibuprofen 　　　　**d.** C_4H_{10}

   e. $Ca(NO_3)_2$ 　　　　　　　　**f.** Mg_3N_2

   g. $FeCO_3$ 　　　　　　　　　　**h.** $(NH_4)_3PO_4$

Answers　　　**a.** 94.2 g　　**b.** 133.5 g　　**c.** 206 g　　**d.** 58.0 g
　　　　　　　　e. 164.1 g　　**f.** 100.9 g　　**g.** 115.9 g　　**h.** 149.0 g

---

Study Note

Use the molar mass as a conversion factor to change the number of moles of a substance to its mass in grams. Find the mass in grams of 0.25 mole of Na_2CO_3.

Molar mass
Grams $\leftrightarrow$ Moles

Example: $0.25 \text{ mole Na}_2\text{CO}_3 \times \dfrac{106.0 \text{ g Na}_2\text{CO}_3}{1 \text{ mole Na}_2\text{CO}_3} = 27 \text{ g of Na}_2\text{CO}_3$

◆ Learning Exercise 5.2B

Find the number of grams in each of the following quantities:

    a. 0.100 mole of SO_2 **b.** 0.100 mole of H_2SO_4

    c. 2.50 moles of NH_3 **d.** 1.25 moles of O_2

    e. 0.500 mole of Mg   **f.** 5.00 moles of H_2

    g. 10.0 moles of PCl_3 **h.** 0.400 mole of S

Answers   **a.** 6.41 g   **b.** 9.81 g   **c.** 42.5 g   **d.** 40.0 g
                        e. 12.2 g   **f.** 10.0 g   **g.** 1380 g   **h.** 12.8 g

---

Study Note

When the grams of a substance are given, the molar mass is used to calculate the number of moles of substance present.

$$\text{grams of substance} \times \frac{1 \text{ mole}}{\text{grams of substance}} = \text{moles}$$

Example: How many moles of NaOH are in 4.0 g of NaOH?

$$4.0 \text{ g NaOH} \times \frac{1 \text{ mole NaOH}}{40.0 \text{ g NaOH}} = 0.10 \text{ mole of NaOH}$$

Molar Mass (inverted)

◆ **Learning Exercise 5.2C**

Calculate the number of moles in each of the following quantities:

    a. 32.0 g of CH_4 **b.** 391 g of K

    c. 8.00 g of C_3H_8 **d.** 25.0 g of Cl_2

    e. 0.220 g of CO_2 **f.** 5.00 g of Al_2O_3

    g. The methane burned in a gas heater has a formula of CH_4. If 725 grams of methane are used in 1 month, how many moles of methane were burned?

    h. There is 18 mg of iron in a vitamin tablet. If there are 100 tablets in a bottle, how many moles of iron are contained in the vitamins in the bottle?

Answers	**a.** 2.00 moles	**b.** 10.0 moles	**c.** 0.182 mole
	d. 0.352 mole	**e.** 0.00500 mole	**f.** 0.0490 mole
	g. 45.3 moles	**h.** 0.032 mole	

5.3 Chemical Changes

- A chemical change occurs when the atoms of the initial substances rearrange to form new substances.
- Chemical change is indicated by a change in properties of the reactants. For example, a rusting nail, souring milk, and a burning match are all chemical changes.
- When new substances form, a chemical reaction has taken place.

◆ **Learning Exercise 5.3**

Identify each of the following as a chemical (C) or a physical (P) change:

    1. _____ tearing a piece of paper   **5.** _____ dissolving salt in water

    2. _____ burning paper   **6.** _____ boiling water

    3. _____ rusting iron   **7.** _____ chewing gum

    4. _____ digestion of food   **8.** _____ removing tarnish with silver polish

Answers   **1.** P   **2.** C   **3.** C   **4.** C   **5.** P   **6.** P   **7.** P   **8.** C

5.4 Chemical Equations

- A chemical equation shows the formulas of the reactants on the left side of the arrow and the formulas of the products on the right side.
- In a balanced equation, *coefficients* in front of the formulas provide the same number of atoms for each kind of element on the reactant and product sides.
- A chemical equation is balanced by placing coefficients in front of the symbols or formulas in the equation.

Example: Balance the following equation:

$$N_2(g) + H_2(g) \rightarrow NH_3(g)$$

1. Count the atoms of N and H on the reactant side and on the product side.

$$N_2(g) + H_2(g) \rightarrow NH_3(g)$$
$$2N, 2H \qquad\qquad 1N, 3H$$

2. Balance the N atoms by placing a coefficient of 2 in front of NH_3. (This increases the H atoms too.) Recheck the number of N atoms and the number of H atoms.

$$N_2(g) + H_2(g) \rightarrow 2NH_3(g)$$

3. Balance the H atoms by placing a coefficient of 3 in front of H_2. Recheck the number of N atoms and the number of H atoms.

$$N_2(g) + 3H_2(g) \rightarrow 2NH_3(g)$$
$$2N, 6H \qquad\qquad 2\,N, 6H$$

The equation is balanced.

◆ Learning Exercise 5.4A

State the number of atoms of each element on the reactant side and on the product side for each of the following balanced equations:

a. $CaCO_3(s) \rightarrow CaO(g) + CO_2(g)$

Element	Atoms on reactant side	Atoms on product side
Ca		
C		
O		

b. $2Na(s) + H_2O(l) \rightarrow Na_2O(s) + H_2(g)$

Element	Atoms on reactant side	Atoms on product side
Na		
H		
O		

c. $C_5H_{12}(g) + 8O_2(g) \rightarrow 5CO_2(g) + 6H_2O(g)$

Element	Atoms on reactant side	Atoms on product side
C		
H		
O		

d. $2AgNO_3(aq) + K_2S(aq) \rightarrow 2KNO_3(aq) + Ag_2S(s)$

Element	Atoms on reactant side	Atoms on product side
Ag		
N		
O		
K		
S		

e. $2Al(OH)_3(aq) + 3H_2SO_4(aq) \rightarrow Al_2(SO_4)_3(s) + 6H_2O(l)$

Element	Atoms on reactant side	Atoms on product side
Al		
O		
H		
S		

Answers

a. $CaCO_3(s) \rightarrow CaO(g) + CO_2(g)$

Element	Atoms on reactant side	Atoms on product side
Ca	1	1
C	1	1
O	3	3

b. $2Na(s) + H_2O(l) \rightarrow Na_2O(s) + H_2(g)$

Element	Atoms on reactant side	Atoms on product side
Na	2	2
H	2	2
O	1	1

c. $C_5H_{12}(g) + 8O_2(g) \rightarrow 5CO_2(g) + 6H_2O(g)$

Element	Atoms on reactant side	Atoms on product side
C	5	5
H	12	12
O	16	16

d. $2AgNO_3(aq) + K_2S(aq) \rightarrow 2KNO_3(aq) + Ag_2S(s)$

Element	Atoms on reactant side	Atoms on product side
Ag	2	2
N	2	2
O	6	6
K	2	2
S	1	1

e. $2Al(OH)_3(aq) + 3H_2SO_4(aq) \rightarrow Al_2(SO_4)_3(s) + 6H_2O(l)$

Element	Atoms on reactant side	Atoms on product side
Al	2	2
O	18	18
H	12	12
S	3	3

◆ Learning Exercise 5.4B

Balance each of the following equations by placing appropriate coefficients in front of the formulas as needed:

a. _____ $MgO(s) \rightarrow$ _____ $Mg(s) +$ _____ $O_2(g)$

b. _____ $Zn(s) +$ _____ $HCl(aq) \rightarrow$ _____ $ZnCl_2(aq) +$ _____ $H_2(g)$

c. _____ $Al(s) +$ _____ $CuSO_4(aq) \rightarrow$ _____ $Cu(s) +$ _____ $Al_2(SO_4)_3(aq)$

d. _____ $Al_2S_3(s) +$ _____ $H_2O(l) \rightarrow$ _____ $Al(OH)_3(aq) +$ _____ $H_2S(g)$

e. _____ $BaCl_2(aq) +$ _____ $Na_2SO_4(aq) \rightarrow$ _____ $BaSO_4(s) +$ _____ $NaCl(g)$

f. _____ $CO(g) +$ _____ $Fe_2O_3(s) \rightarrow$ _____ $Fe(s) +$ _____ $CO_2(g)$

g. _____ $K(s) +$ _____ $H_2O(l) \rightarrow$ _____ $K_2O(s) +$ _____ $H_2(g)$

h. _____ $Fe(OH)_3(s) \rightarrow$ _____ $Fe_2O_3(s) +$ _____ $H_2O(l)$

Answers

a. $2MgO(s) \rightarrow 2Mg(s) + O_2(g)$

b. $Zn(s) + 2HCl(aq) \rightarrow ZnCl_2(aq) + H_2(g)$

c. $2Al(s) + 3CuSO_4(aq) \rightarrow 3Cu(s) + Al_2(SO_4)_3(aq)$

d. $Al_2S_3(s) + 6H_2O(l) \rightarrow 2Al(OH)_3(aq) + 3H_2S(g)$

e. $BaCl_2(aq) + Na_2SO_4(aq) \rightarrow BaSO_4(s) + 2NaCl(aq)$

f. $3CO(g) + Fe_2O_3(s) \rightarrow 2Fe(s) + 3CO_2(g)$

g. $2K(s) + H_2O(l) \rightarrow K_2O(s) + H_2(g)$

h. $2Fe(OH)_3(s) \rightarrow Fe_2O_3(s) + 3H_2O(l)$

5.5 Types of Reactions

- Reactions are classified as combination, decomposition, single replacement, and double replacement.
- In a *combination* reaction, reactants are combined. In a *decomposition* reaction, a reactant splits into simpler products.
- In *single (or double) replacement* reactions, one (or two) elements in the reacting compounds are replaced with element(s) from the other reactant(s).

◆ Learning Exercise 5.5A

Match each of the following reactions with the type of reaction:

a. combination   **b.** decomposition   **c.** single replacement   **d.** double replacement

1. _____ $N_2(g) + 3H_2(g) \rightarrow 2NH_3(g)$

2. _____ $BaCl_2(aq) + K_2CO_3(aq) \rightarrow BaCO_3(s) + 2KCl(aq)$

3. _____ $2H_2O_2(aq) \rightarrow 2H_2O(l) + O_2(g)$

4. _____ $CuO(s) + H_2(g) \rightarrow Cu(s) + H_2O(l)$

5. _____ $N_2(g) + 2O_2(g) \rightarrow 2NO_2(g)$

6. _____ $2NaHCO_3(s) \rightarrow Na_2O(s) + 2CO_2(g) + H_2O(l)$

7. _____ $PbCO_3(s) \rightarrow PbO(s) + CO_2(g)$

8. _____ $Al(s) + Fe_2O_3(s) \rightarrow Fe(s) + Al_2O_3(s)$

Answers   **1.** a   **2.** d   **3.** b   **4.** c

               5. a   **6.** b   **7.** b   **8.** c

◆ Learning Exercise 5.5B

One way to remove tarnish from silver is to place the silver object on a piece of aluminum foil and add boiling water and some baking soda. The unbalanced equation is the following:

$$Al(s) + Ag_2S(s) \rightarrow Ag(s) + Al_2S_3(s)$$

a. What is the balanced equation?

b. What type of reaction takes place?

Answers   **a.** $2Al(s) + 3Ag_2S(s) \rightarrow 6Ag(s) + Al_2S_3(s)$ **b.** single replacement

5.6 Oxidation–Reduction Reactions

- An oxidation–reduction reaction consists of a loss and gain of electrons. In an oxidation, electrons are lost. In a reduction, there is a gain of electrons.
- An oxidation must always be accompanied by a reduction. The number of electrons lost in the oxidation reaction is equal to the number of electrons gained in the reduction reaction.
- In biological systems, the term *oxidation* describes the gain of oxygen or the loss of hydrogen. The term *reduction* is used to describe a loss of oxygen or a gain of hydrogen.

◆ Learning Exercise 5.6

For each of the following reactions, indicate whether the underlined element is *oxidized* or *reduced*.

    a. $4\underline{Al}(s) + 3O_2(g) \rightarrow 2Al_2O_3(s)$ Al is _____

    b. $\underline{Fe}^{3+}(aq) + 1e^- \rightarrow \underline{Fe}^{2+}(aq)$ Fe^{3+} is _____

    c. $\underline{Cu}O(s) + H_2(g) \rightarrow Cu(s) + H_2O(l)$ Cu^{2+} is _____

    d. $2\underline{Cl}^-(aq) \rightarrow Cl_2(g) + 2e^-$ Cl^- is _____

    e. $2H\underline{Br}(aq) + Cl_2(g) \rightarrow 2HCl(aq) + Br_2(g)$ Br^- is _____

    f. $2\underline{Na}(s) + Cl_2(g) \rightarrow 2NaCl(s)$ Na is _____

    g. $\underline{Cu}Cl_2(aq) + Zn(g) \rightarrow ZnCl_2(aq) + Cu(g)$ Cu^{2+} is _____

Answers
    a. Al is oxidized to Al^{3+}; loss of electrons (addition of O)
    b. Fe^{3+} is reduced to Fe^{2+}; gain of electrons
    c. Cu^{2+} is reduced to Cu; gain of electrons (loss of O)
    d. $2Cl^-$ is oxidized to Cl_2; loss of electrons
    e. $2\,Br^-$ is oxidized to Br_2; loss of electrons
    f. Na is oxidized to Na^+; loss of electrons
    g. Cu^{2+} is reduced to Cu; gain of electrons

5.7 Mole Relationships in Chemical Equations

- The coefficients in a balanced chemical equation describe the moles of reactants and products in the reactions.
- Using the coefficients, mole–mole conversion factors are written for any two substances in the equation.
- For the reaction of oxygen forming ozone, $3O_2(g) \rightarrow 2O_3(g)$, the mole–mole conversion factors are the following:

$$\frac{3 \text{ moles } O_2}{2 \text{ moles } O_3} \quad \text{and} \quad \frac{2 \text{ moles } O_3}{3 \text{ moles } O_2}$$

◆ Learning Exercise 5.7A

Write the conversion factors that are possible from the following equation: $N_2(g) + O_2(g) \rightarrow 2NO(g)$

Answers $\dfrac{1 \text{ mole } N_2}{1 \text{ mole } O_2}$ and $\dfrac{1 \text{ mole } O_2}{1 \text{ mole } N_2}$; $\dfrac{1 \text{ mole } N_2}{2 \text{ moles } NO}$ and $\dfrac{2 \text{ moles } NO}{1 \text{ mole } N_2}$

               $\dfrac{1 \text{ mole } O_2}{2 \text{ moles } NO}$ and $\dfrac{2 \text{ moles } NO}{1 \text{ mole } O_2}$

◆ **Learning Exercise 5.7B**

Use the equation below to answer the following questions:

$$C_3H_8(g) + 5O_2(g) \rightarrow 3CO_2(g) + 4H_2O(g)$$

a. How many moles of O_2 are needed to react with 2.00 moles of C_3H_8?

b. How many moles CO_2 are produced when 4.00 moles of O_2 react?

c. How many moles of C_3H_8 react with 3.00 moles of O_2?

d. How many moles of H_2O are produced from 0.50 mole of C_3H_8?

Answers   **a.** 10.0 moles of O_2 **b.** 2.40 moles of CO_2
            c. 0.600 mole of C_3H_8 **d.** 2.0 moles of H_2O

5.8 Mass Calculations for Reactions

- The grams or moles of a substance in an equation are converted to another using molar masses and mole–mole factors.
- Suppose that a problem asks for the number of grams of O_3 (ozone) produced from 8.0 g of O_2. The equation is $3O_2 \rightarrow 2O_3$.

Step 1	**Step 2**	**Step 3**
Use molar mass of O_2	Use mole factor from coefficients	Use molar mass of O_3

8.0 g O_2 $\longrightarrow$ moles O_2 $\longrightarrow$ moles O_3 $\longrightarrow$ g O_3

$$8.0 \text{ g } O_2 \times \frac{1 \text{ mole } O_2}{32.0 \text{ g } O_2} \times \frac{2 \text{ moles } O_3}{3 \text{ moles } O_2} \times \frac{48.0 \text{ g } O_3}{1 \text{ mole } O_3} = 8.0 \text{ g of } O_3$$

◆ **Learning Exercise 5.8**

Consider the equation for the following questions:

$$2C_2H_6(g) + 7O_2(g) \rightarrow 4CO_2(g) + 6H_2O(g)$$

a. How many grams of oxygen (O_2) are needed to react with 4.00 moles of C_2H_6?

b. How many grams of C_2H_6 are needed to react with 115 g of O_2?

c. How many grams of C_2H_6 react if 22.0 g of CO_2 gas is produced?

d. How many grams of CO_2 are produced when 2.00 moles of C_2H_6 react with sufficient oxygen?

e. How many grams of water are produced when 82.5 g of O_2 react with sufficient C_2H_6?

Answers	**a.** 448 g of O_2	**b.** 30.8 g of C_2H_6	**c.** 7.53 g of C_2H_6
	d. 176 g of CO_2	**e.** 39.8 g of H_2O	

5.9 Energy in Chemical Reactions

- In a reaction, molecules (or atoms) must collide with energy equal to or greater than the energy of activation.
- The heat of reaction is the energy difference between the energy of the reactants and the products.
- In exothermic reactions, the heat of reaction is the energy released. In endothermic reactions, the heat of reaction is the energy absorbed.
- Generally, the rate of a reaction (the speed at which products form) can be increased by adding more reacting molecules, raising the temperature of the reaction, or by adding a catalyst.

◆ Learning Exercise 5.9

Indicate whether each of the following is an endothermic or exothermic reaction:

1. $2H_2(g) + O_2(g) \rightarrow 2H_2O(g) + 582 \text{ kJ}$ _____

2. $C_2H_4(g) + 176 \text{ kJ} \rightarrow H_2(g) + C_2H_2(g)$ _____

3. $2C(s) + O_2(g) \rightarrow 2CO(g) + 220 \text{ kJ}$ _____

4. $C_6H_{12}O_6(s) + 6O_2(g) \rightarrow 6CO_2(g) + 6H_2O(l) + 1350 \text{ kcal}$ _____
   glucose

5. $C_2H_4(g) + H_2O(g) \rightarrow C_2H_5OH(l) + 21 \text{ kcal}$ _____

Answers: **1.** exothermic **2.** endothermic **3.** exothermic **4.** exothermic **5.** exothermic

Checklist for Chapter 5

You are ready to take the practice test for Chapter 5. Be sure that you have accomplished the following learning goals for this chapter. If you are not sure, review the section listed at the end of the goal. Then apply your new skills and understanding to the practice test.

After studying Chapter 5, I can successfully:

_____ Calculate the number of particles in a mole of a substance (5.1).

_____ Calculate the molar mass given the formula of a substance (5.2).

_____ Convert the grams of a substance to moles; moles to grams (5.2).

_____ Identify a chemical and physical change (5.3).

_____ State a chemical equation in words and calculate the total atoms of each element in the reactants and products (5.4).

_____ Write a balanced equation for a chemical reaction from the formulas of the reactants and products (5.4).

_____ Identify a reaction as a combination, decomposition, and single or double replacement (5.5).

_____ Identify an oxidation and reduction reaction (5.6).

_____ Use mole–mole factors for the mole relationships in an equation to calculate the moles of another substance in an equation for a chemical reaction (5.7).

_____ Calculate the mass of a substance in an equation using mole factors and molar masses (5.8).

_____ Identify exothermic and endothermic reactions from the heat in a chemical reaction (5.9).

Practice Test for Chapter 5

Indicate whether each change is a (A) physical change or a (B) chemical change:

1. _____ a melting ice cube

4. _____ a burning candle

2. _____ breaking glass

5. _____ milk turning sour

3. _____ bleaching a stain

For each of the unbalanced equations in questions 6 through 10, balance and indicate the correct coefficient for the component in the equation written in boldface type.

    a. 1   **b.** 2   **c.** 3   **d.** 4   **e.** 5

6. _____ $Sn(s) + \mathbf{Cl_2}(g) \rightarrow SnCl_4(s)$

7. _____ $Al(s) + H_2O(l) \rightarrow Al_2O_3(s) + \mathbf{H_2}(g)$

8. _____ $C_3H_8(g) + \mathbf{O_2}(g) \rightarrow CO_2(g) + H_2O(g)$

9. _____ $\mathbf{NH_3}(g) + O_2(g) \rightarrow N_2(g) + H_2O(g)$

10. _____ $N_2O(g) \rightarrow N_2(g) + \mathbf{O_2}(g)$

For questions 11 through 15, classify each reaction as one of the following:

    a. combination   **b.** decomposition   **c.** single replacement   **d.** double replacement

11. _____ $S(s) + O_2(g) \rightarrow SO_2(g)$

12. _____ $Fe_2O_3(s) + 3C(s) \rightarrow 2Fe(s) + 3CO(g)$

13. _____ $CaCO_3(s) \rightarrow CaO(s) + CO_2(g)$

14. _____ $Mg(s) + 2AgNO_3(aq) \rightarrow Mg(NO_3)_2(aq) + 2Ag(s)$

15. _____ $Na_2S(aq) + Pb(NO_3)_2(aq) \rightarrow PbS(s) + 2NaNO_3(aq)$

For questions 16 through 20, identify as an (A) oxidation or a (B) reduction.

16. $Ca \rightarrow Ca^{2+} + 2e^-$ _____

19. $Br_2 + 2e^- \rightarrow 2Br^-$ _____

17. $Fe^{3+} + 3e^- \rightarrow Fe$ _____

20. $Sn^{2+} \rightarrow Sn^{4+} + 2e^-$ _____

18. $Al^{3+} + 3e^- \rightarrow Al$ _____

21. The moles of oxygen (O) in 2.0 moles of $Al(OH)_3$ is
    A. 1   **B.** 2   **C.** 3   **D.** 4   **E.** 6

22. What is the molar mass of Li_2SO_4?
    A. 55.1 g   **B.** 62.1 g   **C.** 100.1 g   **D.** 109.9 g   **E.** 103.1 g

23. What is the molar mass of $NaNO_3$?
    A. 34.0 g   **B.** 37.0 g   **C.** 53.0 g   **D.** 75.0 g   **E.** 85.0 g

24. The number of grams in 0.600 mole of Cl_2 is
    A. 71.0 g   **B.** 118 g   **C.** 42.6 g   **D.** 84.5 g   **E.** 4.30 g

25. How many grams are in 4.00 moles of NH_3?
    A. 4.00 g   **B.** 17.0 g   **C.** 34.0 g   **D.** 68.0 g   **E.** 0.240 g

26. How many moles is 8.0 g of NaOH?

    A. 0.10 mole   **B.** 0.20 mole   **C.** 0.40 mole   **D.** 2.0 moles   **E.** 4.0 moles

27. The number of moles of aluminum in 54 g of Al is

    A. 0.50 mole   **B.** 1.0 mole   **C.** 2.0 moles   **D.** 3.0 moles   **E.** 4.0 moles

28. The number of moles of water in 36 g of H_2O is

    A. 0.50 mole   **B.** 1.0 mole   **C.** 2.0 moles   **D.** 3.0 moles   **E.** 4.0 moles

29. What is the number of moles in 2.2 g of CO_2?

    A. 2.0 moles   **B.** 1.0 mole   **C.** 0.20 mole   **D.** 0.050 mole   **E.** 0.010 mole

30. 0.20 g H_2 = _____ mole H_2

    A. 0.10 mole   **B.** 0.20 mole   **C.** 0.40 mole   **D.** 0.040 mole   **E.** 0.010 mole

For questions 31 to 35, use the reaction: $C_2H_5OH(g) + 3O_2(g) \rightarrow 2CO_2(g) + 3H_2O(g)$
                                           Ethanol

31. How many grams of oxygen are needed to react with 1.0 mole of ethanol?

    A. 8.0 g   **B.** 16 g   **C.** 32 g   **D.** 64 g   **E.** 96 g

32. How many moles of water are produced when 12 moles of oxygen react?

    A. 3.0 moles   **B.** 6.0 moles   **C.** 8.0 moles   **D.** 12.0 moles   **E.** 36.0 moles

33. How many grams of carbon dioxide are produced when 92 g of ethanol react?

    A. 22 g   **B.** 44 g   **C.** 88 g   **D.** 92 g   **E.** 176 g

34. How many moles of oxygen would be needed to produce 44 g of CO_2?

    A. 0.67 mole   **B.** 1.0 mole   **C.** 1.5 moles   **D.** 2.0 moles   **E.** 3.0 moles

35. How many grams of water will be produced if 23 g of ethanol react?

    A. 54 g   **B.** 27 g   **C.** 18 g   **D.** 9.0 g   **E.** 6.0 g

For questions, 36 to 38, indicate if the reactions are (1) exothermic or (2) endothermic:

36. $N_2(g) + 3H_2(g) \rightarrow 2NH_3(g) + 22$ kcal

37. $2HCl(g) + 44$ kcal $\rightarrow H_2(g) + Cl_2(g)$

38. $C_3H_8(g) + 5O_2(g) \rightarrow 3CO_2(g) + 4H_2O(g) + 531$ kcal

Answers to the Practice Test

1. A	**2.** A	**3.** B	**4.** B	**5.** B
6. B	**7.** C	**8.** E	**9.** D	**10.** A
11. A	**12.** C	**13.** B	**14.** C	**15.** D
16. A	**17.** B	**18.** B	**19.** B	**20.** A
21. E	**22.** D	**23.** E	**24.** C	**25.** D
26. B	**27.** C	**28.** C	**29.** D	**30.** A
31. E	**32.** D	**33.** E	**34.** C	**35.** B
36. 1	**37.** 2	**38.** 1		

Answers and Solutions to Text Problems

5.1 One mole is the amount of a substance that contains 6.02×10^{23} items. For example, one mole of water contains 6.02×10^{23} molecules of water.

5.3 **a.** $0.500 \ \text{mole C} \times \dfrac{6.02 \times 10^{23} \ \text{atoms C}}{1 \ \text{mole C}} = 3.01 \times 10^{23} \ \text{C atoms}$

   b. $1.28 \ \text{moles SO}_2 \times \dfrac{6.02 \times 10^{23} \ \text{molecules SO}_2}{1 \ \text{mole SO}_2} = 7.71 \times 10^{23} \ \text{SO}_2 \ \text{molecules}$

   c. $5.22 \times 10^{22} \ \text{atoms Fe} \times \dfrac{1 \ \text{mole Fe}}{6.02 \times 10^{23} \ \text{atoms Fe}} = 0.0867 \ \text{mole of Fe}$

   d. $8.50 \times 10^{24} \ \text{atoms C}_2\text{H}_5\text{OH} \times \dfrac{1 \ \text{mole C}_2\text{H}_5\text{OH}}{6.02 \times 10^{23} \ \text{atoms C}_2\text{H}_5\text{OH}} = 14.1 \ \text{moles of C}_2\text{H}_5\text{OH}$

5.5 1 mole of H_3PO_4 molecules contains 3 moles of H atoms, 1 mole of P atoms, and 4 moles of O atoms.

   a. $2.00 \ \text{moles H}_3\text{PO}_4 \times \dfrac{3 \ \text{moles H}}{1 \ \text{moles H}_3\text{PO}_4} = 6.00 \ \text{moles of H}$

   b. $2.00 \ \text{moles H}_3\text{PO}_4 \times \dfrac{4 \ \text{moles O}}{1 \ \text{mole H}_3\text{PO}_4} = 8.00 \ \text{moles of O}$

   c. $2.00 \ \text{moles H}_3\text{PO}_4 \times \dfrac{1 \ \text{mole P}}{1 \ \text{mole H}_3\text{PO}_4} \times \dfrac{6.02 \times 10^{23} \ \text{atoms P}}{1 \ \text{mole P}} = 1.20 \times 10^{24} \ \text{P atoms}$

   d. $2.00 \ \text{moles H}_3\text{PO}_4 \times \dfrac{4 \ \text{moles O}}{1 \ \text{mole H}_3\text{PO}_4} \times \dfrac{6.02 \times 10^{23} \ \text{atoms O}}{1 \ \text{mole O}} = 4.82 \times 10^{24} \ \text{O atoms}$

5.7 The subscripts indicate the moles of each element in one mole of that compound.

   a. $1.0 \ \text{mole quinine} \times \dfrac{24 \ \text{moles H}}{1 \ \text{mole quinine}} = 24 \ \text{moles of H}$

   b. $5.0 \ \text{moles quinine} \times \dfrac{20 \ \text{moles C}}{1 \ \text{mole quinine}} = 1.0 \times 10^2 \ \text{moles of C}$

   c. $0.020 \ \text{mole quinine} \times \dfrac{2 \ \text{moles N atoms}}{1 \ \text{mole quinine}} = 0.040 \ \text{mole of N atoms}$

5.9 **a.** 1 mole of Na and 1 mole of Cl: $23.0 \ \text{g} + 35.5 \ \text{g} = 58.5 \ \text{g/mole NaCl}$
   b. 2 moles of Fe and 3 moles of O: $111.8 \ \text{g} + 48.0 \ \text{g} = 159.8 \ \text{g/mole Fe}_2\text{O}_3$
   c. 2 moles of Li and 1 mole of C and 3 moles of O: $13.8 \ \text{g} + 12.0 \ \text{g} + 48.0 \ \text{g}$
      $= 73.8 \ \text{g/mole Li}_2\text{CO}_3$
   d. 2 moles of Al and 3 moles of S and 12 moles of O: $54.0 \ \text{g} + 96.3 \ \text{g} + 192.0 \ \text{g}$
      $= 342.3 \ \text{g/mole Al}_2(\text{SO}_4)_3$
   e. 1 mole of Mg and 2 moles of O and 2 moles of H: $24.3 \ \text{g} + 32.0 \ \text{g} + 2.0 \ \text{g}$
      $= 58.3 \ \text{g/mole Mg(OH)}_2$
   f. 16 moles of C and 19 moles of H and 3 moles of N and 5 moles of O and 1 mole of S:
      $192.0 \ \text{g} + 19.0 \ \text{g} + 42.0 \ \text{g} + 80.0 \ \text{g} + 32.1 \ \text{g} = 365.1 \ \text{g/mole C}_{16}\text{H}_{19}\text{N}_3\text{O}_5\text{S}$

5.11 **a.** $2.00 \ \text{moles Na} \times \dfrac{23.0 \ \text{g Na}}{1 \ \text{mole Na}} = 46.0 \ \text{g of Na}$

   b. $2.80 \ \text{moles Ca} \times \dfrac{40.1 \ \text{g Ca}}{1 \ \text{mole Ca}} = 112 \ \text{g of Ca}$

c. $0.125 \ \text{mole Sn} \times \dfrac{118.7 \ \text{g Sn}}{1 \ \text{mole Sn}} = 14.8 \ \text{g of Sn}$

d. $1.76 \ \text{mole Cu} \times \dfrac{63.6 \ \text{g Cu}}{1 \ \text{mole Cu}} = 112 \ \text{g of Cu}$

5.13 a. $0.500 \ \text{moles NaCl} \times \dfrac{58.5 \ \text{g NaCl}}{1 \ \text{mole NaCl}} = 29.3 \ \text{g of NaCl}$

b. $1.75 \ \text{moles Na}_2\text{O} \times \dfrac{62.0 \ \text{g Na}_2\text{O}}{1 \ \text{mole Na}_2\text{O}} = 109 \ \text{g of Na}_2\text{O}$

c. $0.225 \ \text{mole H}_2\text{O} \times \dfrac{18.0 \ \text{g H}_2\text{O}}{1 \ \text{mole H}_2\text{O}} = 4.05 \ \text{g of H}_2\text{O}$

d. $4.42 \ \text{mole CO}_2 \times \dfrac{44.0 \ \text{g CO}_2}{1 \ \text{mole CO}_2} = 194 \ \text{g of CO}_2$

5.15 a. $5.00 \ \text{moles MgSO}_4 \times \dfrac{120.4 \ \text{g MgSO}_4}{1 \ \text{mole MgSO}_4} = 602 \ \text{g of MgSO}_4$

b. $0.25 \ \text{mole CO}_2 \times \dfrac{44.0 \ \text{g CO}_2}{1 \ \text{mole CO}_2} = 11 \ \text{g of CO}_2$

5.17 a. $50.0 \ \text{g Ag} \times \dfrac{1 \ \text{mole Ag}}{107.9 \ \text{g Ag}} = 0.463 \ \text{mole of Ag}$

b. $0.200 \ \text{g C} \times \dfrac{1 \ \text{mole C}}{12.0 \ \text{g C}} = 0.0167 \ \text{of mole C}$

c. $15.0 \ \text{g NH}_3 \times \dfrac{1 \ \text{mole NH}_3}{17.0 \ \text{g NH}_3} = 0.882 \ \text{mole of NH}_3$

d. $75.0 \ \text{g SO}_2 \times \dfrac{1 \ \text{mole SO}_2}{64.1 \ \text{g SO}_2} = 1.17 \ \text{moles of SO}_2$

5.19 a. $25.0 \ \text{g Ne} \times \dfrac{1 \ \text{mole Ne}}{20.2 \ \text{g Ne}} = 1.24 \ \text{moles of Ne}$

b. $25.0 \ \text{g O}_2 \times \dfrac{1 \ \text{mole O}_2}{32.0 \ \text{g O}_2} = 0.781 \ \text{mole of O}_2$

c. $25.0 \ \text{g Al(OH)}_3 \times \dfrac{1 \ \text{mole Al(OH)}_3}{78.0 \ \text{g Al(OH)}_3} = 0.321 \ \text{mole of Al(OH)}_3$

d. $25.0 \ \text{g Ga}_2\text{S}_3 \times \dfrac{1 \ \text{mole Ga}_2\text{S}_3}{235.7 \ \text{g Ga}_2\text{S}_3} = 0.106 \ \text{mole of Ga}_2\text{S}_3$

5.21 a. $25 \ \text{g S} \times \dfrac{1 \ \text{mole S}}{32.1 \ \text{g S}} = 0.78 \ \text{mole of S}$

b. $125 \ \text{g SO}_2 \times \dfrac{1 \ \text{mole SO}_2}{64.1 \ \text{g SO}_2} \times \dfrac{1 \ \text{mole S}}{1 \ \text{mole SO}_2} = 1.95 \ \text{moles of S}$

c. $2.0 \ \text{moles Al}_2\text{S}_3 \times \dfrac{3 \ \text{moles S}}{1 \ \text{mole Al}_2\text{SO}_3} = 6.0 \ \text{moles of S}$

5.23 A chemical change occurs when the atoms of the initial substances rearrange to form new substances. Chemical change is indicated by a change in properties of the reactants. For example, a rusting nail, souring milk, and a burning match are all chemical changes.
a. physical: the shape changes, but not the substance
b. chemical: new substances form

    c. physical: water evaporates forming gaseous water

    d. chemical: the composition of the substances change to give new substances

    e. physical: water freezes

    f. chemical: new substances form

5.25 An equation is balanced when there are equal numbers of atoms of each element on the reactant as on the product side.

    a. not balanced   **b.** balanced   **c.** not balanced   **d.** balanced

5.27 Place coefficients in front of formulas until you make the atoms of each element equal on each side of the equation. Try starting with the formula that has subscripts.

    a. $N_2(g) + O_2(g) \rightarrow 2NO(g)$ **b.** $2HgO(s) \rightarrow 2Hg(l) + O_2(g)$

    c. $4Fe(s) + 3O_2(g) \rightarrow 2Fe_2O_3(s)$ **d.** $2Na(s) + Cl_2(g) \rightarrow 2NaCl(s)$

5.29 **a.** There are two NO_3 in the product. Balance by placing a 2 before $AgNO_3$.

       $Mg(s) + 2AgNO_3(aq) \rightarrow Mg(NO_3)_2(aq) + 2Ag(s)$

    b. Start with the formula $Al_2(SO_4)_3$. Balance the Al by writing 2 Al and balance the SO_4 by writing 3 $CuSO_4$.

       $Al(s) + 3CuSO_4(aq) \rightarrow 3Cu(s) + Al_2(SO_4)_3(aq)$

    c. $Pb(NO_3)_2(aq) + 2NaCl(aq) \rightarrow PbCl_2(s) + 2NaNO_3(aq)$

    d. $2Al(s) + 6HCl(aq) \rightarrow 2AlCl_3(aq) + 3H_2(g)$

5.31 **a.** This is a decomposition reaction because a single reactant splits into two simpler substances.

    b. This is a single replacement reaction because I_2 in BaI_2 is replaced by Br_2.

5.33 **a.** combination   **b.** single replacement   **c.** decomposition

    d. double replacement   **e.** decomposition   **f.** double replacement

    g. combination

5.35 **a.** $Mg(s) + Cl_2(g) \rightarrow MgCl_2(s)$

    b. $2HBr(g) \rightarrow H_2(g) + Br_2(g)$

    c. $Mg(s) + Zn(NO_3)_2(aq) \rightarrow Zn(s) + Mg(NO_3)_2(aq)$

    d. $K_2S(aq) + Pb(NO_3)_2(aq) \rightarrow 2KNO_3(aq) + PbS(s)$

5.37 **a.** reduction   **b.** oxidation   **c.** reduction   **d.** reduction

5.39 **a.** Zinc (Zn) is oxidized because it loses electrons to form Zn^{2+}; chlorine (Cl_2) is reduced.

    b. Bromide ion $(2Br^-)$ is oxidized to Br_2; chlorine (Cl_2) is reduced to $2\,Cl^-$ (gains electrons).

    c. Oxide ion (O^{2-}) is oxidized to O_2 (loses electrons); lead(II) ion (Pb^{2+}) is reduced.

    d. Sn^{2+} ion is oxidized to Sn^{4+} (loses electrons); Fe^{3+} ion is reduced to Fe^{2+} (gains electrons).

5.41 **a.** $Fe^{3+} + e^- \rightarrow Fe^{2+}$ is a reduction.

    b. $Fe^{2+} \rightarrow Fe^{3+} + e^-$ is an oxidation.

5.43 Because the linoleic acid adds hydrogen, the acid has been reduced.

5.45 **a.** Two SO_2 molecules react with one O_2 molecule to produce two SO_3 molecules. Two moles of SO_2 react with one mole of O_2 to produce two moles of SO_3.

    b. Four P atoms react with five O_2 molecules to produce two P_2O_5 molecules. Four moles of P react with five moles of O_2 to produce two moles of P_2O_5.

5.47 **a.** $\dfrac{2 \text{ moles } SO_2}{1 \text{ mole } O_2}$ and $\dfrac{1 \text{ mole } O_2}{2 \text{ moles } SO_2}$ $\dfrac{2 \text{ moles } SO_2}{2 \text{ moles } SO_3}$ and $\dfrac{2 \text{ moles } SO_3}{2 \text{ moles } SO_2}$

        $\dfrac{1 \text{ mole } O_2}{2 \text{ moles } SO_3}$ and $\dfrac{2 \text{ moles } SO_3}{1 \text{ mole } O_2}$

b. $\dfrac{4 \text{ moles P}}{5 \text{ moles O}_2}$ and $\dfrac{5 \text{ moles O}_2}{4 \text{ moles P}}$ $\dfrac{4 \text{ moles P}}{2 \text{ moles P}_2\text{O}_5}$ and $\dfrac{2 \text{ moles P}_2\text{O}_5}{4 \text{ moles P}}$

   $\dfrac{5 \text{ moles O}_2}{2 \text{ moles P}_2\text{O}_5}$ and $\dfrac{2 \text{ moles P}_2\text{O}_5}{5 \text{ moles O}_2}$

5.49 **a.** $2.0 \text{ moles H}_2 \times \dfrac{1 \text{ mole O}_2}{2 \text{ moles H}_2} = 1.0$ mole of O_2

   b. $5.0 \text{ moles O}_2 \times \dfrac{2 \text{ moles H}_2}{1 \text{ mole O}_2} = 10.$ moles of H_2

   c. $2.5 \text{ moles O}_2 \times \dfrac{2 \text{ moles H}_2\text{O}}{1 \text{ mole O}_2} = 5.0$ moles of H_2O

5.51 **a.** $0.500 \text{ mole SO}_2 \times \dfrac{5 \text{ moles C}}{2 \text{ moles SO}_2} = 1.25$ moles of C

   b. $1.2 \text{ moles C} \times \dfrac{4 \text{ moles CO}}{5 \text{ moles C}} = 0.96$ mole of CO

   c. $0.50 \text{ mole CS}_2 \times \dfrac{2 \text{ moles SO}_2}{1 \text{ mole CS}_2} = 1.0$ mole of SO_2

   d. $2.5 \text{ moles C} \times \dfrac{1 \text{ mole CS}_2}{5 \text{ moles C}} = 0.50$ mole of CS_2

5.53 **a.** $2.50 \text{ moles Na} \times \dfrac{2 \text{ moles Na}_2\text{O}}{4 \text{ moles Na}} \times \dfrac{62.0 \text{ g Na}_2\text{O}}{1 \text{ mole Na}_2\text{O}} = 77.5$ g of Na_2O

   b. $18.0 \text{ g Na} \times \dfrac{1 \text{ mole Na}}{23.0 \text{ g Na}} \times \dfrac{1 \text{ mole O}_2}{4 \text{ moles Na}} \times \dfrac{32.0 \text{ g O}_2}{1 \text{ mole O}_2} = 6.26$ g of O_2

   c. $75.0 \text{ g Na}_2\text{O} \times \dfrac{1 \text{ mole Na}_2\text{O}}{62.0 \text{ g Na}_2\text{O}} \times \dfrac{1 \text{ mole O}_2}{2 \text{ moles Na}_2\text{O}} \times \dfrac{32.0 \text{ g O}_2}{1 \text{ mole O}_2} = 19.4$ g of O_2

5.55 **a.** $8.00 \text{ moles NH}_3 \times \dfrac{3 \text{ moles O}_2}{4 \text{ moles NH}_3} \times \dfrac{32.0 \text{ g O}_2}{1 \text{ mole O}_2} = 192$ g of O_2

   b. $6.50 \text{ g O}_2 \times \dfrac{1 \text{ mole O}_2}{32.0 \text{ g O}_2} \times \dfrac{2 \text{ moles N}_2}{3 \text{ moles O}_2} \times \dfrac{28.0 \text{ g N}_2}{1 \text{ mole N}_2} = 3.79$ g of N_2

   c. $34.0 \text{ g NH}_3 \times \dfrac{1 \text{ mole NH}_3}{17.0 \text{ g NH}_3} \times \dfrac{6 \text{ moles H}_2\text{O}}{4 \text{ moles NH}_3} \times \dfrac{18.0 \text{ g H}_2\text{O}}{1 \text{ mole H}_2\text{O}} = 54.0$ g of H_2O

5.57 **a.** $28.0 \text{ g NO}_2 \times \dfrac{1 \text{ mole NO}_2}{46.0 \text{ g NO}_2} \times \dfrac{1 \text{ mole H}_2\text{O}}{3 \text{ moles NO}_2} \times \dfrac{18.0 \text{ g H}_2\text{O}}{1 \text{ mole H}_2\text{O}} = 3.65$ g of H_2O

   b. $15.8 \text{ g NO}_2 \times \dfrac{1 \text{ mole NO}_2}{46.0 \text{ g NO}_2} \times \dfrac{1 \text{ mole NO}}{3 \text{ moles NO}_2} \times \dfrac{30.0 \text{ g NO}}{1 \text{ mole NO}} = 3.43$ g of NO

   c. $8.25 \text{ g NO}_2 \times \dfrac{1 \text{ mole NO}_2}{46.0 \text{ g NO}_2} \times \dfrac{2 \text{ moles HNO}_3}{3 \text{ moles NO}_2} \times \dfrac{63.0 \text{ g HNO}_3}{1 \text{ mole HNO}_3} = 7.53$ g of HNO_3

5.59 **a.** $2\text{PbS}(s) + 3\text{O}_2(g) \rightarrow 2\text{PbO}(s) + 2\text{SO}_2(g)$

   b. $0.125 \text{ mole PbS} \times \dfrac{3 \text{ moles O}_2}{2 \text{ moles PbS}} \times \dfrac{32.0 \text{ g O}_2}{1 \text{ mole O}_2} = 6.00$ g of O_2

   c. $65.0 \text{ g PbS} \times \dfrac{1 \text{ mole PbS}}{239.3 \text{ g PbS}} \times \dfrac{2 \text{ moles SO}_2}{2 \text{ moles PbS}} \times \dfrac{64.1 \text{ g SO}_2}{1 \text{ moles SO}_2} = 17.4$ g of SO_2

   d. $128 \text{ g PbO} \times \dfrac{1 \text{ mole PbO}}{223.2 \text{ g PbO}} \times \dfrac{2 \text{ moles PbS}}{2 \text{ moles PbO}} \times \dfrac{239.3 \text{ g PbS}}{1 \text{ mole PbS}} = 137$ g of PbS

5.61 **a.** The energy of activation is the energy required to break the bonds in the reacting molecules.

b. A catalyst lowers the activation energy by providing an alternate path, which increases the rate of a reaction.

c. In exothermic reactions, the energy of the products is lower than the reactants.

d.

5.63 **a.** exothermic; heat loss   **b.** endothermic; heat gain   **c.** exothermic; heat loss

5.65 **a.** exothermic   **b.** endothermic   **c.** exothermic

5.67 **a.** The rate of a reaction relates the speed at which reactants are transformed into products.

b. Because more reactants will have the energy necessary to proceed to products (the activation energy) at room temperature than at refrigerator temperatures, the rate of formation of mold will be higher at room temperature.

5.69 **a.** Addition of a reactant increases the reaction rate.

b. Increasing the temperature increases the number of collisions with the energy of activation. The rate of reaction will be increased.

c. Addition of a catalyst increases the reaction rate.

d. Removal of reactant decreases the reaction rate.

5.71 **1. a.** S_2Cl_2 **b.** 135.2 g/mole   **c.** 0.0740 mole

2. a. C_6H_6 **b.** 78.0 g/mole   **c.** 0.0128 mole

5.73 **a.** $10 \times C(12.0) + 8 \times H(1.0) + 2 \times N(14.0) + 2 \times O(16.0) + 2 \times S(32.1) = 252.2$ g/mole

b. $25.0 \text{ g } C_{10}H_8N_2O_2S_2 \times \dfrac{1 \text{ mole}}{252.2 \text{ g } C_{10}H_8N_2O_2S_2} = 0.0991$ mole

c. $25.0 \text{ g } C_{10}H_8N_2O_2S_2 \times \dfrac{1 \text{ mole } C_{10}H_8N_2O_2S_2}{252.2 \text{ g } C_{10}H_8N_2O_2S_2} \times \dfrac{10 \text{ moles C}}{1 \text{ mole } C_{10}H_8N_2O_2S_2} = 0.991$ mole of C

d. $8.2 \times 10^{24} \text{ N atoms} \times \dfrac{1 \text{ mole N}}{6.02 \times 10^{23} \text{ N atoms}} \times \dfrac{1 \text{ mole } C_{10}H_8O_2N_2S_2}{2 \text{ moles N}}$
$= 6.8$ moles of $C_{10}H_8O_2N_2S_2$

5.75 **a.** 1, 1, 2 combination   **b.** 2, 2, 1 decomposition

5.77 Physical: solid candle wax melts (changes state), candle height is shorter, melted wax turns solid (changes state), shape of the wax changes, the wick becomes shorter.

Chemical: wax burns in oxygen, heat and light are emitted, wick burns in the presence of oxygen.

5.79 **a.** $2NO(g) + O_2(g) \rightarrow 2NO_2(g)$ **b.** combination

5.81 **a.** $2NI_3(g) \rightarrow N_2(g) + 3I_2(g)$ **b.** decomposition

5.83 **a.** $2Cl_2(g) + O_2(g) \rightarrow 2OCl_2(g)$ **b.** combination

5.85 **a.** $1.50 \text{ moles } C_3H_8 \times \dfrac{44.0 \text{ g } C_3H_8}{1 \text{ mole } C_3H_8} = 66.0$ g of C_3H_8

b. $34.0 \text{ g } C_3H_8 \times \dfrac{1 \text{ mole } C_3H_8}{44.0 \text{ g } C_3H_8} = 0.773$ mole of C_3H_8

c. $0.771 \text{ moles } C_3H_8 \times \dfrac{3 \text{ moles C}}{1 \text{ mole } C_3H_8} \times \dfrac{12.0 \text{ g C}}{1 \text{ mole C}} = 27.8$ g of C

5.87 **a.** $1 \text{ Fe atom} \times \dfrac{55.9 \text{ amu}}{1 \text{ Fe atom}} = 55.9 \text{ amu}$

$1 \text{ S atom} \times \dfrac{32.1 \text{ amu}}{1 \text{ S atom}} = 32.1 \text{ amu}$

$4 \text{ O atoms} \times \dfrac{16.0 \text{ amu}}{1 \text{ O atom}} = \dfrac{64.0 \text{ amu}}{152.0 \text{ amu}}$

b. $1 \text{ Ca atom} \times \dfrac{40.1 \text{ amu}}{1 \text{ Ca atom}} = 40.1 \text{ amu}$

$2 \text{ I atoms} \times \dfrac{126.9 \text{ amu}}{1 \text{ I atom}} = 253.8 \text{ amu}$

$6 \text{ O atoms} \times \dfrac{16.0 \text{ amu}}{1 \text{ O atom}} = \dfrac{96.0 \text{ amu}}{389.9 \text{ amu}}$

c. $5 \text{ C atoms} \times \dfrac{12.0 \text{ amu}}{1 \text{ C atom}} = 60.0 \text{ amu}$

$8 \text{ H atoms} \times \dfrac{1.0 \text{ amu}}{1 \text{ H atom}} = 8.0 \text{ amu}$

$1 \text{ N atom} \times \dfrac{14.0 \text{ amu}}{1 \text{ N atom}} = 14.0 \text{ amu}$

$1 \text{ Na atom} \times \dfrac{23.0 \text{ amu}}{1 \text{ Na atom}} = 23.0 \text{ amu}$

$4 \text{ O atoms} \times \dfrac{16.0 \text{ amu}}{1 \text{ O atom}} = \dfrac{64.0 \text{ amu}}{169.0 \text{ amu}}$

d. $6 \text{ C atoms} \times \dfrac{12.0 \text{ amu}}{1 \text{ C atom}} = 72.0 \text{ amu}$

$12 \text{ H atoms} \times \dfrac{1.0 \text{ amu}}{1 \text{ H atom}} = 12.0 \text{ amu}$

$2 \text{ O atoms} \times \dfrac{16.0 \text{ amu}}{1 \text{ O atom}} = \dfrac{32.0 \text{ amu}}{116.0 \text{ amu}}$

5.89 **a.** $0.150 \text{ mole K} \times \dfrac{39.1 \text{ g K}}{1 \text{ mole K}} = 5.87 \text{ g of K}$

b. $0.150 \text{ mole Cl}_2 \times \dfrac{71.0 \text{ g Cl}_2}{1 \text{ mole Cl}_2} = 10.7 \text{ g of Cl}_2$

c. $0.150 \text{ mole Na}_2\text{CO}_3 \times \dfrac{106.0 \text{ g Na}_2\text{CO}_3}{1 \text{ mole Na}_2\text{CO}_3} = 15.9 \text{ g of Na}_2\text{CO}_3$

5.91 **a.** $25.0 \text{ g CO}_2 \times \dfrac{1 \text{ mole CO}_2}{44.0 \text{ g CO}_2} = 0.568 \text{ mole of CO}_2$

b. $25.0 \text{ g Al(OH)}_3 \times \dfrac{1 \text{ mole Al(OH)}_3}{78.0 \text{ g Al(OH)}_3} = 0.321 \text{ mole of Al(OH)}_3$

c. $25.0 \text{ g MgCl}_2 \times \dfrac{1 \text{ mole MgCl}_2}{95.3 \text{ g MgCl}_2} = 0.262 \text{ mole of MgCl}_2$

5.93 **a.** $0.500 \text{ mole C}_3\text{H}_6\text{O}_3 \times \dfrac{6.02 \times 10^{23} \text{ molecules}}{1 \text{ mole C}_3\text{H}_6\text{O}_3} = 3.01 \times 10^{23} \text{ molecules}$

b. $1.50 \text{ mole C}_3\text{H}_6\text{O}_3 \times \dfrac{3 \text{ moles C}}{1 \text{ mole C}_3\text{H}_6\text{O}_3} \times \dfrac{6.02 \times 10^{23} \text{ atoms}}{1 \text{ mole C}} = 2.71 \times 10^{24} \text{ C atoms}$

c. 4.5×10^{24} ~~O atoms~~ $\times \dfrac{1 \text{ mole O}}{6.02 \times 10^{23} \text{ O atoms}} \times \dfrac{1 \text{ mole } C_3H_6O_3}{3 \text{ moles O}} = 2.5$ moles of $C_3H_6O_3$

d. $3 \times C(12.0) + 6 \times H(1.0) + 3 \times O(16.0) = 90.0$ g/mole

5.95 a. $NH_3(g) + HCl(g) \rightarrow NH_4Cl(s)$ combination

b. $Fe_3O_4(s) + 4H_2(g) \rightarrow 3Fe(s) + 4H_2O(g)$ single replacement

c. $2Sb(s) + 3Cl_2(g) \rightarrow 2SbCl_3(s)$ combination

d. $2\,NI_3(s) \rightarrow N_2(g) + 3I_2(g)$ decomposition

e. $2\,KBr(aq) + Cl_2(aq) \rightarrow 2KCl(aq) + Br_2(l)$ single replacement

f. $2Fe(s) + 3H_2SO_4(aq) \rightarrow Fe_2(SO_4)_3(s) + 3H_2(g)$ single replacement

g. $Al_2(SO_4)_3(aq) + 6NaOH(aq) \rightarrow 3Na_2SO_4(aq) + 2Al(OH)_3(s)$ double replacement

5.97 a. $Zn^{2+} + 2e^- \rightarrow Zn$ reduction   **b.** $Al \rightarrow Al^{3+} + 3e^-$ oxidation

c. $Pb \rightarrow Pb^{2+} + 2e^-$ oxidation   **d.** $Cl_2 + 2e^- \rightarrow 2Cl^-$ reduction

5.99 a. 124 ~~g C_2H_6O~~ $\times \dfrac{1 \text{ mole } C_2H_6O}{46.0 \text{ g } C_2H_6O} \times \dfrac{1 \text{ mole } C_6H_{12}O_6}{2 \text{ moles } C_2H_6O} = 1.35$ moles of $C_6H_{12}O_6$

b. 0.240 ~~kg $C_6H_{12}O_6$~~ $\times \dfrac{1000 \text{ g}}{1 \text{ kg}} \times \dfrac{1 \text{ mole } C_6H_{12}O_6}{180.0 \text{ g } C_6H_{12}O_6} \times \dfrac{2 \text{ moles } C_2H_6O}{1 \text{ mole } C_6H_{12}O_6} \times \dfrac{46.0 \text{ g } C_2H_6O}{1 \text{ mole } C_2H_6O}$

$= 123$ g of C_2H_6O

5.101 $2NH_3(g) + 5F_2(g) \rightarrow N_2F_4(g) + 6HF(g)$

a. 4.00 ~~moles HF~~ $\times \dfrac{2 \text{ moles } NH_3}{6 \text{ moles HF}} = 1.33$ moles of NH_3

4.00 ~~moles HF~~ $\times \dfrac{5 \text{ moles } F_2}{6 \text{ moles HF}} = 3.33$ moles of F_2

b. 1.50 ~~moles NH_3~~ $\times \dfrac{5 \text{ moles } F_2}{2 \text{ moles } NH_3} \times \dfrac{38.0 \text{ g } F_2}{1 \text{ mole } F_2} = 143$ g of F_2

c. 3.40 ~~g NH_3~~ $\times \dfrac{1 \text{ mole } NH_3}{17.0 \text{ g } NH_3} \times \dfrac{1 \text{ mole } N_2F_4}{2 \text{ moles } NH_3} \times \dfrac{104.0 \text{ g } N_2F_4}{1 \text{ mole } N_2F_4} = 10.4$ g of N_2F_4

5.103 a. 4.0 ~~moles H_2O~~ $\times \dfrac{1 \text{ mole } C_5H_{12}}{6 \text{ moles } H_2O} \times \dfrac{72.0 \text{ g } C_5H_{12}}{1 \text{ mole } C_5H_{12}} = 48$ g of C_5H_{12}

b. 32.0 ~~g O_2~~ $\times \dfrac{1 \text{ mole } O_2}{32.0 \text{ g } O_2} \times \dfrac{5 \text{ moles } CO_2}{8 \text{ moles } O_2} \times \dfrac{44.0 \text{ g } CO_2}{1 \text{ moles } CO_2} = 27.5$ g of CO_2

5.105 a. Heat is a product; exothermic

b. Heat is given off; the energy of the products is lower than the energy of the reactants.

5.107 a. 5.00 g of gold   **b.** 1.52×10^{22} Au atoms

c. 0.038 moles of oxygen   **d.** Au_2O_3

5.109 a. $3Pb(NO_3)_2(aq) + 2Na_3PO_4(aq) \rightarrow Pb_3(PO_4)_2(s) + 6NaNO_3(aq)$ double replacement

b. $4Ga(s) + 3O_2(g) \rightarrow 2Ga_2O_3(s)$ combination

c. $2NaNO_3(s) \rightarrow 2NaNO_2(s) + O_2(g)$ decomposition

d. $Bi_2O_3(s) + 3C(s) \rightarrow 2Bi(s) + 3CO(g)$ single replacement

5.111 a. $4Al(s) + 3O_2(g) \rightarrow 2Al_2O_3(s)$ **b.** This is a combination reaction.
    c. 5.63 moles of oxygen   **d.** 94.9 g of aluminum oxide
    e. 17.0 g of aluminum oxide

Answers to Combining Ideas from Chapters 3 to 5

CI.7 a. X is a metal; Y is a nonmetal.   **b.** Y has the higher electronegativity.
     c. X^{2+}, Y^-
     d. 1. X = 2, 8, 2   Y = 2, 8, 7
        2. X^{2+} = 2, 8 Y^- = 2, 8, 8
        3. X^{2+} has the same electron arrangement as Ne.
          Y^- has the same electron arrangement as Ar.
        4. $MgCl_2$, magnesium chloride
     e. Li is D; Na is A; K is C, and Rb is B.

CI.9 a.

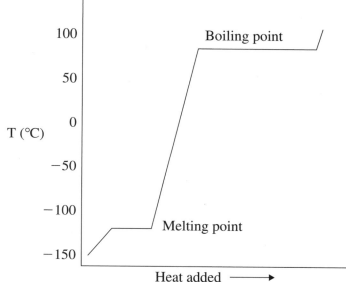

b. $20.0 \cancel{g} \times 92 \, \cancel{°C} \times \dfrac{0.588 \text{ cal}}{\cancel{g \cdot °C}} = 1.1 \times 10^3 \text{ cal}$

c. $1000 \, \cancel{mL} \times \dfrac{0.796 \, \cancel{g}}{\cancel{mL}} \times \dfrac{201 \, \cancel{cal}}{\cancel{g}} \times \dfrac{4.184 \, \cancel{J}}{\cancel{cal}} \times \dfrac{1 \text{ kJ}}{1000 \, \cancel{J}} = 669 \text{ kJ}$

d. $15 \, \cancel{gal} \times \dfrac{4 \, \cancel{qt}}{1 \, \cancel{gal}} \times \dfrac{1 \, \cancel{L \, E10 \, fuel}}{1.06 \, \cancel{qt}} \times \dfrac{10 \text{ L ethanol}}{100 \, \cancel{L \, E10 \, fuel}} = 5.7 \text{ L of ethanol}$

e. $C_2H_5OH(l) + 3O_2(g) \rightarrow 2CO_2(g) + 3H_2O(g)$

f. $5.7 \, \cancel{L \, ethanol} \times \dfrac{1000 \, \cancel{mL}}{1 \, \cancel{L}} \times \dfrac{0.796 \, \cancel{g}}{1 \, \cancel{mL}} \times \dfrac{1 \, \cancel{mole \, ethanol}}{46.0 \, \cancel{g}} \times \dfrac{2 \, \cancel{moles \, CO_2}}{1 \, \cancel{mole \, ethanol}}$

     $\times \dfrac{44.0 \, \cancel{g}}{1 \, \cancel{mole \, CO_2}} \times \dfrac{1 \text{ kg}}{1000 \, \cancel{g}} = 8.7 \text{ kg of } CO_2$

CI.11 a. $C_7H_{10}O_5$

     b. $7 \times C(12.0) + 10 \times H(1.0) + 5 \times O(16.0) = 174 \text{ g/mole}$

c. $1.3 \ \cancel{g} \ SA \times \dfrac{1 \text{ mole SA}}{174 \ \cancel{g \ SA}} = 0.0075$ mole of shikimic acid

d. $154 \ \cancel{g \ anise} \times \dfrac{0.13 \ \cancel{g \ SA}}{2.6 \ \cancel{g \ anise}} \times \dfrac{1 \text{ capsule Tamiflu}}{0.13 \ \cancel{g \ SA}} = 59$ capsules of Tamiflu

e. $C_{16}H_{28}N_2O_4 = 16 \times C(12.0) + 28 \times H(1.0) + 2 \times N(14.0) + 4 \times O(16.0)$
   $= 312.0$ g/mole

f. $1 \ \cancel{capsule} \times \dfrac{75 \ \cancel{mg \ Tamiflu}}{1 \ \cancel{capsule}} \times \dfrac{1 \ \cancel{g}}{1000 \ \cancel{mg}} \times \dfrac{1 \ \cancel{mole \ Tamiflu}}{312 \ \cancel{g}}$

   $\times \dfrac{16 \ \cancel{moles \ C}}{1 \ \cancel{mole \ Tamiflu}} \times \dfrac{12.0 \text{ g C}}{1 \ \cancel{mole \ C}} = 0.046$ g of C

g. $500 \ 000 \ \cancel{people} \times \dfrac{2 \text{ capsules}}{1 \ \cancel{day} \cdot 1 \ \cancel{person}} \times 5 \ \cancel{days} \times \dfrac{75 \ \cancel{mg \ Tamiflu}}{1 \ \cancel{capsule}} \times \dfrac{1 \ \cancel{g}}{1000 \ \cancel{mg}}$

   $\times \dfrac{1 \text{ kg}}{1000 \ \cancel{g}} = 380$ kg of Tamiflu

Study Goals

- Describe the kinetic molecular theory of gases.
- Describe the units of measurement used for pressure, and change from one unit to another.
- Use the gas laws to determine the new pressure, volume, or temperature of a specific amount of gas.
- Describe the relationship between the amount of a gas and its volume.
- Use the combined gas law to find the new pressure, volume, or temperature of a gas when changes in two of these properties are given.
- Use partial pressure to calculate the total pressure of a gas mixture.

Think About It

1. How does a barometer work?

2. What happens to the pressure on a person who is scuba diving?

3. Why are airplanes pressurized?

4. Why does a bag of chips expand when you take it to a higher altitude?

Key Terms

Match the key terms with the statements shown below.

  a. kinetic theory **b.** pressure   **c.** Boyle's law
  d. Charles' law **e.** partial pressure

1. _____ The volume of a gas varies directly with the Kelvin temperature when pressure and amount of gas remain constant.

2. _____ Force exerted by gas particles that collide with the sides of a container.

3. _____ The pressure exerted by the individual gases in a gas mixture.

4. _____ The volume of a gas varies inversely with the pressure of a gas when temperature and amount of gas are constant.

5. _____ A model that explains the behavior of gaseous particles.

Answers   **1.** d   **2.** b   **3.** e   **4.** c   **5.** a

6.1 Properties of Gases

- In a gas, particles are so far apart and moving so fast that they are not attracted to each other.
- A gas is described by the physical properties of pressure (P), volume (V), temperature (T), and amount in moles (n).

◆ Learning Exercise 6.1

True or false:

a. _____ Gases are composed of small particles.

b. _____ Gas molecules are usually close together.

c. _____ Gas molecules move rapidly because they are strongly attracted.

d. _____ The distances between gas molecules are great.

e. _____ Gas molecules travel in straight lines until they collide.

Answers   **a.** T   **b.** F   **c.** F   **d.** T   **e.** T

6.2 Gas Pressure

- A gas exerts pressure, which is the force of the gas particles on the surface of a container.
- Units of gas pressure include torr, mmHg, and atmosphere.

◆ Learning Exercise 6.2

Complete the following:

a. 1.50 atm = _____ mmHg

b. 550 mmHg = _____ atm

c. 725 mmHg = _____ torr

d. 1520 mmHg = _____ atm

Answers   **a.** 1140 mmHg   **b.** 0.72 atm   **c.** 725 torr   **d.** 2.00 atm

6.3 Pressure and Volume (Boyle's Law)

- According to Boyle's law, pressure increases if volume decreases; pressure decreases if volume increases.
- The volume (V) of a gas changes inversely with the pressure (P) of the gas when T and n are held constant: $P_1V_1 = P_2V_2$.

◆ Learning Exercise 6.3A

Complete with *increases* or *decreases:*

1. Gas pressure increases (T constant) when volume _____.

2. Gas volume increases at constant T when pressure _____.

Answers   **1.** decreases   **2.** decreases

◆ Learning Exercise 6.3B

Calculate the variable in each of the following gas problems using Boyle's law.

a. Four (4.0) liters of helium gas has a pressure of 800. mmHg. What will be the new pressure if the volume is reduced to 1.0 liter (n and T constant)?

b. A gas occupies a volume of 360 mL at 750 mmHg. What volume does it occupy at a pressure of (1) 1500 mmHg? (2) 375 mmHg? (*n* and *T* constant)

c. A gas sample at a pressure of 5.0 atm has a volume of 3.00 L. If the gas pressure is changed to 760 mmHg, what volume will the gas occupy (*n* and *T* constant)?

d. A sample of 250. mL of nitrogen is initially at a pressure of 2.50 atm. If the pressure changes to 825 mmHg, what is the new volume in mL (*n* and *T* constant)?

Answers   **a.** 3200 mmHg   **b.** (1) 180 mL (2) 720 mL   **c.** 15 L   **d.** 576 mL

6.4 Temperature and Volume (Charles' Law)

- The volume (*V*) of a gas is directly related to its Kelvin temperature (*T*) when there is no change in the pressure of the gas:

$$\frac{V_1}{T_1} = \frac{V_2}{T_2}$$

- According to Charles' law, temperature increases if the volume of the gas increases; temperature decreases if volume decreases.

◆ **Learning Exercise 6.4A**

Complete with *increases* or *decreases:*

a. When the temperature of a gas increases at constant pressure, its volume _____.
b. When the volume of a gas decreases at constant pressure, its temperature _____.

Answers: **a.** increases   **b.** decreases

◆ **Learning Exercise 6.4B**

Use Charles' law to solve the following gas problems.

a. A large balloon has a volume of 2.5 L at a temperature of 0 °C. What is the new volume of the balloon when the temperature rises to 120 °C and the pressure remains constant?

b. Consider a balloon filled with helium to a volume of 6600 L at a temperature of 223 °C. To what temperature must the gas be cooled to decrease the volume to 4800 L? (*P* constant)

c. A sample of 750 mL of neon is heated from 120 °C to 350 °C. If pressure is kept constant, what is the new volume?

d. What is the final temperature of 355 mL of oxygen gas at 22 °C if its volume increases to 0.840 L? (*P* constant)

Answers   **a.** 3.6 L   **b.** 88 °C   **c.** 1200 mL   **d.** 425 °C

6.5 Temperature and Pressure (Gay-Lussac's Law)

- The pressure (P) of a gas is directly related to its Kelvin temperature (T).

$$\frac{P_1}{T_1} = \frac{P_2}{T_2}$$

This means that an increase in temperature increases the pressure of a gas, or a decrease in temperature decreases the pressure, as long as the volume stays constant.

◆ Learning Exercise 6.5

Solve the following gas law problems using Gay-Lussac's law.

a. A sample of helium gas has a pressure of 860 mmHg at a temperature of 225 K. At what pressure (mmHg) will the helium sample reach a temperature of 675 K?
(V constant)

b. A balloon contains a gas with a pressure of 580 mmHg and a temperature of 227 °C. What is the new pressure (mmHg) of the gas when the temperature drops to 27 °C?
(V constant)

c. A spray can contains a gas with a pressure of 3.0 atm at a temperature of 17 °C. What is the pressure (atm) in the container if the temperature inside the can rises to 110 °C?
(V constant)

d. A gas has a pressure of 1200 mmHg at 300. °C. What will the temperature (°C) be when the pressure falls to 1.10 atm?
(V constant)

Answers   a. 2580 mmHg   b. 348 mmHg   c. 4.0 atm   d. 126 °C

6.6 The Combined Gas Law

- The gas laws can be combined into a relationship of pressure (P), volume (V), and temperature (T).

$$\frac{P_1V_1}{T_1} = \frac{P_2V_2}{T_2}$$

This expression is used to determine the effect of changes in two of the variables on the third.

◆ Learning Exercise 6.6

Solve the following using the combined gas laws:

a. A 5.0-L sample of nitrogen gas has a pressure of 1200 mmHg at 220 K. What is the pressure of the sample when the volume increases to 20.0 L at 440 K?

b. A 10.0-L sample of gas is emitted from a volcano with a pressure of 1.20 atm and a temperature of 150.0 °C. What is the volume of the gas when its pressure is 0.900 atm and the temperature is −40.0 °C?

c. A 25.0-mL bubble forms at the ocean depths where the pressure is 10.0 atm and the temperature is 5.0 °C. What is the volume of that bubble at the ocean surface where the pressure is 760.0 mmHg and the temperature is 25 °C?

d. A 35.0-mL sample of argon gas has a pressure of 1.0 atm and a temperature of 15 °C. What is the final volume if the pressure goes to 2.0 atm and the temperature to 45 °C?

e. A 315-L weather balloon is launched at Earth's surface where the temperature is 12 °C and the pressure is 0.93 atm. What is the volume of the balloon in the upper atmosphere if the pressure is 116 mmHg and the temperature is −35 °C?

Answers   **a.** 600 mmHg   **b.** 7.34 L   **c.** 268 mL   **d.** 19 mL   **e.** 1600 L

6.7 Volume and Moles (Avogadro's Law)

● Avogadro's law states that equal volumes of gases at the same temperature and pressure contain the same number of moles. The volume (*V*) of a gas is directly related to the number of moles of the gas when the pressure and temperature of the gas do not change.

$$\frac{V_1}{n_1} = \frac{V_2}{n_2}$$

● If the moles of gas increase, the volume must increase; if the moles of gas decrease, the volume decreases.
● At STP conditions, standard pressure (1 atm) and temperature (0 °C), one mole of a gas occupies a volume of 22.4 L.

Study Note

At STP, the molar volume factor 22.4 L/mole converts between volume and moles of a gas.
Example: How many liters would 2.00 moles of N_2 occupy at STP?

Solution: $2.00 \ \text{moles N}_2 \times \dfrac{22.4 \ \text{L (STP)}}{1 \ \text{mole N}_2} = 44.8 \ \text{L (STP)}$

◆ **Learning Exercise 6.7A**

Use Avogadro's law to solve the following gas problems:

a. A gas containing 0.50 mole of helium has a volume of 4.00 L. What is the new volume when 1.0 mole of nitrogen is added to the container when pressure and temperature remain constant?

b. A balloon containing 1.00 mole of oxygen has a volume of 15 L. What is the new volume of the balloon when 2.00 moles of helium are added (*T* and *P* constant)?

c. What is the volume occupied by 28.0 g of nitrogen (N_2) at STP?

d. What is the volume (in liters) of a container that holds 6.40 g of O_2 at STP?

Answers   **a.** 12 L   **b.** 45 L   **c.** 22.4 L   **d.** 4.48 L

◆ Learning Exercise 6.7B

Calculate the amount reacted or produced in each of the following equations involving gases:

a. How many liters of Cl_2 gas at STP are required to completely react 8.50 g of K?
$$2K(s) + Cl_2(g) \rightarrow 2KCl(s)$$

b. How many grams of $KClO_3$ must be heated to produce 12.6 L of O_2 gas at STP?
$$2KClO_3(s) \rightarrow 2KCl(s) + 3O_2(g)$$

c. How many liters of NO gas at STP can be produced when 55.8 L of O_2 react at STP?
$$4NH_3(g) + 5O_2(g) \rightarrow 4NO(g) + 6H_2O(l)$$

Answers

a. $8.50 \text{ g K} \times \dfrac{1 \text{ mole K}}{39.1 \text{ g K}} \times \dfrac{1 \text{ mole Cl}_2}{2 \text{ moles K}} \times \dfrac{22.4 \text{ L}}{1 \text{ mole Cl}_2} = 2.43 \text{ L of Cl}_2$

b. $12.6 \text{ L} \times \dfrac{1 \text{ mole O}_2}{22.4 \text{ L}} \times \dfrac{2 \text{ moles KClO}_3}{3 \text{ moles O}_2} \times \dfrac{122.6 \text{ g KClO}_3}{1 \text{ mole KClO}_3} = 46.0 \text{ g of KClO}_3$

c. $55.8 \text{ L O}_2 \times \dfrac{1 \text{ mole O}_2}{22.4 \text{ L O}_2} \times \dfrac{4 \text{ moles NO}}{5 \text{ moles O}_2} \times \dfrac{22.4 \text{ L NO}}{1 \text{ mole NO}} = 44.6 \text{ L of NO}$

6.8 Partial Pressures (Dalton's Law)

- In a mixture of two or more gases, the total pressure is the sum of the partial pressures of the individual gases.

$$P_{total} = P_1 + P_2 + P_3 + \ldots$$

- The partial pressure of a gas in a mixture is the pressure it would exert if it were the only gas in the container.

◆ Learning Exercise 6.8A

Use Dalton's law to solve the following problems about gas mixtures:

a. What is the pressure in mmHg of a sample of gases containing oxygen at 0.500 atm, nitrogen (N_2) at 132 torr, and helium at 224 mmHg?

b. What is the pressure (atm) of a gas sample containing helium at 285 mmHg and oxygen (O_2) at 1.20 atm?

104

c. A gas sample containing nitrogen (N_2) and oxygen (O_2) has a pressure of 1500 mmHg. If the partial pressure of the nitrogen is 0.900 atm, what is the partial pressure (mmHg) of the oxygen gas in the mixture?

Answers　　**a.** 736 mmHg　　**b.** 1.58 atm　　**c.** 816 mmHg

◆ Learning Exercise 6.8B

Fill in the blanks by writing I (increases) or D (decreases) for a gas in a closed container.

	Pressure	Volume	Moles	Temperature
a.	____	increases	constant	constant
b.	increases	constant	____	constant
c.	constant	decreases	____	constant
d.	____	constant	constant	increases
e.	constant	____	constant	decreases
f.	____	constant	decreases	constant

Answers　　**a.** D　　**b.** I　　**c.** D
　　　　　　d. I　　**e.** D　　**f.** D

◆ Learning Exercise 6.8C

Complete the table for typical blood gas values for partial pressures.

Gas	Alveoli	Oxygenated blood	Deoxygenated blood	Tissues
CO_2				
O_2				

Answers

Gas	Alveoli	Oxygenated blood	Deoxygenated blood	Tissues
CO_2	40 mmHg	40 mmHg	46 mmHg	50 mmHg or greater
O_2	100 mmHg	100 mmHg	40 mmHg	30 mmHg or less

Checklist for Chapter 6

You are ready to take the practice test for Chapter 6. Be sure that you have accomplished the following learning goals for this chapter. If you are not sure, review the section listed at the end of the goal. Then apply your new skills and understanding to the practice test.

After studying Chapter 6, I can successfully:

_____ Describe the kinetic theory of gases (6.1).

_____ Change the units of pressure from one to another (6.2).

_____ Use the pressure–volume relationship (Boyle's law) to determine the new pressure or volume of a fixed amount of gas at constant temperature (6.3).

_____ Use the temperature–volume relationship (Charles' law) to determine the new temperature or volume of a fixed amount of gas at a constant pressure (6.4).

_____ Use the temperature–pressure relationship (Gay-Lussac's law) to determine the new temperature or pressure of a certain amount of gas at a constant volume (6.5).

_____ Use the combined gas law to find the new pressure, volume, or temperature of a gas when changes in two of these properties are given (6.6).

_____ Describe the relationship between the amount of a gas and its volume, and use this relationship in calculations with gases (6.7).

_____ Calculate the total pressure of a gas mixture from the partial pressures (6.8).

Practice Test for Chapter 6

Answer questions 1 through 5 using T (true) or F (false):

1. _____ A gas does not have its own volume or shape.

2. _____ The molecules of a gas are moving extremely fast.

3. _____ The collisions of gas molecules with the walls of their container create pressure.

4. _____ Gas molecules are close together and move in straight-line patterns.

5. _____ We consider gas molecules to have no attractions between them.

6. When a gas is heated in a closed metal container, the
   A. pressure increases.   B. pressure decreases.   C. volume increases.
   D. volume decreases.   E. number of molecules increases.

7. The pressure of a gas will increase when
   A. the volume increases (n and T constant).
   B. the temperature decreases (n and V constant).
   C. more molecules of gas are added (V and T constant).
   D. molecules of gas are removed (V and T constant).
   E. None of these

8. If the temperature of a gas is increased,
   A. the pressure will decrease (V and n constant).
   B. the volume will increase (P and n constant).
   C. the volume will decrease (P and n constant).
   D. the number of molecules will increase (P and V constant).
   E. None of these

9. The relationship that the volume of a gas is inversely related to its pressure at constant temperature is known as
   A. Boyle's law   **B.** Charles' law   **C.** Gay-Lussac's law
   D. Dalton's law   **E.** Avogadro's law

10. What is the pressure (atm) of a gas with a pressure of 1200 mmHg?
    A. 0.63 atm   **B.** 0.79 atm   **C.** 1.2 atm
    D. 1.6 atm   **E.** 2.0 atm

11. A 6.00-L sample of oxygen has a pressure of 660 mmHg. When the volume is reduced to 2.00 liters at constant temperature, it will have a new pressure of
    A. 1980 mmHg   **B.** 1320 mmHg   **C.** 330. mmHg
    D. 220. mmHg   **E.** 110. mmHg

12. A sample of nitrogen gas at 180 K has a pressure of 1.0 atm. When the temperature is increased to 360 K at constant volume, the new pressure will be
    A. 0.50 atm   **B.** 1.0 atm   **C.** 1.5 atm
    D. 2.0 atm   **E.** 4.0 atm

13. If two gases have the same volume, temperature, and pressure, they also have the same
    A. density.   **B.** number of particles.   **C.** molar mass.
    D. speed.   **E.** size molecules.

14. A gas sample has a volume of 4.00 L, a pressure of 750 mmHg, and a temperature of 77 °C. What is its new volume at 277 °C and 250 mmHg?
    A. 7.6 L   **B.** 19 L   **C.** 2.1 L
    D. 0.00056 L   **E.** 3.3 L

15. If the temperature of a gas does not change, but its volume doubles, its pressure will
    A. double.
    B. triple.
    C. decrease to one-half the original pressure.
    D. decrease to one-fourth the original pressure.
    E. not change.

16. A sample of oxygen with a pressure of 400 mmHg contains 2.0 moles of gas and has a volume of 4.0 L. What will the new pressure be when the volume expands to 5.0 L, and 3.0 moles of helium gas is added while temperature is constant?
    A. 160 mmHg   **B.** 250 mmHg   **C.** 800 mmHg
    D. 1000 mmHg   **E.** 1560 mmHg

17. A sample of 2.00 moles of gas initially at STP is converted to a volume of 5.0 L and a temperature of 27 °C. What is its new pressure in atm?
    A. 0.12 atm   **B.** 5.5 atm   **C.** 7.5 atm
    D. 8.9 atm   **E.** 9.8 atm

18. The conditions for standard temperature and pressure (STP) are
    A. 0 K, 1 atm   **B.** 0 °C, 10 atm   **C.** 25 °C, 1 atm
    D. 273 K, 1 atm   **E.** 273 K, 0.5 atm

19. The volume occupied by 1.50 moles of CH_4 at STP is
    A. 44.8 L   **B.** 33.6 L   **C.** 22.4 L
    D. 11.2 L   **E.** 5.60 L

20. How many grams of oxygen gas (O_2) are present in 44.8 L of oxygen at STP?
    A. 8.0 g   **B.** 16.0 g   **C.** 32.0 g
    D. 48.0 g   **E.** 64.0 g

21. What is the volume in liters of 0.500 mole of nitrogen gas (N_2) at STP?
    A. 0.500 L   **B.** 1.00 L   **C.** 11.2 L
    D. 22.4 L   **E.** 44.8 L

22. A gas mixture contains helium with a partial pressure of 0.80 atm, oxygen with a partial pressure of 450 mmHg, and nitrogen with a partial pressure of 230 mmHg. What is the total pressure in atm for the gas mixture?

    A. 1.10 atm   **B.** 1.39 atm   **C.** 1.69 atm
    D. 2.00 atm   **E.** 8.00 atm

23. A mixture of oxygen and nitrogen has a total pressure of 840 mmHg. If the oxygen has a partial pressure of 510 mmHg, what is the partial pressure of the nitrogen?

    A. 240 mmHg   **B.** 330 mmHg   **C.** 1.69 mmHg
    D. 1040 mmHg   **E.** 1350 mmHg

24. The exchange of gases between the alveoli, blood, and tissues of the body is a result of

    A. pressure gradients.   **B.** differences in molar mass.   **C.** shapes of molecules.
    D. altitude.   **E.** All of these.

25. Oxygen moves into the tissues from the blood because its partial pressure

    A. in arterial blood is higher than in the tissues.
    B. in venous blood is higher than in the tissues.
    C. in arterial blood is lower than in the tissues.
    D. in venous blood is lower than in the tissues.
    E. is equal in the blood and in the tissues.

Answers to the Practice Test

1. T	**2.** T	**3.** T	**4.** F	**5.** T
6. A	**7.** C	**8.** B	**9.** A	**10.** D
11. A	**12.** D	**13.** B	**14.** B	**15.** C
16. C	**17.** E	**18.** D	**19.** B	**20.** E
21. C	**22.** C	**23.** B	**24.** A	**25.** A

Answers and Solutions to Selected Text Problems

6.1   **a.** Gaseous particles have greater kinetic energies at higher temperatures. Because kinetic energy is a measure of the energy of motion, the gaseous particles must be moving faster at higher temperatures than at lower values.

      b. Because particles in a gas are very far apart, gases can be easily compressed without the particles bumping into neighboring gas particles. Neighboring particles are much closer together in solids and liquids, and they will "bump" into each other and repel each other if the sample is compressed.

6.3   **a.** temperature   **b.** volume   **c.** amount of gas   **d.** pressure

6.5   Some units used to describe the pressure of a gas are pounds per square inch (lb/in.^{2} which is also abbreviated as psi), atmospheres (abbreviated atm), torr, mmHg, in. Hg, and pascals.

6.7   **a.** $2.00 \; \cancel{atm} \times \dfrac{760 \text{ torr}}{1 \; \cancel{atm}} = 1520 \text{ torr}$

      b. $2.00 \; \cancel{atm} \times \dfrac{760 \text{ mmHg}}{1 \; \cancel{atm}} = 1520 \text{ mmHg}$

6.9   The gases in the diver's lungs (and dissolved in the blood) will expand because pressure decreases as the diver ascends. Unless the diver exhales, the expanding gases could rupture the membranes in the lung tissues. In addition, the formation of gas bubbles in the bloodstream could cause "the bends."

6.11 **a.** According to Boyle's law, for the pressure to increase while temperature and quantity of gas remains constant, the gas volume must decrease. Thus, cylinder A would represent the final volume.

b.

	Conditions 1	Conditions 2	Know	Predict
P	650 mmHg	1.2 atm (910 mmHg)	*P* increases	
V	220 mL	160 mL		*V* decreases

Because $P_1V_1 = P_2V_2$, then $V_2 = P_1V_1/P_2$

$$V_2 = 220 \text{ mL} \times \frac{650 \text{ mmHg}}{1.2 \text{ atm}} \times \frac{1 \text{ atm}}{760 \text{ mmHg}} = 160 \text{ mL}$$

6.13 **a.** The pressure doubles when the volume is halved.
b. The pressure falls to one-third the initial pressure when the volume expands to three times its initial volume.
c. The pressure increases to ten times the original pressure when the volume decreases to 1/10 of its initial volume.

6.15 From Boyle's law we know that pressure is inversely related to volume. (For example, the pressure increases when the volume decreases.)

a. Volume increases; pressure must decrease.

$$655 \text{ mmHg} \times \frac{10.0 \text{ L}}{20.0 \text{ L}} = 328 \text{ mmHg}$$

b. Volume decreases; pressure must increase.

$$655 \text{ mmHg} \times \frac{10.0 \text{ L}}{2.50 \text{ L}} = 2620 \text{ mmHg}$$

c. The mL units must be converted to L for unit cancellation in the calculation, and because the volume decreases; pressure must increase.

$$655 \text{ mmHg} \times \frac{10.0 \text{ L}}{1500 \text{ mL}} \times \frac{1000 \text{ mL}}{1 \text{ L}} = 4400 \text{ mmHg}$$

6.17 From Boyle's law we know that pressure is inversely related to volume.

a. Pressure increases; volume must decrease.

$$50.0 \text{ L} \times \frac{760 \text{ mmHg}}{1500 \text{ mmHg}} = 25 \text{ L}$$

b. The mmHg units must be converted to atm for unit cancellation in the calculation.

$$50.0 \text{ L} \times \frac{760 \text{ mmHg}}{2.0 \text{ atm}} \times \frac{1 \text{ atm}}{760 \text{ mmHg}} = 25 \text{ L}$$

c. The mmHg units must be converted to atm for unit cancellation in the calculation.

$$50.0 \text{ L} \times \frac{760 \text{ mmHg}}{0.500 \text{ atm}} \times \frac{1 \text{ atm}}{760 \text{ mmHg}} = 100. \text{ L}$$

6.19 Pressure decreases; volume must increase.

$$5.0 \text{ L} \times \frac{5.0 \text{ atm}}{1.0 \text{ atm}} = 25 \text{ L}$$

6.21 **a.** Inspiration begins when the diaphragm flattens causing the lungs to expand. The increased volume reduces the pressure in the lungs such that air flows into the lungs.

b. Expiration occurs as the diaphragm relaxes causing a decrease in the volume of the lungs. The pressure of the air in the lungs increases and air flows out of the lungs.

c. Inspiration occurs when the pressure in the lungs is less than the pressure of the air in the atmosphere.

6.23 According to Charles' law, there is a direct relationship between temperature and volume. For example, volume increases when temperature increases while the pressure and amount of gas remains constant.

a. Diagram C describes an increased volume corresponding to an increased temperature.
b. Diagram A describes a decreased volume corresponding to a decrease in temperature.
c. Diagram B shows no change in volume, which corresponds to no change in temperature.

6.25 $15\,°C + 273 = 288\,K$

a. $288\,K \times \dfrac{5.00\,\cancel{L}}{2.50\,\cancel{L}} = 576\,K - 273 = 303\,°C$

b. $288\,K \times \dfrac{1.25\,\cancel{L}}{2.50\,\cancel{L}} = 144\,K - 273 = -129\,°C$

c. $288\,K \times \dfrac{7.50\,\cancel{L}}{2.50\,\cancel{L}} = 864\,K - 273 = 591\,°C$

d. $288\,K \times \dfrac{3.55\,\cancel{L}}{2.50\,\cancel{L}} = 409\,K - 273 = 136\,°C$

6.27 According to Charles' law, gas volume is directly proportional to Kelvin temperature when P and n are constant. In all gas law computations, temperatures must be in Kelvin units. (Temperatures in $°C$ are converted to K by the addition of 273.)

a. When temperature decreases, volume must also decrease.

$75\,°C + 273 = 348\,K \quad 55\,°C + 273 = 328\,K$

$2500\,mL \times \dfrac{328\,K}{348\,K} = 2400\,mL$

b. When temperature increases, volume must also increase.

$2500\,mL \times \dfrac{680\,K}{348\,K} = 4900\,mL$

c. $-25\,°C + 273 = 248\,K$

$2500\,mL \times \dfrac{248\,K}{348\,K} = 1800\,mL$

d. $2500\,mL \times \dfrac{240\,K}{348\,K} = 1700\,mL$

6.29 Because gas pressure increases with an increase in temperature, the gas pressure in an aerosol container might exceed the tolerance of the can when it is heated and cause it to explode.

6.31 **a.** $25\,°C + 273 = 298\,°C \qquad T_2 = T_1 \times P_2/P_1$

Prediction: A decrease in pressure will make T_1 lower than T_2

$298\,K \times \dfrac{620\,\cancel{mmHg}}{740\,\cancel{mmHg}} = 250.\,K - 273 = -23\,°C$

b. $-18\,°C + 273 = 255\,K \qquad 0.950\,\cancel{atm} \times 760\,torr/1\,\cancel{atm} = 722\,torr$

Prediction: An increase in pressure will make T_1 higher than T_2

$255\,K \times \dfrac{1250\,\cancel{torr}}{722\,\cancel{torr}} = 441\,K - 273 = 168\,°C$

6.33 According to Gay-Lussac's law, temperature is directly related to pressure. For example, temperature increases when the pressure increases. In all gas law computations, temperatures must be in Kelvin units. (Temperatures in °C are converted to K by the addition of 273.)

   a. When temperature decreases, pressure must also decrease.

$$155\,°C + 273 = 428\,K \qquad\qquad 0\,°C + 273 = 273\,K$$

$$1200\text{ torr} \times \frac{273\,K}{428\,K} = 700\text{ torr}$$

   b. When temperature increases, pressure must also increase.

$$12\,°C + 273 = 285\,K \qquad\qquad 35\,°C + 273 = 308\,K$$

$$1.40\text{ atm} \times \frac{308\,K}{285\,K} = 1.51\text{ atm}$$

6.35 $T_1 = 325\,K$; $V_1 = 6.50\,L$; $P_1 = 845\text{ mmHg}$ (1.11 atm)

   a. $T_2 = 125\,°C + 273 = 398\,K$; $V_2 = 1.85\,L$

$$1.11\text{ atm} \times \frac{6.50\,L}{1.85\,L} \times \frac{325\,K}{298\,K} = 4.25\text{ atm}$$

   b. $T_2 = 12\,°C + 273 = 285\,K$; $V_2 = 2.25\,L$

$$1.11\text{ atm} \times \frac{6.50\,L}{2.25\,L} \times \frac{285\,K}{298\,K} = 3.07\text{ atm}$$

   c. $T_2 = 47.\,°C + 273 = 320\,K$; $V_2 = 12.8\,L$

$$1.11\text{ atm} \times \frac{6.50\,L}{12.8\,L} \times \frac{320\,K}{298\,K} = 0.605\text{ atm}$$

6.37 $T_1 = 225\,°C + 273 = 498\,K$; $P_1 = 1.80\text{ atm}$; $V_1 = 100.0\text{ mL}$

$T_2 = -25\,°C + 273 = 248\,K$; $P_2 = 0.80\text{ atm}$

$$100.0\text{ atm} \times \frac{1.80\,L}{0.80\,L} \times \frac{248\,K}{498\,K} = 110\text{ mL}$$

6.39 Addition of more air molecules to a tire or basketball will increase its volume.

6.41 According to Avogadro's law, a change in a gas's volume is directly proportional to the change in the number of moles of gas.

   a. When moles decrease, volume must also decrease. (One half of 4.00 moles = 2.00 moles)

$$8.00\,L \times \frac{2.00\text{ moles}}{4.00\text{ moles}} = 4.00\,L$$

   b. When moles increase, volume must also increase.

$$25.0\text{ g neon} \times \frac{1\text{ mole neon}}{20.2\text{ g neon}} = 1.24\text{ moles of Ne added}$$

   (1.50 moles + 1.24 moles = 2.74 moles)

$$8.00\,L \times \frac{2.74\text{ moles}}{1.50\text{ moles}} = 14.6\,L$$

   c. 1.50 moles + 3.50 moles = 5.00 moles of gases

$$8.0\,L \times \frac{5.00\text{ moles}}{1.50\text{ moles}} = 26.7\,L$$

6.43 At STP, the molar volume of any gas is 22.4 L per mole.

    a. $44.8 \, \cancel{L} \times \dfrac{1 \text{ mole O}_2}{22.4 \, \cancel{L}} = 2.00$ moles of O_2

    b. $4.00 \, \cancel{L} \times \dfrac{1 \text{ mole CO}_2}{22.4 \, \cancel{L}} = 0.179$ mole of CO_2

    c. $6.40 \, \cancel{\text{g O}_2} \times \dfrac{1 \text{ mole } \cancel{O_2}}{32.0 \, \cancel{\text{g O}_2}} \times \dfrac{22.4 \text{ L}}{1 \text{ mole } \cancel{O_2}} = 4.48$ L

    d. $50.0 \, \cancel{\text{g Ne}} \times \dfrac{1 \text{ mole } \cancel{Ne}}{20.2 \, \cancel{\text{g Ne}}} \times \dfrac{22.4 \, \cancel{L}}{1 \text{ mole } \cancel{Ne}} \times \dfrac{1000 \text{ mL}}{1 \, \cancel{L}} = 55\,400$ mL

6.45 $8.25 \, \cancel{\text{g Mg}} \times \dfrac{1 \text{ mole } \cancel{Mg}}{24.3 \, \cancel{\text{g Mg}}} \times \dfrac{1 \text{ mole H}_2}{1 \text{ mole } \cancel{Mg}} \times \dfrac{22.4 \text{ L}}{1 \text{ mole } \cancel{H_2}} = 7.60$ L of H_2

6.47 Each gas particle in a gas mixture exerts a pressure as it strikes the walls of the container. The total gas pressure for any gaseous sample is the sum of all of the individual pressures. When the pressure of a particular type of gas is discussed, it is only part of the total. Thus, it is referred to as a partial pressure.

6.49 To obtain the total pressure in a gaseous mixture, add up all of the partial pressures using the same pressure unit.

$$P_{\text{total}} = P_{\text{Nitrogen}} + P_{\text{oxygen}} + P_{\text{Helium}}$$
$$= 425 \text{ torr} + 115 \text{ torr} + 225 \text{ torr} = 765 \text{ torr}$$

6.51 Because the total pressure in a gaseous mixture is the sum of the partial pressures using the same pressure unit, addition and subtraction is used to obtain the unknown partial pressure.

$$P_{\text{Nitrogen}} = P_{\text{total}} - (P_{\text{oxygen}} + P_{\text{Helium}})$$
$$= 925 \text{ torr} - (425 \text{ torr} + 75 \text{ torr}) = 425 \text{ torr}$$

6.53 **a.** If oxygen cannot readily cross from the lungs into the bloodstream, then the partial pressure of oxygen will be lower in the blood of an emphysema patient.
    b. An increase in the partial pressure of oxygen in the air supplied to the lungs will result in an increase in the partial pressure of oxygen in the bloodstream (addition of reactant causes the formation of more product). Because an emphysema patient has a lower partial pressure of oxygen in the blood, the use of a portable oxygen tank helps to bring the oxygenation of the patient's blood to a more desirable level.

6.55 **a.** 2 Fewest number of gas particles exerts the lowest pressure.
    b. 1 Greatest number of gas particles exerts the highest pressure.

6.57 **a.** A Volume decreases when temperature decreases.
    b. C Volume increases when pressure decreases.
    c. A Volume decreases when the moles of gas decrease.
    d. B Doubling temperature doubles the volume, but losing half the gas particles decreases the volume by half. The two effects cancel and no change in volume occurs.
    e. C Increasing the moles increases the volume to keep T and P constant.

6.59 **a.** The volume of the chest and lungs will decrease when compressed during the Heimlich maneuver.
    b. A decrease in volume causes the pressure to increase. A piece of food would be dislodged with a sufficiently high pressure.

6.61 $31\,000 \, \cancel{L} \times \dfrac{1 \text{ mole } \cancel{H_2}}{22.4 \, \cancel{L}} \times \dfrac{2.0 \text{ g H}_2}{1 \text{ mole } \cancel{H_2}} = 2.8 \times 10^3$ g of H_2

6.63 All temperatures *must* be in Kelvin in calculations involving gas laws.

$$10. \text{ atm} \times \frac{348 \text{ K}}{298 \text{ K}} = 12 \text{ atm}$$

6.65 Remember to use Kelvin temperature units in the calculation and convert to Celsius degrees after completing the calculation.

$$400. \text{ K} \times \frac{0.25 \text{ atm}}{2.00 \text{ atm}} = 50. \text{ K} \qquad 50. \text{ K} - 273 = -223 \,°\text{C}$$

$$P_{\text{Helium}} = 2400 \text{ torr} \times \frac{2.00 \text{ moles}}{8.0 \text{ moles}} = 600 \text{ torr (or } 6.0 \times 10^2 \text{ torr to show two sig figs)}$$

$$P_{\text{Oxygen}} = 2400 \text{ torr} \times \frac{6.0 \text{ moles}}{8.0 \text{ moles}} = 1800 \text{ torr}$$

6.67 **a.** $25.0 \text{ L} \times \dfrac{1 \text{ mole He}}{22.4 \text{ L}} \times \dfrac{4.00 \text{ g He}}{1 \text{ mole He}} = 4.46 \text{ g of He}$

b. $P_2 = P_1 \times V_1/V_2 \times T_2/T_1$

Conditions 1	Conditions 2
$P_1 = 760$ mmHg	$P_2 = ?$
$V_1 = 25.0$ L	$V_2 = 2460$ mmHg
$T_1 = 273$ K	$T_2 = 238$ K

$$760 \text{ mmHg} \times \frac{25.0 \text{ L}}{2460 \text{ L}} \times \frac{238 \text{ K}}{273 \text{ K}} = 6.73 \text{ mmHg}$$

6.69 Because the partial pressure of nitrogen is to be reported in torr, the atm and mmHg units must be converted to torr.

$$0.60 \text{ atm} \times \frac{760 \text{ torr}}{1 \text{ atm}} = 460 \text{ torr}$$

$$425 \text{ mmHg} \times \frac{1 \text{ torr}}{1 \text{ mmHg}} = 425 \text{ torr}$$

$$\text{and } P_{\text{Nitrogen}} = P_{\text{total}} - (P_{\text{oxygen}} + P_{\text{Argon}})$$
$$= 1250 \text{ torr} - (460 \text{ torr} + 425 \text{ torr}) = 370 \text{ torr}$$

6.71 $D = \dfrac{\text{mass}}{\text{volume}} = \dfrac{32.0 \text{ g O}_2}{1 \text{ mole}} \times \dfrac{1 \text{ mole}}{22.4 \text{ L (STP)}} = 1.43 \text{ g/L}$

6.73 **a.** The partial pressure of oxygen in the lungs is greater than what is present in blood in the alveoli.
   b. Arterial blood picks up oxygen in the lungs and delivers oxygen to body tissues. Arterial blood has a higher partial pressure of oxygen than venous blood.
   c. Because carbon dioxide is generated in body tissues, the partial pressure of CO_2 will be greater in the tissues than in arterial blood.
   d. The venous blood is returning to the lungs from body tissues. Venous blood, which has a higher partial pressure of carbon dioxide than the lungs, transports the excess CO_2 to the lungs to be exhaled.

6.75 The mole–mole conversion factor is obtained from the reaction and used to convert to moles of gas, and the STP molar volume conversion factor is used to convert moles of gas into liters of gas, as shown below:

$$2.00 \text{ moles CaCO}_3 \times \frac{1 \text{ mole CO}_2}{1 \text{ mole CaCO}_3} \times \frac{22.4 \text{ L}}{1 \text{ mole CO}_2} = 44.8 \text{ L of CO}_2$$

6.77 $425 \text{ mL} \times \dfrac{745 \text{ mmHg}}{0.115 \text{ atm}} \times \dfrac{1 \text{ atm}}{760 \text{ mmHg}} \times \dfrac{178 \text{ K}}{297 \text{ K}} = 2170 \text{ mL}$

6.79 $5.4 \text{ g Al} \times \dfrac{1 \text{ mole Al}}{27.0 \text{ g Al}} \times \dfrac{3 \text{ moles O}_2}{4 \text{ moles Al}} \times \dfrac{22.4 \text{ L O}_2(\text{STP})}{1 \text{ mole H}_2} = 3.4 \text{ L of O}_2$

6.81 **a.** False. The flask containing helium has more atoms because one gram of helium contains more atoms than one gram of neon.
b. False. There are different numbers of moles in the flasks, which means the pressures are different.
c. True. There are more moles of helium, which makes the pressure of helium greater than that of neon.
d. True. Density is mass divided by volume. Because there is the same mass of gas in flasks of equal volume, the densities are the same.

6.83 $T_1 = 15\,°\text{C} + 273 = 288 \text{ K}; P_1 = 745 \text{ mmHg}; V_1 = 4250 \text{ mL}$

$P_2 = 1.20 \text{ atm} \times \dfrac{760 \text{ mmHg}}{1 \text{ atm}} = 912 \text{ mmHg}; V_2 = 2.50 \text{ L} \times \dfrac{1000 \text{ mL}}{1 \text{ L}} = 2.50 \times 10^3 \text{ mL}$

$288 \text{ K} \times \dfrac{2500 \text{ mL}}{4250 \text{ mL}} \times \dfrac{912 \text{ mmHg}}{745 \text{ mmHg}} = 207 \text{ K} - 273 = -66\,°\text{C}$

6.85 $413 \text{ K} \times \dfrac{1.30 \text{ atm}}{0.900 \text{ atm}} = 597 \text{ K} - 273 = 324\,°\text{C}$

6.87 $132 \text{ g NaN}_3 \times \dfrac{1 \text{ mole NaN}_3}{65.0 \text{ g NaN}_3} \times \dfrac{3 \text{ moles N}_2}{2 \text{ moles NaN}_3} \times \dfrac{22.4 \text{ L}}{1 \text{ mole N}_2} = 68.2 \text{ L at STP}$

6.89 **a.** Year 2000:

$780 \text{ Tg} \times \dfrac{1 \times 10^{12} \text{ g}}{1 \text{ Tg}} \times \dfrac{1 \text{ kg}}{1 \times 10^3 \text{ g}} = 7.8 \times 10^{11} \text{ kg}$

Year 2020:

$990 \text{ Tg} \times \dfrac{1 \times 10^{12} \text{ g}}{1 \text{ Tg}} \times \dfrac{1 \text{ kg}}{1 \times 10^3 \text{ g}} = 9.9 \times 10^{11} \text{ kg}$

b. Year 2000:

$780 \text{ Tg CO}_2 \times \dfrac{1 \times 10^{12} \text{ g CO}_2}{1 \text{ Tg CO}_2} \times \dfrac{1 \text{ mole CO}_2}{44.0 \text{ g CO}_2} = 1.8 \times 10^{13} \text{ moles of CO}_2$

Year 2020:

$990 \text{ Tg CO}_2 \times \dfrac{1 \times 10^{12} \text{ g CO}_2}{1 \text{ Tg CO}_2} \times \dfrac{1 \text{ mole CO}_2}{44.0 \text{ g CO}_2} = 2.3 \times 10^{13} \text{ moles of CO}_2$

c. $990 \text{ Tg} - 780 \text{ Tg} = 210 \text{ Tg}$

$210 \text{ Tg} \times \dfrac{1 \times 10^{12} \text{ g}}{1 \text{ Tg}} \times \dfrac{1 \text{ Mg}}{1 \times 10^6 \text{ g}} = 2.1 \times 10^8 \text{ Mg}$

Study Goals

- Identify the solute and solvent in a solution.
- Describe hydrogen bonding in water.
- Describe electrolytes in a solution.
- Define solubility.
- Calculate the percent concentrations and molarity of a solution.
- Use the molarity of a solution in a chemical reaction to calculate the volume or quantity of a reactant or product.
- Distinguish between a solution, a colloid, and a suspension.
- Describe osmosis and dialysis.
- Describe the behavior of a red blood cell in hypotonic, isotonic, and hypertonic solutions.

Think About It

1. Why is salt used to preserve foods?

2. Why do raisins or dried prunes swell when placed in water?

3. Why are pickles made in a brine solution with a high salt concentration?

4. Why can't you drink seawater?

5. How do your kidneys remove toxic substances from the blood but retain the usable substances?

Key Terms

Match the following terms with the correct statement shown below:

    a. solution   **b.** concentration   **c.** molarity
    d. osmosis   **e.** electrolyte

1. _____ A substance that dissociates into ions when it dissolves in water.

2. _____ The flow of solvent through a semipermeable membrane into a solution with a higher solute concentration.

3. _____ The amount of solute that is dissolved in a specified amount of solution.

4. _____ The number of moles of solute in one liter of solution.

5. _____ A mixture of at least two components called a solute and a solvent.

Answers   **1.** e   **2.** d   **3.** b   **4.** c   **5.** a

7.1 Solutions

- A polar solute is soluble in a polar solvent; a nonpolar solute is soluble in a nonpolar solvent; "like dissolves like."
- A solution forms when a solute dissolves in a solvent.
- The partial positive charge of hydrogen and the partial negative charge of oxygen permits water molecules to hydrogen bond to other water molecules.
- An ionic solute dissolves in water, a polar solvent, because the polar water molecules attract and pull the positive and negative ions into solution. In solution, water molecules surround the ions in a process called hydration.

◆ Learning Exercise 7.1A

Indicate the solute and solvent in each of the following: **Solute**   **Solvent**

    a. 10 g of KCl dissolved in 100 g of water   _____   _____

    b. Soda water: $CO_2(g)$ dissolved in water   _____   _____

    c. An alloy composed of 80% Zn and 20% Cu   _____   _____

    d. A mixture of O_2 (200 mm Hg) and He (500 mm Hg)   _____   _____

    e. A solution of 40 mL of CCl_4 and 2 mL of Br_2 _____   _____

Answers   **a.** KCl; water   **b.** CO_2; water   **c.** Cu; Zn
             d. oxygen; helium   **e.** Br_2; CCl_4

◆ Learning Exercise 7.1B

Essay: How does the polarity of the water molecule allow it to hydrogen bond?

Answer   The O—H bonds in water molecules are polar because the hydrogen atoms are partially positive and the oxygen atoms are partially negative. Hydrogen bonding occurs because the partially positive hydrogen atoms in one water molecule are attracted to partially negative oxygen atoms of other water molecules.

◆ Learning Exercise 7.1C

Water is polar and hexane is nonpolar. In which solvent is each of the following soluble?

    a. Bromine, Br_2, nonpolar _____   **d.** Vitamin D, nonpolar   _____

    b. HCl, polar   _____   **e.** Vitamin C, polar   _____

    c. Cholesterol, nonpolar _____

Answers   **a.** hexane   **b.** water   **c.** hexane   **d.** hexane   **e.** water

7.2 Electrolytes and Nonelectrolytes

- Electrolytes conduct an electrical current because they produce ions in aqueous solutions.
- Strong electrolytes are nearly completely ionized, whereas weak electrolytes are slightly ionized. Nonelectrolytes do not form ions in solution but dissolve as molecules.
- An equivalent is the amount of an electrolyte that carries 1 mole of electrical charge.
- The number of equivalents per mole of a positive or negative ion is equal to the charge on the ion.

◆ **Learning Exercise 7.2A**

Write an equation for the formation of an aqueous solution of each of the following strong electrolytes:

  a. LiCl　$\rule{8cm}{0.4pt}$

  b. $Mg(NO_3)_2$　$\rule{7cm}{0.4pt}$

  c. Na_3PO_4　$\rule{7cm}{0.4pt}$

  d. K_2SO_4　$\rule{7cm}{0.4pt}$

  e. $MgCl_2$　$\rule{7cm}{0.4pt}$

Answers　　**a.** $LiCl(s) \xrightarrow{H_2O} Li^+(aq) + Cl^-(aq)$

　　　　　　b. $Mg(NO_3)_2(s) \xrightarrow{H_2O} Mg^{2+}(aq) + 2\,NO_3^-(aq)$

　　　　　　c. $Na_3PO_4(s) \xrightarrow{H_2O} 3\,Na^+(aq) + PO_4^{3-}(aq)$

　　　　　　d. $K_2SO_4(s) \xrightarrow{H_2O} 2\,K^+(aq) + SO_4^{2-}(aq)$

　　　　　　e. $MgCl_2(s) \xrightarrow{H_2O} Mg^{2+}(aq) + 2\,Cl^-(aq)$

◆ **Learning Exercise 7.2B**

Indicate whether an aqueous solution of each of the following contains mostly ions, molecules only, or mostly molecules with some ions. Write an equation for the formation of the solution:

  a. glucose, $C_6H_{12}O_6$, a nonelectrolyte　$\rule{6cm}{0.4pt}$

  b. NaOH, a strong electrolyte　$\rule{6cm}{0.4pt}$

  c. K_2SO_4, a strong electrolyte　$\rule{6cm}{0.4pt}$

  d. HF, a weak electrolyte　$\rule{6cm}{0.4pt}$

Answers　　**a.** $C_6H_{12}O_6(s) \rightarrow C_6H_{12}O_6(aq)$ molecules only

　　　　　　b. $NaOH(s) \rightarrow Na^+(aq) + OH^-(aq)$　　　mostly ions

　　　　　　c. $Li_2SO_4(s) \rightarrow 2\,Li^+(aq) + SO_4^{2-}(aq)$　　mostly ions

　　　　　　d. $HF(g) + H_2O(l) \rightleftharpoons H_3O^+(aq) + F^-(aq)$　mostly molecules and a few ions

◆ **Learning Exercise 7.2C**

Calculate the following:

  a. Number of equivalents in 1 mole of Mg^{2+}

  b. Number of equivalents of Cl^- in 2.5 moles of Cl^-

  c. Number of equivalents of Ca^{2+} in 2.0 moles of Ca^{2+}

Answers　　**a.** 2 Eq　　　**b.** 2.5 Eq　　　**c.** 4.0 Eq

7.3 Solubility

- The amount of solute that dissolves depends on the nature of the solute and solvent.
- Solubility describes the maximum amount of a solute that dissolves in 100 g of solvent at a given temperature.
- A saturated solution contains the maximum amount of dissolved solute at a certain temperature.
- An increase in temperature increases the solubility of most solids, but decreases the solubility of gases in water.

◆ Learning Exercise 7.3A

Identify each of the following as a saturated solution (S) or an unsaturated solution (U):

1. A sugar cube dissolves when added to a cup of coffee. _____

2. A KCl crystal added to a KCl solution does not change in size. _____

3. A layer of sugar forms in the bottom of a glass of iced tea. _____

4. The rate of crystal formation equals the rate of solution. _____

5. Upon heating, all the sugar dissolves. _____

Answers   **1.** U   **2.** S   **3.** S   **4.** S   **5.** U

◆ Learning Exercise 7.3B

Use the KNO_3 solubility chart for the following problems:

Solubility of KNO_3

Temperature (°C)	g KNO_3/100 g H_2O
0	15
20	30
40	65
60	110
80	170
100	250

a. How many grams of KNO_3 will dissolve in 100 g of water at 40 °C?

b. How many grams of KNO_3 will dissolve in 300 g of water at 60 °C?

c. A solution is prepared using 200 g of water and 350 g of KNO_3 at 80 °C. Will any solute remain undissolved? If so, how much?

d. Will 200 g of KNO_3 dissolve when added to 100 g of water at 100 °C?

Answers   **a.** 65 g   **b.** 330 g
                 c. Yes, 10 g of KNO_3 will not dissolve.
                 d. Yes, all 200 g of KNO_3 will dissolve.

7.4 Percent Concentration

- The concentration of a solution is the relationship between the amount of solute (g or mL) and the amount (g or mL) of solution.
- A mass percent (mass/mass) expresses the ratio of the mass of solute to the mass of solution multiplied by 100.

$$\text{Percent (m/m)} = \frac{\text{grams of solute}}{\text{grams of solution}} \times 100$$

- Percent concentrations can also be expressed as a mass/volume ratio.

$$\text{Percent (m/v)} = \frac{\text{grams of solute}}{\text{volume (mL) of solution}} \times 100$$

Study Note

Calculate mass percent concentration (% m/m) as

$$\frac{\text{grams of solute}}{\text{grams of solution}} \times 100$$

Example: What is the percent (m/m) when 2.4 g of $NaHCO_3$ dissolves in 120 g of solution?

Solution: $\quad \dfrac{2.4 \text{ g } NaHCO_3}{120 \text{ g solution}} \times 100 = 2.0\% \text{ (m/m)}$

◆ **Learning Exercise 7.4A**

Determine the percent concentration of the following solutions:

    a. The mass/mass % of 18.0 g of NaCl in 90.0 g of solution.

    b. The mass/volume % of 5.0 g of KCl in 2.0 liters of solution.

    c. The mass/mass % of 4.0 g of KOH in 50.0 g of solution.

    d. The mass/volume % of 0.25 kg of glucose in 5.0 liters of solution.

Answers   **a.** 20.0%   **b.** 0.25%   **c.** 8.0%   **d.** 5.0%

Study Note

In solution problems, the percent concentration is useful as a conversion factor. The factor is obtained by rewriting the % as g of solute/100 g (or mL) solution.

Example 1: How many g of KI are needed to prepare 250 mL of a 4% (m/v) KI solution?

Solution: $\qquad 250 \text{ mL solution} \times \underset{\%\,(m/v)\,factor}{\frac{4 \text{ g KI}}{100 \text{ mL solution}}} = 10 \text{ g of KI}$

Example 2: How many g of a 25% (m/m) NaOH solution can be prepared from 75 g of NaOH?

Solution: $\qquad 75 \text{ g NaOH} \times \underset{\%\,(m/m)\,factor\,(inverted)}{\frac{100 \text{ g NaOH solution}}{25 \text{ g NaOH}}} = 300 \text{ g of NaOH solution}$

◆ **Learning Exercise 7.4B**

Calculate the number of grams of solute needed to prepare each of the following solutions:

    a. How many grams of glucose are needed to prepare 400. mL of a 10.0% (m/v) solution?

    b. How many grams of lidocaine hydrochloride are needed to prepare 50.0 g of a 2.0% (m/m) solution?

    c. How many grams of KCl are needed to prepare 0.80 liter of a 0.15% (m/v) KCl solution?

    d. How many grams of NaCl are needed to prepare 250 mL of a 1.0% (m/v) solution?

Answers   **a.** 40.0 g   **b.** 1.0 g   **c.** 1.2 g   **d.** 2.5 g

◆ **Learning Exercise 7.4C**

Use percent-concentration factors to calculate the volume (mL) of each solution that contains the amount of solute stated in each problem.

    a. 2.00 g of NaCl from a 1.00% (m/v) NaCl solution

    b. 25 g of glucose from a 5% (m/v) glucose solution

    c. 1.5 g of KCl from a 0.50% (m/v) KCl solution

    d. 75.0 g of NaOH from a 25.0% (m/v) NaOH solution

Answers   **a.** 200. mL   **b.** 500 mL   **c.** 300 mL   **d.** 300. mL

7.5 Molarity and Dilution

- Molarity is a concentration term that describes the number of moles of solute dissolved in 1 L (1000 mL) of solution.

$$M = \frac{\text{moles of solute}}{\text{L solution}}$$

- *Dilution* is the process of mixing a solution with solvent to obtain a lower concentration.
- For dilutions, use the expression $C_1V_1 = C_2V_2$ and solve for the unknown value.

◆ **Learning Exercise 7.5A**

Calculate the molarity of the following solutions:

    a. 2.0 moles of HCl in 1.0 liter of HCl solution

    b. 10.0 moles of glucose $(C_6H_{12}O_6)$ in 2.0 liters of glucose solution

    c. 80.0 g of NaOH in 4.0 liters of NaOH solution (Hint: Find moles NaOH.)

Answers   **a.** 2.0 M HCl   **b.** 5.0 M glucose   **c.** 0.50 M NaOH

Study Note

Molarity can be used as a conversion factor to convert between the amount of solute and the volume of solution.

Example 1: How many g of NaOH are in 0.20 L of a 4.0 M NaOH solution?

Solution: The concentration 4.0 M can be expressed as the conversion factors:

$$\frac{4.0 \text{ moles NaOH}}{1 \text{ L NaOH}} \text{ and } \frac{1 \text{ L NaOH}}{4.0 \text{ moles NaOH}}$$

$$0.20 \cancel{\text{ L NaOH}} \times \frac{4.0 \cancel{\text{ moles NaOH}}}{1 \cancel{\text{ L NaOH}}} \times \frac{40.0 \text{ g NaOH}}{1 \cancel{\text{ mole NaOH}}} = 32 \text{ g NaOH}$$

Example 2: How many mL of a 6 M HCl solution will provide 0.36 mole of HCl?

Solution: $$0.36 \cancel{\text{ mole HCl solution}} \times \underset{\text{Molarity factor(inverted)}}{\frac{1 \cancel{L}}{6 \cancel{\text{ moles HCl}}}} \times \frac{1000 \text{ mL}}{1 \cancel{L}} = 60 \text{ mL of HCl solution}$$

◆ **Learning Exercise 7.5B**

Calculate the quantity of solute in the following solutions:

a. How many moles of HCl are in 1.50 liter of a 2.50 M HCl solution?

b. How many moles of KOH are in 125 ml of a 2.40 M KOH solution?

c. How many grams of NaOH are needed to prepare 225 mL of a 3.00 M NaOH solution? (Hint: Find moles of NaOH.)

d. How many grams of NaCl are in 415 mL of a 1.30 M NaCl solution?

Answers   **a.** 3.75 moles of HCl   **b.** 0.300 mole of KOH
              c. 27.0 g of NaOH   **d.** 31.6 g of NaCl

◆ **Learning Exercise 7.5C**

Calculate the milliliters needed of each solution to obtain each of the following:

a. 0.200 mole of $Mg(OH)_2$ from a 2.50 M $Mg(OH)_2$ solution

b. 0.125 mole of glucose from a 5.00 M glucose solution

c. 0.250 mole of KI from a 4.00 M KI solution

d. 16.0 g of NaOH from a 3.20 M NaOH solution

Answers   **a.** 80.0 mL   **b.** 25.0 mL   **c.** 62.5 mL   **d.** 125 mL

◆ Learning Exercise 7.5D

Solve each of the following dilution problems (assume the volumes add):

a. What is the final molar concentration after 100 mL of a 5.0 M KCl solution is diluted with water to give a final volume of 200 mL?

b. What is the final %(m/v) concentration of the diluted solution if 5.0 mL of a 15% (m/v) KCl solution is diluted to 25 mL?

c. What is the final %(m/v) concentration after 250 mL of an 8% (m/v) NaOH is diluted with 750 mL of water?

d. 48 mL of water is added to 12 mL of a 1.0 M NaCl solution. What is the final molar concentration?

e. What volume(L) of water must be added to 2.0 L of 12% (m/v) KCl to obtain a 4.0% (m/v) KCl solution? What is the total volume(L) of the solution?

f. What volume of 6.0 M HCl is needed to prepare 300. mL of 1.0 M HCl? How much water must be added?

Answers

a. 2.5 M	**b.** 3.0% (m/v)	**c.** 2% (m/v)
d. 0.20 M	**e.** add 4.0 L; V_2 = 6.0 L	**f.** V_1 = 50 mL; add 250 mL water

7.6 Solutions in Chemical Reactions

- For a balanced equation, the molarity and volume of one solution can be used to determine the moles of a reacting substance or one that is produced.
- For a balanced equation, the molarity and the number of moles of a solute can be used to determine the volume of a solution of another reactant.

◆ Learning Exercise 7.6A

For the following reaction,

$$2\ AgNO_3(aq) + H_2SO_4(aq) \rightarrow Ag_2SO_4(s) + 2\ H_2O(l)$$

a. How many milliliters of 1.5 M $AgNO_3$ will react with 40.0 mL of 1.0 M H_2SO_4?

b. How many grams of Ag_2SO_4 will be produced?

Answers

a. $40.0\ \text{mL}\ H_2SO_4 \times \dfrac{1\ L\ H_2SO_4}{1000\ \text{mL}\ H_2SO_4} \times \dfrac{1.0\ \text{mole}\ H_2SO_4}{1\ L\ H_2SO_4} \times \dfrac{2\ \text{moles}\ AgNO_3}{1\ \text{mole}\ H_2SO_4} \times$

$\dfrac{1000\ \text{mL}\ AgNO_3}{1.5\ \text{moles}\ AgNO_3} = 53\ \text{mL}$

b. $40.0\ \text{mL}\ H_2SO_4 \times \dfrac{1\ L\ H_2SO_4}{1000\ \text{mL}\ H_2SO_4} \times \dfrac{1.0\ \text{moles}\ H_2SO_4}{1\ L\ H_2SO_4} \times \dfrac{1\ \text{mole}\ Ag_2SO_4}{1\ \text{mole}\ H_2SO_4} \times \dfrac{311.9\ g\ Ag_2SO_4}{1\ \text{mole}\ Ag_2SO_4}$

$= 12\ g\ \text{of}\ Ag_2SO_4$

◆ **Learning Exercise 7.6B**

For the following reaction, calculate the milliliters of 1.8 M KOH that react with 18.5 mL of 2.2 M HCl.

$$HCl(aq) + KOH(aq) \rightarrow KCl(aq) + H_2O(l)$$

Answers

b. $18.5 \text{ mL HCl} \times \dfrac{1 \text{ L HCl}}{1000 \text{ mL HCl}} \times \dfrac{2.2 \text{ moles HCl}}{1 \text{ L HCl}} \times \dfrac{1 \text{ mole KOH}}{1 \text{ mole HCl}} \times \dfrac{1000 \text{ mL KOH}}{1.8 \text{ moles KOH}}$

$= 22.6$ mL of KOH

7.7 Properties of Solutions

- Colloids contain particles that do not settle out and pass through filters but not through semipermeable membranes.
- Suspensions are composed of large particles that settle out of solution.
- In the process of osmosis, water (solvent) moves through a semipermeable membrane from the solution that has a lower solute concentration to a solution where the solute concentration is higher.
- Osmotic pressure is the pressure that prevents the flow of water into a more concentrated solution.
- Isotonic solutions have osmotic pressures equal to that of body fluids. A hypotonic solution has a lower osmotic pressure than body fluids; a hypertonic solution has a higher osmotic pressure.
- A red blood cell maintains its volume in an isotonic solution, but it swells (hemolysis) in a hypotonic solution and shrinks (crenation) in a hypertonic solution.
- In dialysis, water and small solute particles can pass through a dialyzing membrane, while larger particles are retained.

◆ **Learning Exercise 7.7A**

Identify each of the following as a solution, colloid, or suspension:

1. _____ Contains single atoms, ions, or small molecules.

2. _____ Settles out with gravity.

3. _____ Retained by filters.

4. _____ Cannot diffuse through a cellular membrane.

5. _____ Aggregates of atoms, molecules, or ions larger in size than solution particles.

6. _____ Large particles that are visible.

Answers   **1.** solution   **2.** suspension   **3.** suspension   **4.** colloid   **5.** colloid   **6.** suspension

◆ **Learning Exercise 7.7B**

Fill in the blanks:

In osmosis, the direction of solvent flow is from the (1) _____ solute concentration to the

(2) _____ solute concentration. A semipermeable membrane separates 5% and 10% sucrose

solutions. The (3) _____ % solution has the greater osmotic pressure. Water will move from the

(4) _____ % solution into the (5) _____ % solution. The compartment that contains the

(6) _____ % solution increases in volume.

Answers   **1.** lower   **2.** higher   **3.** 10   **4.** 5   **5.** 10   **6.** 10

◆ **Learning Exercise 7.7C**

What occurs when 2% (A) and 10% (B) starch solutions are separated by a semipermeable membrane?

Semipermeable
membrane

2% starch	10% starch
A	**B**

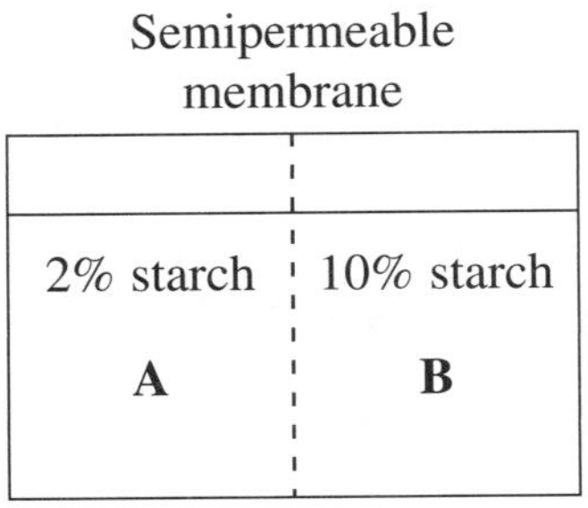

a. Water will flow from side _____ to side _____.

b. The volume in compartment _____ will increase and decrease in compartment _____.

c. The final concentration of the solutions in both compartments will be _____.

Answers   **a.** A, B   **b.** B, A   **c.** 6%

◆ **Learning Exercise 7.7D**

Fill in the blanks:

A (1) _____ % NaCl solution and a (2) _____ % glucose solution are isotonic to the body fluids. A red blood cell placed in these solutions does not change in volume because these solutions are (3) _____ tonic. When a red blood cell is placed in water, it undergoes (4) _____ because water is (5) _____ tonic. A 20% glucose solution will cause a red blood cell to undergo (6) _____ because the 20% glucose solution is (7) _____ tonic.

Answers   **1.** 0.9   **2.** 5   **3.** iso   **4.** hemolysis
             5. hypo   **6.** crenation   **7.** hyper

◆ **Learning Exercise 7.7E**

Indicate whether the following solutions are

1. hypotonic   2. hypertonic   3. isotonic

a. _____ 5% (m/v) glucose   d. _____ water

b. _____ 3% (m/v) NaCl   e. _____ 0.9% (m/v) NaCl

c. _____ 2% (m/v) glucose   f. _____ 10% (m/v) glucose

Answers   **a.** 3   **b.** 2   **c.** 1   **d.** 1   **e.** 3   **f.** 2

◆ Learning Exercise 7.7F

Indicate whether the following solutions will cause a red blood cell to undergo

1. crenation   **2.** hemolysis   **3.** no change (stays the same)

    a. _____ 10% (m/v) NaCl   **d.** _____ 0.5% (m/v) NaCl

    b. _____ 1% (m/v) glucose   **e.** _____ 10% (m/v) glucose

    c. _____ 5% (m/v) glucose   **f.** _____ water

Answers   **a.** 1   **b.** 2   **c.** 3   **d.** 2   **e.** 1   **f.** 2

◆ Learning Exercise 7.7G

A dialysis bag contains starch, glucose, NaCl, protein, and urea.

    a. When the dialysis bag is placed in water, which of the components would you expect to dialyze through the bag? Why?

    b. Which components will stay inside the dialysis bag? Why?

Answers   **a.** Glucose, NaCl, urea; they are solution particles.
                        b. Starch, protein; colloids are retained by semipermeable membranes.

Checklist for Chapter 7

You are ready to take the practice test for Chapter 7. Be sure that you have accomplished the following learning goals for this chapter. If you are not sure, review the section listed at the end of the goal. Then apply your new skills and understanding to the practice test.

After studying Chapter 7, I can successfully:

_____ Describe hydrogen bonding in water (7.1).

_____ Identify the solute and solvent in a solution (7.1).

_____ Describe the process of dissolving an ionic solute in water (7.1).

_____ Identify the components in solutions of electrolytes and nonelectrolytes (7.2).

_____ Calculate the number of equivalents for an electrolyte (7.2).

_____ Identify a saturated and an unsaturated solution (7.3).

_____ Describe the effects of temperature and nature of the solute on its solubility in a solvent (7.3).

_____ Calculate the percent concentration of a solute in a solution and use percent concentration to calculate the amount of solute or solution (7.4).

_____ Calculate the diluted volume of a solution (7.4).

_____ Calculate the molarity of a solution (7.5).

_____ Use molarity as a conversion factor to calculate between the mole (or grams) of a solute and the volume of the solution (7.6).

_____ Identify a mixture as a solution, a colloid, or a suspension (7.7).

_____ Explain the processes of osmosis and dialysis (7.7).

Practice Test for Chapter 7

Indicate if the following are more soluble in (A) water (polar solvent) or (B) benzene (nonpolar solvent).

1. _____ $I_2(g)$, nonpolar   3. _____ $KI(s)$, polar

2. _____ $NaBr(s)$, polar   4. _____ C_6H_{12}, nonpolar

5. When dissolved in water, $Ca(NO_3)_2$ dissociates into
   - **A.** $Ca^{2+} + (NO_3)_2^{2-}$
   - **B.** $Ca^+ + NO_3^-$
   - **C.** $Ca^{2+} + 2\,NO_3^-$
   - **D.** $Ca^{2+} + 2\,N^{5+} + 2\,O_3^{6-}$
   - **E.** $CaNO_3^+ + NO_3^-$

6. What is the number of equivalents in 2 moles of Mg^{2+}?
   - **A.** 0.50 equiv **B.** 1 equiv **C.** 1.5 equiv **D.** 2 equiv **E.** 4 equiv

7. CH_3CH_2OH, ethyl alcohol, is a nonelectrolyte. When placed in water it
   - **A.** dissociates completely. **B.** dissociates partially. **C.** does not dissociate.
   - **D.** makes the solution acidic. **E.** makes the solution basic.

8. The solubility of NH_4Cl is 46 g in 100 g of water at 40 °C. How much NH_4Cl can dissolve in 500 g of water at 40 °C?
   - **A.** 9.2 g **B.** 46 g **C.** 100 g **D.** 184 g **E.** 230 g

9. A solution containing 1.20 g of sucrose in 50.0 mL of solution has a mass–volume percent concentration of
   - **A.** 0.600% **B.** 1.20% **C.** 2.40% **D.** 30.0% **E.** 41.6%

10. The amount of lactose in 250 mL of a 3.0% (m/v) lactose solution of infant formula is
    - **A.** 0.15 g **B.** 1.2 g **C.** 6.0 g **D.** 7.5 g **E.** 30 g

11. The volume needed to obtain 0.40 g of glucose from a 5.0% (m/v) glucose solution is
    - **A.** 1.0 mL **B.** 2.0 mL **C.** 4.0 mL **D.** 5.0 mL **E.** 8.0 mL

12. The amount of NaCl needed to prepare 50.0 mL of a 4.00% (m/v) NaCl solution is
    - **A.** 20.0 g **B.** 15.0 g **C.** 10.0 g **D.** 4.00 g **E.** 2.00 g

13. A solution containing 6.0 g of NaCl in 1500 mL of solution has a mass–volume percent concentration of
    - **A.** 0.40 (m/v)% **B.** 0.25 (m/v)% **C.** 4.0 (m/v)% **D.** 0.90 (m/v)% **E.** 2.5 (m/v)%

For questions 14 through 18, indicate whether each statement describes a
   - **A.** solution **B.** colloid **C.** suspension

14. Contains single atoms, ions, or small molecules of solute.

15. Settles out upon standing.

16. Can be separated by filtering.

17. Can be separated by semipermeable membranes.

18. Passes through semipermeable membranes.

19. The separation of colloids from solution particles by use of a membrane is called
    - **A.** osmosis **B.** dispersion **C.** dialysis **D.** hemolysis **E.** collodian

20. Any two solutions that have identical osmotic pressures are
    - **A.** hypotonic **B.** hypertonic **C.** isotonic **D.** isotopic **E.** blue

21. In osmosis, water flows
    A. between solutions of equal concentrations.
    B. from higher solute concentrations to lower solute concentrations.
    C. from lower solute concentrations to higher solute concentrations.
    D. from colloids to solutions of equal concentrations.
    E. from lower solvent concentrations to higher solvent concentrations.

22. A normal red blood cell will shrink when placed in a solution that is
    A. isotonic   B. hypotonic   C. hypertonic   D. colloidal   E. semitonic

23. A red blood cell undergoes hemolysis when placed in a solution that is
    A. isotonic   B. hypotonic   C. hypertonic   D. colloidal   E. semitonic

24. A solution that has the same osmotic pressure as body fluids is
    A. 0.1% NaCl   B. 0.9% NaCl   C. 5% NaCl   D. 10% glucose   E. 15% glucose

25. Which of the following is hypertonic to red blood cells?
    A. 0.5% NaCl   B. 0.9% NaCl   C. 1% glucose   D. 5% glucose   E. 10% glucose

26. Which of the following is hypotonic to red blood cells?
    A. 2.0% NaCl   B. 0.9% NaCl   C. 1% glucose   D. 5% glucose   E. 10% glucose

For questions 27 through 31, select the correct term from the following:
    A. isotonic   B. hypertonic   C. hypotonic   D. osmosis   E. dialysis

27. _____ A solution with a higher osmotic pressure than the blood.

28. _____ A solution of 10% NaCl surrounding a red blood cell.

29. _____ A 1% glucose solution.

30. _____ The cleansing process of the artificial kidney.

31. _____ The flow of water up the stem of a plant.

32. _____ In dialysis,
    A. dissolved salts and small molecules are separated from colloids.
    B. nothing but water passes through the membrane.
    C. only ions pass through a membrane.
    D. two kinds of colloids are separated.
    E. colloids are separated from suspensions.

33. A dialyzing membrane
    A. is a semipermeable membrane.
    B. allows only water and true solution particles to pass through.
    C. does not allow colloidal particles to pass through.
    D. All of the above
    E. None of the above

34. Which substance will remain inside a dialysis bag?
    A. water   B. NaCl   C. starch   D. glucose   E. Mg^{2+}

35. Waste removal in hemodialysis is based on
    A. concentration gradients between the bloodstream and the dialysate.
    B. a pH difference between the bloodstream and the dialysate.
    C. use of an osmotic membrane.
    D. greater osmotic pressure in the bloodstream.
    E. renal compensation.

36. The moles of KOH needed to prepare 2400 mL of a 2.0 M KOH solution is
    A. 1.2 moles   B. 2.4 moles   C. 4.8 moles   D. 12 moles   E. 48 moles

37. The amount in grams of NaOH needed to prepare 7.5 mL of a 5.0 M NaOH is
    A. 1.5 g   B. 3.8 g   C. 6.7 g   D. 15 g   E. 38 g

For questions 38 through 40, consider a 20.0-mL sample of a solution that contains 2.0 g of NaOH.

38. The % (m/v) concentration of the solution is
    A. 1.0%(m/v)   **B.** 4.0%(m/v)   **C.** 5%(m/v)   **D.** 10%(m/v)   **E.** 20%(m/v)

39. The number of moles of NaOH in the sample is
    A. 0.050 mole   **B.** 0.40 mole   **C.** 1.0 mole   **D.** 2.5 moles   **E.** 4.0 moles

40. The molarity of the sample is
    A. 0.10 M   **B.** 0.5 M   **C.** 1.0 M   **D.** 1.5 M   **E.** 2.5 M

Answers to the Practice Test

1. B	**2.** A	**3.** A	**4.** B	**5.** C
6. E	**7.** C	**8.** E	**9.** C	**10.** D
11. E	**12.** C	**13.** A	**14.** A	**15.** E
16. C	**17.** B	**18.** A	**19.** C	**20.** C
21. C	**22.** C	**23.** B	**24.** B	**25.** E
26. C	**27.** B	**28.** B	**29.** C	**30.** E
31. D	**32.** A	**33.** D	**34.** C	**35.** A
36. C	**37.** A	**38.** D	**39.** A	**40.** E

Answers and Solutions to Selected Text Problems

7.1   The component present in the smaller amount is the solute; the larger amount is the solvent.

    a. Sodium chloride, solute; water, solvent
    b. Water, solute; ethanol, solvent
    c. Oxygen, solute; nitrogen, solvent

7.3   The K^+ and I^- ions at the surface of the solid are pulled into solution by the polar water molecules where the hydration process surrounds separate ions with water molecules.

7.5   **a.** Potassium chloride, an ionic solute, would be soluble in water (a polar solvent).
    b. Iodine, a nonpolar solute, would be soluble in carbon tetrachloride (a nonpolar solvent).
    c. Sugar, a polar solute, would be soluble in water, which is a polar solvent.
    d. Gasoline, a nonpolar solute, would be soluble in carbon tetrachloride, which is a nonpolar solvent.

7.7   The salt KF dissociates into ions when it dissolves in water. The weak acid HF exists as mostly molecules along with some ions when it dissolves in water.

7.9   Strong electrolytes dissociate into ions.

    a. $KCl(s) \xrightarrow{H_2O} K^+(aq) + Cl^-(aq)$
    b. $CaCl_2(s) \xrightarrow{H_2O} Ca^{2+}(aq) + 2Cl^-(aq)$
    c. $K_3PO_4(s) \xrightarrow{H_2O} K^+(aq) + PO_4^{3-}(aq)$
    d. $Fe(NO_3)_3(s) \xrightarrow{H_2O} Fe^{3+}(aq) + 3 NO_3^-(aq)$

7.11   **a.** In solution, a weak electrolyte exists mostly as molecules with a few ions.
    b. Sodium bromide is a strong electrolyte and forms ions in solution.
    c. A nonelectrolyte does not dissociate and forms only molecules in solution.

7.13   **a.** Strong electrolyte; only ions are present in the K_2SO_4 solution.
    b. Weak electrolyte; both ions and molecules are present in the NH_4OH solution.
    c. Nonelectrolyte; only molecules are present in the $C_6H_{12}O_6$ solution.

7.15 **a.** $1 \text{ mole } K^+ \times \dfrac{1 \text{ Eq } K^+}{1 \text{ mole } K^+} = 1 \text{ Eq of } K^+$

b. $2 \text{ moles } OH^+ \times \dfrac{1 \text{ Eq } OH^-}{1 \text{ mole } OH^-} = 2 \text{ Eq of } OH^-$

c. $1 \text{ mole } Ca^{2+} \times \dfrac{2 \text{ Eq } Ca^{2+}}{1 \text{ mole } Ca^{2+}} = 2 \text{ Eq of } Ca^{2+}$

d. $3 \text{ moles } CO_3^{2-} \times \dfrac{2 \text{ Eq } CO_3^{2-}}{1 \text{ moles } CO_3^{2-}} = 6 \text{ Eq of } CO_3^{2-}$

7.17 $1.0 \text{ L} \times \dfrac{154 \text{ mEq}}{1 \text{ L}} \times \dfrac{1 \text{ Eq}}{1000 \text{ mEq}} \times \dfrac{1 \text{ mole } Na^+}{1 \text{ Eq}} = 0.154 \text{ mole of } Na^+$

$1.0 \text{ L} \times \dfrac{154 \text{ mEq}}{1 \text{ L}} \times \dfrac{1 \text{ Eq}}{1000 \text{ mEq}} \times \dfrac{1 \text{ mole } Cl^-}{1 \text{ Eq}} = 0.154 \text{ mole of } Cl^-$

7.19 The total equivalents of anions must be equal to the equivalents of cations in any solution.

mEq of anions $= 40 \text{ mEq } Cl^-/L + 15 \text{ mEq } HPO_4^{2-}/L = 55 \text{ mEq/L}$

mEq $Na^+ =$ mEq anions $= 55 \text{ mEq } Na^+/L$

7.21 **a.** The solution must be saturated because no additional solute dissolves.
b. The solution was unsaturated because the sugar cube dissolves.

7.23 **a.** It is unsaturated because 34 g of KCl is the maximum that dissolves in 100 g of H_2O at 20 °C.
b. 11 g of $NaNO_3$ in 25 g of H_2O is 44 g in 100 g of H_2O. At 20 °C, 88 g of $NaNO_3$ can dissolve so the solution is unsaturated.
c. Adding 400. g of sugar to 125 g of H_2O is 320 g in 100 g of H_2O at 20 °C, only 204 g of sugar can dissolve, which is less than 320 g. The sugar solution is saturated and excess undissolved sugar is present.

7.25 **a.** $\dfrac{34 \text{ g KCl}}{100 \text{ g } H_2O} \times 200 \text{ g } H_2O = 68 \text{ g of KCl}$ (This will dissolve at 20 °C)

At 20 °C 68 g of KCl can dissolve in 200 g of H_2O.

b. Since 80. g of KCl dissolves at 50 °C and 68 g is in solution at 20 °C, the mass of solid is 80. g $-$ 68 g $= 12$ g of KCl.

7.27 **a.** In general, the solubility of solid ionic solutes increases as temperature is increased.
b. The solubility of a gaseous solute (CO_2) decreases as the temperature is increased.
c. The solubility of a gaseous solute is lowered as temperature increases. When the can of warm soda is opened, more CO_2 is released producing more spray.

7.29 A 5% (m/m) glucose solution contains 5 g of glucose in 100 g of solution (5 g of glucose + 95 g of water), while a 5% (m/v) glucose solution contains 5 g of glucose in 100 mL solution.

7.31 **a.** $\dfrac{25 \text{ g of KCl}}{150 \text{ g solution}} \times 100 = 17\%(\text{m/m})$

b. $\dfrac{12 \text{ g sugar}}{225 \text{ g solution}} \times 100 = 5.3\%(\text{m/m})$

7.33 **a.** $\dfrac{75 \text{ g } Na_2SO_4}{250 \text{ mL solution}} \times 100 = 30.\%(\text{m/v})$

b. $\dfrac{39 \text{ g sucrose}}{355 \text{ mL solution}} \times 100 = 11\%(\text{m/v})$

7.35 **a.** $50.0 \text{ mL solution} \times \dfrac{5.0 \text{ g KCl}}{100 \text{ mL solution}} = 2.5 \text{ g of KCl}$

b. $1250 \text{ mL solution} \times \dfrac{4.0 \text{ g NH}_4\text{Cl}}{100 \text{ mL solution}} = 50.\text{g of NH}_4\text{Cl}$

7.37 $355 \text{ mL solution} \times \dfrac{22.5 \text{ mL alcohol}}{100 \text{ mL solution}} = 79.9 \text{ mL of alcohol}$

7.39 **a.** $1 \text{ L} \times \dfrac{100 \text{ mL solution}}{1 \text{ L}} \times \dfrac{20. \text{ g mannitol}}{100 \text{ mL solution}} = 20. \text{ g of mannitol}$

b. $12 \text{ L} \times \dfrac{100 \text{ mL solution}}{1 \text{ L}} \times \dfrac{20. \text{ g mannitol}}{100 \text{ mL solution}} = 240 \text{ g of mannitol}$

7.41 $100 \text{ g glucose} \times \dfrac{100 \text{ mL solution}}{5 \text{ g glucose}} \times \dfrac{1 \text{ L}}{1000 \text{ mL}} = 2 \text{ L solution}$

7.43 Molarity = moles of solute/L of solution

a. $\dfrac{2.0 \text{ moles glucose}}{4.0 \text{ L solution}} = 0.50 \text{ M glucose}$

b. $\dfrac{4.0 \text{ g KOH}}{2.0 \text{ L solution}} \times \dfrac{1 \text{ mole KOH}}{56.1 \text{ g KOH}} = 0.036 \text{ M KOH}$

c. $\dfrac{5.85 \text{ g NaCl}}{0.400 \text{ L}} \times \dfrac{1 \text{ mole NaCl}}{58.5 \text{ g NaCl}} = 0.250 \text{ mole/L} = 0.250 \text{ M}$

7.45 **a.** $1.0 \text{ L solution} \times \dfrac{3.0 \text{ moles NaCl}}{1 \text{ L solution}} = 3.0 \text{ moles of NaCl}$

b. $0.40 \text{ L solution} \times \dfrac{1.0 \text{ mole KBr}}{1 \text{ L solution}} = 0.40 \text{ mole of KBr}$

c. $0.125 \text{ L solution} \times \dfrac{2.0 \text{ moles MgCl}_2}{1 \text{ L solution}} = 0.30 \text{ mole of MgCl}_2$

7.47 **a.** $2.0 \text{ L} \times \dfrac{1.5 \text{ moles NaOH}}{1 \text{ L}} \times \dfrac{40.0 \text{ g NaOH}}{1 \text{ mole NaOH}} = 120 \text{ g of NaOH}$

b. $4.0 \text{ L} \times \dfrac{0.20 \text{ mole KCl}}{1 \text{ L}} \times \dfrac{74.6 \text{ g KCl}}{1 \text{ mole KCl}} = 60. \text{ g of KCl}$

c. $0.025 \text{ L} \times \dfrac{6.0 \text{ moles HCl}}{1 \text{ L}} \times \dfrac{36.5 \text{ g HCl}}{1 \text{ mole HCl}} = 5.5 \text{ g of HCl}$

7.49 **a.** $3.0 \text{ moles NaOH} \times \dfrac{1 \text{ L}}{2.0 \text{ moles NaOH}} = 1.5 \text{ L NaOH}$

b. $15 \text{ moles NaCl} \times \dfrac{1 \text{ L}}{1.5 \text{ moles NaCl}} = 10. \text{ L NaCl}$

c. $0.0500 \text{ mole Ca(NO}_3)_2 \times \dfrac{1 \text{ L}}{0.800 \text{ mole Ca(NO}_3)_2} \times \dfrac{1000 \text{ mL}}{1 \text{ L}} = 62.5 \text{ mL}$

7.51 Adding water (solvent) to the soup increases the volume and dilutes the tomato concentration.

7.53 The concentration of a diluted solution can be calculated using the relationship:

$$\%(m/v) \text{ of dilute solution} = \frac{\text{grams of solute}}{\text{volume dilute solution}} \times 100$$

$$\text{or Molarity of dilute solution} = \frac{\text{moles of solute}}{\text{volume dilute solution in L}}$$

a. From the initial solution: moles solute is $2.0 \, \cancel{L} \times 6.0 \text{ moles HCl}/\cancel{L} = 12$ moles of HCl

Molarity of the dilute solution $= \dfrac{12 \text{ moles HCl}}{6.0 \text{ L}} = 2.0$ M HCl

b. The moles of solute is $12 \text{ moles NaOH}/\cancel{L} \times 0.50 \, \cancel{L} = 6.0$ moles of NaOH

Final molarity is $\dfrac{6.0 \text{ moles NaOH}}{3.0 \text{ L}} = 2.0$ M NaOH

c. Initial grams of solute is $10.0 \, \cancel{\text{mL solution}} \times \dfrac{25 \text{ g KOH}}{100 \, \cancel{\text{mL solution}}} = 2.5$ g of KOH

d. Initial grams of solute is $50.0 \, \cancel{\text{mL}} \times \dfrac{15 \text{ g H}_2\text{SO}_4}{100 \, \cancel{\text{mL}}} = 7.5$ g of H_2SO_4

Final $\%(m/v)$ is $\dfrac{7.5 \text{ g H}_2\text{SO}_4}{50.0 \text{ mL}} \times 100 = 3.0\%(m/v)$ H_2SO_4

7.55 The final volume of a diluted solution can be found by using the relationship: $C_1V_1 = C_2V_2$ where C_1 is the concentration (M or %) of the initial (concentrated) and C_2 is the concentration (M or %) of the final (dilute) solution and V_1 is the volume of the initial solution. Solving for V_2 gives the volume of the dilute solution.

a. $V_2 = \dfrac{M_1V_1}{M_2} = \dfrac{6.0 \, \cancel{\text{moles/L}}}{0.20 \, \cancel{\text{mole/L}}} \times 0.0200 \text{ L} = 0.60$ L

b. $V_2 = \dfrac{\%_1V_1}{\%_2} = \dfrac{10.0 \, \cancel{\%}}{2.0 \, \cancel{\%}} \times 50.0 \text{ mL} = 250$ mL

c. $V_2 = \dfrac{M_1V_1}{M_2} = \dfrac{6.0 \, \cancel{\text{moles/L}}}{0.50 \, \cancel{\text{mole/L}}} \times 0.500 \text{ L} = 6.0$ L

d. $V_2 = \dfrac{\%_1V_1}{\%_2} = \dfrac{12 \, \cancel{\%}}{5.0 \, \cancel{\%}} \times 75 \text{ mL} = 180$ mL

7.57 a. $50.0 \, \cancel{\text{mL}} \times \dfrac{1 \text{ L}}{1000 \, \cancel{\text{mL}}} = 0.0500 \, \cancel{L} \times \dfrac{1.50 \text{ moles KCl}}{1 \, \cancel{L}} = 0.0750$ mole of KCl

$0.0750 \, \cancel{\text{mole KCl}} \times \dfrac{1 \, \cancel{\text{mole PbCl}_2}}{2 \, \cancel{\text{moles KCl}}} \times \dfrac{278.1 \text{ g}}{1 \, \cancel{\text{mole PbCl}_2}} = 10.4$ g of $PbCl_2$

b. $50.0 \, \cancel{\text{mL}} \times \dfrac{1 \text{ L}}{1000 \, \cancel{\text{mL}}} = 0.0500 \, \cancel{L} \times \dfrac{1.50 \text{ moles KCl}}{1 \, \cancel{L}} = 0.0750$ mole of KCl

$0.0750 \, \cancel{\text{mole KCl}} \times \dfrac{1 \, \cancel{\text{mole Pb(NO}_3)_2}}{2 \, \cancel{\text{mole KCl}}} \times \dfrac{1 \text{ L solution}}{2.00 \, \cancel{\text{mole Pb(NO}_3)_2}} = 0.0188$ L of solution

$0.0188 \, \cancel{L} \times \dfrac{1000 \text{ mL}}{1 \, \cancel{L}} = 18.8$ mL of solution

7.59 a. $15.0 \, \text{g Mg} \times \dfrac{1 \text{ mole Mg}}{24.3 \text{ g Mg}} = 0.617 \text{ mole of Mg}$

$0.617 \, \text{mole Mg} \times \dfrac{2 \text{ moles HCl}}{1 \text{ mole Mg}} = 1.23 \, \text{moles HCl} \times \dfrac{1 \text{ L}}{6.00 \text{ moles HCl}} = 0.206 \text{ L}$

$0.206 \, \text{L} \times \dfrac{1000 \text{ mL}}{1 \text{ L}} = 206 \text{ mL}$

b. $\dfrac{2.00 \text{ moles HCl}}{1 \text{ L solution}} \times 0.500 \, \text{L} = 1.00 \, \text{mole HCl} \times \dfrac{1 \text{ mole H}_2}{2 \text{ moles HCl}} = 0.500 \text{ mole of H}_2 \text{ gas}$

7.61 a. A solution cannot be separated by a semipermeable membrane.
b. A suspension settles as time passes.

7.63 a. Water in the soil diffuses through the plant's root membranes to dilute the solutions in these cells. Because the plant's cells above these root cells contain more concentrated solutions than the root cells, water moves up from the roots to dilute the more concentrated cell solutions.
b. The pickling (brine) solution contains more solutes and less solvent than the cucumber's cells. Thus, solvent flows out of the cells of the cucumber and into the brine solution, and the cucumber shrivels and becomes a pickle.

7.65 a. The 10% (m/v) starch solution has a higher osmotic pressure than pure water.
b. The water will initially flow into the starch solution to dilute solute concentration.
c. The volume of the starch solution will increase due to inflow of water.

7.67 Water flows out of the solution with the higher solvent concentration (which corresponds to a lower solute concentration) to the solution with a lower solvent concentration (which corresponds to a higher solute concentration).

a. Water flows into compartment B, which contains the 10% (m/v) glucose solution.
b. Water flows into compartment B, which contains the 8% (m/v) albumin solution.
c. Water flows into compartment B, which contains the 10% (m/v) NaCl solution.

7.69 A red blood cell has osmotic pressure of a 5% (m/v) glucose solution or a 0.90% (m/v) NaCl solution. In a hypotonic solution (lower osmotic pressure), solvent flows from the hypotonic solution into the red blood cell. When a red blood cell is placed in a hypertonic solution (higher osmotic pressure), solvent (water) flows from the red blood cell to the hypertonic solution. Isotonic solutions have the same osmotic pressure, and a red blood cell in an isotonic solution will not change volume because the flow of solvent into and out of the cell is equal.

a. Distilled water is a hypotonic solution when compared to a red blood cell's contents.
b. A 1% (m/v) glucose solution is a hypotonic solution.
c. A 0.90% (m/v) NaCl solution is isotonic with a red blood cell's contents.
d. A 5% (m/v) glucose solution is an isotonic solution.

7.71 Colloids cannot pass through the semipermeable dialysis membrane; water and solutions freely pass through semipermeable membranes.

a. Sodium and chloride ions will both pass through the membrane into the distilled water.
b. The amino acid alanine can pass through a dialysis membrane; the colloid starch will not.
c. Sodium and chloride ions will both be present in the water surrounding the dialysis bag; the colloid starch will not.
d. Urea will diffuse through the dialysis bag into the water.

7.73 a. (1) Polar solute and a polar solvent combine to form a homogeneous solution.
b. (2) Nonpolar solute and a polar solvent do not form a solution.
c. (1) Nonpolar solute and a nonpolar solvent combine to form a homogeneous solution.

7.75 a. 3 (no dissociation) **b.** 1 (some dissociation, a few ions) **c.** 2 (all ionized)

7.77 A "brine" salt-water solution has a high concentration of $Na^+ Cl^-$, which is hypertonic to the cucumber. Therefore, water flows from the cucumber into the hypertonic salt solution that surrounds it. The loss of water causes the cucumber to become a wrinkled pickle.

7.79 **a.** 2 Water will flow into the B (8%) side.
b. 1 Water will continue to flow equally in both directions; no change in volumes.
c. 3 Water will flow into the A (5%) side.
d. 2 Water will flow into the B (1%) side.

7.81 Iodine is a nonpolar molecule and needs a nonpolar solvent such as hexane. Iodine does not dissolve in water because water is a polar solvent.

7.83 $$\frac{15.5 \text{ g } Na_2SO_4}{15.5 \text{ g } Na_2SO_4 + 75.5 \text{ g water}} \times 100 = 17.0\% \text{ m/m}$$

7.85 **a.** $24 \text{ h} \times \dfrac{750 \text{ mL solution}}{12 \text{ h}} \times \dfrac{4 \text{ g amino acids}}{100 \text{ mL solution}} = 60 \text{ g of amino acids}$

$24 \text{ h} \times \dfrac{750 \text{ mL solution}}{12 \text{ h}} \times \dfrac{25 \text{ g glucose}}{100 \text{ mL solution}} = 380 \text{ g of glucose}$

$24 \text{ h} \times \dfrac{500 \text{ mL solution}}{12 \text{ h}} \times \dfrac{10 \text{ g lipid}}{100 \text{ mL solution}} = 100 \text{ g of lipid}$

b. $60 \text{ g amino acids (protein)} \times \dfrac{4 \text{ kcal}}{1 \text{ g protein}} = 240 \text{ kcal}$

$380 \text{ g glucose (carb)} \times \dfrac{4 \text{ kcal}}{1 \text{ g carb}} = 1520 \text{ kcal}$

$100 \text{ g lipid (fat)} \times \dfrac{9 \text{ kcal}}{1 \text{ g fat}} = 900 \text{ kcal}$

Sum: 240 kcal + 1520 kcal + 900 kcal = (2660) = 2700 kcal/day

7.87 $4.5 \text{ mL propyl alcohol} \times \dfrac{100 \text{ mL solution}}{12 \text{ mL propyl alcohol}} = 38 \text{ mL of solution}$

7.89 $250 \text{ mL} \times \dfrac{1 \text{ L}}{1000 \text{ mL}} \times \dfrac{2 \text{ moles KCl}}{1 \text{ L}} \times \dfrac{74.6 \text{ g KCl}}{1 \text{ moles KCl}} = 37.3 \text{ g of KCl}$

To make a 2.00 M KCl solution, weigh out 37.3 g of KCl (0.500 mole) and place in a volumetric flask. Add enough water to dissolve the KCl and give a final volume of 0.250 L.

7.91 Mass of solution: 70.0 g of solute + 130.0 g of solvent = 200.0 g

a. $\dfrac{70.0 \text{ g } HNO_3}{200.0 \text{ g solution}} \times 100\% = 35.5\% \text{ (m/m) } HNO_3$

b. $200.0 \text{ g solution} \times \dfrac{1 \text{ mL solution}}{1.21 \text{ g solution}} = 165 \text{ mL of solution}$

c. $\dfrac{70.0 \text{ g } HNO_3}{165 \text{ mL solution}} \times 100 = 42.4\% \text{ (m/v) } HNO_3$

d. $\dfrac{70.0 \text{ g } HNO_3}{0.165 \text{ L solution}} \times \dfrac{1 \text{ mole } HNO_3}{63.0 \text{ g } HNO_3} = 6.73 \text{ M } HNO_3$

7.93 **a.** $2.5 \, \cancel{L} \times \dfrac{3.0 \text{ moles } \cancel{Al(NO_3)_3}}{1 \, \cancel{L}} \times \dfrac{213 \text{ g } Al(NO_3)_3}{1 \text{ mole } \cancel{Al(NO_3)_3}} = 1600 \text{ g of } Al(NO_3)_3$

b. $75 \, \cancel{mL} \times \dfrac{1 \, \cancel{L}}{1000 \, \cancel{mL}} \times \dfrac{0.50 \text{ mole } \cancel{C_6H_{12}O_6}}{1 \, \cancel{L}} \times \dfrac{180. \text{ g } C_6H_{12}O_6}{1 \text{ mole } \cancel{C_6H_{12}O_6}} = 6.8 \text{ g of } C_6H_{12}O_6$

7.95 $60.0 \, \cancel{mL \, Al(OH)_3} \times \dfrac{1 \, \cancel{L}}{1000 \, \cancel{mL \, Al(OH)_3}} \times \dfrac{1.00 \text{ mole } \cancel{Al(OH)_3}}{1 \, \cancel{L}} \times \dfrac{3 \text{ moles } \cancel{HCl}}{1 \text{ mole } \cancel{Al(OH)_3}} \times$

$\dfrac{1000 \text{ mL HCl}}{6.00 \text{ moles } \cancel{HCl}} = 30.0 \text{ mL of HCl solution}$

7.97 A solution with a high salt (solute) concentration will dry flowers because water (solvent) flows out of the flowers' cells and into the salt solution to dilute the salt concentration.

7.99 Drinking seawater, which is hypertonic, will cause water to flow out of the body cells and dehydrate the body's cells.

7.101 **a.** Mass of NaCl is $25.50 \text{ g} - 24.10 \text{ g} = 1.40 \text{ g}$

Mass of solution is $36.15 \text{ g} - 24.10 \text{ g} = 12.05 \text{ g}$

$\dfrac{1.40 \text{ g NaCl}}{12.05 \text{ g solution}} \times 100 = 11.6\% \text{ (m/m)}$

b. $1.40 \, \cancel{g \, NaCl} \times \dfrac{1 \text{ mole NaCl}}{58.5 \, \cancel{g \, NaCl}} = 0.0239 \text{ mole of NaCl}$

$\dfrac{0.0239 \text{ mole NaCl}}{0.0100 \text{L}} = 2.39 \text{ M}$

c. $M_1V_1 = M_2V_2 \quad M_2 = \dfrac{M_1V_1}{V_2} = \dfrac{(2.39 \text{ M})(10.0 \, \cancel{mL})}{60.0 \, \cancel{mL}} = 0.398 \text{ M}$

7.103 **a.** The solution is saturated: 35 g of KF and 25 g of H_2O is 140 g of KF in 100 g of H_2O, which exceeds the solubility of KF at 18 °C.
b. The solution is unsaturated: 42 g of KF and 50. g of H_2O is 84 g of KF in 100 g of H_2O, which is less than the solubility of 92 g of KF per 100 g of H_2O.
c. The solution is saturated: 145 g of KF and 150. g of H_2O is 97 g of KF per 100 g of H_2O, which exceeds the solubility of 92 g of KF per 100 g of H_2O.

7.105 $15.2 \, \cancel{g \, LiCl} \times \dfrac{1 \text{ mole LiCl}}{42.4 \, \cancel{g \, LiCl}} = 0.358 \text{ mole of LiCl}$

$0.358 \, \cancel{mole \, LiCl} \times \dfrac{1 \, \cancel{L}}{1.75 \, \cancel{moles}} \times \dfrac{1000 \text{ mL}}{1 \, \cancel{L}} = 205 \text{ mL}$

7.107 $4.20 \, \cancel{L} \times \dfrac{1 \text{ mole}}{22.4 \, \cancel{L}} = 0.188 \, \cancel{mole \, H_2} \times \dfrac{2 \text{ moles HCl}}{1 \text{ mole } \cancel{H_2}} = 0.376 \text{ mole of HCl}$

$\dfrac{0.376 \text{ mole HCl}}{0.250 \text{ L}} = 1.50 \text{ M}$

8
Acids and Bases

Study Goals

- Describe the characteristics of acids and bases.
- Identify conjugate acid–base pairs in Brønsted–Lowry acids and bases.
- Use the ion product of water to calculate $[H_3O^+]$, $[OH^-]$, and pH.
- Write balanced equations for reactions of an acid with metals, carbonates, and bases.
- Calculate the concentration of an acid solution from titration data.
- Describe the function of a buffer.

Think About It

1. Why do a lemon, grapefruit, and vinegar taste sour?

2. What do antacids do? What are some bases listed on the labels of antacids?

3. Why are some aspirin products buffered?

Key Terms

    a. acid   **b.** base   **c.** pH   **d.** neutralization   **e.** buffer

1. _____ A substance that forms hydroxide ions (OH^-) in water and/or accepts protons (H^+).

2. _____ A reaction between an acid and a base to form a salt and water.

3. _____ A substance that forms hydrogen ions (H^+) in water.

4. _____ A mixture of a weak acid (or base) and its salt that maintains the pH of a solution.

5. _____ A measure of the acidity of a solution.

Answers   **1.** b   **2.** d   **3.** a   **4.** e   **5.** c

8.1 Acids and Bases

- In water, an Arrhenius acid produces H_3O^+, and an Arrhenius base produces OH^-.
- Protons (H^+) form hydronium ions, H_3O^+, in water when they bond to polar water molecules.
- According to the Brønsted–Lowry theory, acids donate protons (H^+) to bases.
- Conjugate acid–base pairs are molecules or ions linked by the loss and gain of a proton.

◆ **Learning Exercise 8.1A**

Indicate if the following characteristics describe an (A) acid or (B) base.

1. _____ Turns blue litmus red

2. _____ Tastes sour

3. _____ Has more OH^- ions than H_3O^+ ions

4. _____ Neutralizes bases

5. _____ Tastes bitter

6. _____ Turns red litmus blue

7. _____ Has more H_3O^+ ions than OH^- ions

8. _____ Neutralizes acids

Answers   **1.** A   **2.** A   **3.** B   **4.** A   **5.** B   **6.** B   **7.** A   **8.** B

◆ **Learning Exercise 8.1B**

Fill in the blanks with the formula or name of an acid or base:

1. HCl   _____

2. _____   sodium hydroxide

3. _____   sulfurous acid

4. _____   nitric acid

5. $Ca(OH)_2$ _____

6. H_2CO_3 _____

7. $Al(OH)_3$ _____

8. _____   potassium hydroxide

Answers   **1.** hydrochloric acid   **2.** NaOH   **3.** H_2SO_3
             4. HNO_3 **5.** calcium hydroxide   **6.** carbonic acid
             7. aluminum hydroxide   **8.** KOH

Study Note

Identify the conjugate acid–base pairs in the following equation:

$$HCl + H_2O \rightarrow H_3O^+ + Cl^-$$

Solution: HCl (proton donor) and Cl^- (proton acceptor)
             H_2O (proton acceptor) and H_3O^+ (proton donor)

◆ **Learning Exercise 8.1C**

Complete the following conjugate acid–base pairs:

Conjugate Acid	**Conjugate Base**
1. H_2O	_____
2. HSO_4^-	_____
3. _____	F^-
4. _____	CO_3^{2-}
5. HNO_3	_____
6. NH_4^+	_____
7. _____	HS^-
8. _____	$H_2PO_4^-$

Answers   1. OH^- 2. SO_4^{2-} 3. HF   4. HCO_3^-
          5. NO_3^- 6. NH_3 7. H_2S 8. H_3PO_4

◆ **Learning Exercise 8.1D**

Identify the conjugate acid–base pairs in each of the following:

1. $HF + H_2O \rightleftharpoons H_3O^+ + F^-$

2. $NH_4^+ + SO_4^{2-} \rightleftharpoons NH_3 + HSO_4^-$

3. $NH_3 + H_2O \rightleftharpoons NH_4^+ + OH^-$

4. $HNO_3 + OH^- \rightleftharpoons H_2O + NO_3^-$

Answers   1. HF/F^- and H_2O/H_3O^+ 2. NH_4^+/NH_3 and SO_4^{2-}/HSO_4^-
          3. NH_3/NH_4^+ and H_2O/OH^- 4. HNO_3/NO_3^- and OH^-/H_2O

8.2 Strengths of Acids and Bases

- In aqueous solution, a strong acid donates nearly all of its protons to water, whereas a weak acid donates only a small percentage of protons to water.
- Most hydroxides of Groups 1A (1) and 2A (2) are strong bases, which dissociate nearly completely in water. In an aqueous ammonia solution, NH_3, which is a weak base, accepts only a small percentage of protons to form NH_4^+.

Study Note

Only six common acids are strong acids: other acids are considered as weak acids.

HCl	HNO_3
HBr	H_2SO_4(first H)
HI	$HClO_4$

Example: Is H_2S a strong or weak acid?
Solution: H_2S is a weak acid because it is not one of the six strong acids.

◆ **Learning Exercise 8.2A**

Identify each of the following as a strong or weak acid or base:

1. HNO_3 _____   **4.** NH_3 _____   **7.** $Ca(OH)_2$ _____

2. H_2CO_3 _____   **5.** LiOH _____   **8.** H_2SO_4 _____

3. $H_2PO_4^-$ _____   **6.** H_3BO_3 _____

Answers   **1.** strong acid   **2.** weak acid   **3.** weak acid or weak base   **4.** weak base
        5. strong base   **6.** weak acid   **7.** strong base   **8.** strong acid

◆ **Learning Exercise 8.2B**

Using Table 8.3, identify the stronger acid in each of the following pairs of acids:

1. HCl or H_2CO_3 _____   **4.** H_2SO_4 or HSO_4^- _____

2. HNO_2 or HCN _____   **5.** HF or H_3PO_4 _____

3. H_2S or HBr _____

Answers   **1.** HCl   **2.** HNO_2 **3.** HBr   **4.** H_2SO_4 **5.** H_3PO_4

8.3 Ionization of Water

● In pure water, a few water molecules transfer a proton to other water molecules producing small but equal amounts of $[H_3O^+]$ and $[OH^-] = 1 \times 10^{-7}$ moles/L.

● K_w, the ion product, $[H_3O^+][OH^-] = [1 \times 10^{-7}][1 \times 10^{-7}] = 1 \times 10^{-14}$, applies to all aqueous solutions.

● In acidic solutions, the $[H_3O^+]$ is greater than the $[OH^-]$. In basic solutions, the $[OH^-]$ is greater than the $[H_3O^+]$.

Study Note

Example: What is the $[H_3O^+]$ in a solution that has $[OH^-] = 2.0 \times 10^{-9}$ M?

Solution: $[H_3O^+] = \dfrac{1.0 \times 10^{-14}}{2.0 \times 10^{-9}} = 5.0 \times 10^{-6}$ M

◆ **Learning Exercise 8.3**

Write the $[H_3O^+]$ when the $[OH^-]$ has the following values:

a. $[OH^-] = 1.0 \times 10^{-10}$ M $[H_3O^+] =$

b. $[OH^-] = 2.0 \times 10^{-5}$ M $[H_3O^+] =$

c. $[OH^-] = 4.5 \times 10^{-7}$ M $[H_3O^+] =$

d. $[OH^-] = 8.0 \times 10^{-4}$ M $[H_3O^+] =$

e. $[OH^-] = 5.5 \times 10^{-8}$ M $[H_3O^+] =$

Answers   **a.** 1.0×10^{-4} M   **b.** 5.0×10^{-10} M   **c.** 2.2×10^{-8} M
        d. 1.3×10^{-11} M   **e.** 1.8×10^{-7} M

8.4 The pH Scale

- The pH scale is a range of numbers from 0 to 14 related to the $[H_3O^+]$ of the solution.
- A neutral solution has a pH of 7. In acidic solutions, the pH is below 7, and in basic solutions the pH is above 7.
- Mathematically, pH is the negative logarithm of the hydronium ion concentration:

$$pH = -\log[H_3O^+]$$

◆ Learning Exercise 8.4A

State whether the following pH values are acidic, basic, or neutral:

1. _____ plasma, pH = 7.4
2. _____ soft drink, pH = 2.8
3. _____ maple syrup, pH = 6.8
4. _____ beans, pH = 5.0
5. _____ tomatoes, pH = 4.2

6. _____ lemon juice, pH = 2.2
7. _____ saliva, pH = 7.0
8. _____ eggs, pH = 7.8
9. _____ lime, pH = 12.4
10. _____ strawberries, pH = 3.0

Answers 1. basic 2. acidic 3. acidic 4. acidic 5. acidic
6. acidic 7. neutral 8. basic 9. basic 10. acidic

◆ Learning Exercise 8.4B

Calculate the pH of each of the following solutions.

a. $[H_3O^+] = 1 \times 10^{-8}$ M _____
b. $[OH^-] = 1 \times 10^{-12}$ M _____

c. $[H_3O^+] = 1 \times 10^{-3}$ M _____
d. $[OH^-] = 1 \times 10^{-10}$ M _____

Answers a. 8.0 b. 2.0 c. 3.0 d. 4.0

◆ Learning Exercise 8.4C

Calculate the pH of each of the following solutions.

a. $[H_3O^+] = 1 \times 10^{-3}$ M _____
b. $[OH^-] = 1 \times 10^{-6}$ M _____

c. $[H_3O^+] = 1 \times 10^{-8}$ M _____
d. $[OH^-] = 1 \times 10^{-10}$ M _____

Answers a. 3.0 b. 8.0 c. 8.0 d. 4.0

◆ Learning Exercise 8.4D

Complete the following table:

	$[H_3O^+]$	$[OH^-]$	pH
a.		1×10^{-12} M	
b.			8.0
c.	1×10^{-10} M		
d.			7.0
e.			1.0

Answers		$[H_3O^+]$	$[OH^-]$	pH
	a.	1×10^{-2} M	1×10^{-12} M	2.0
	b.	1×10^{-8} M	1×10^{-6} M	8.0
	c.	1×10^{-10} M	1×10^{-4} M	10.0
	d.	1×10^{-7} M	1×10^{-7} M	7.0
	e.	1×10^{-1} M	1×10^{-13} M	1.0

8.5 Reactions of Acids and Bases

- Acids react with many metals to yield hydrogen gas (H_2) and the salt of the metal.
- Acids react with carbonates and bicarbonates to yield CO_2, H_2O, and the salt of the metal.
- Acids neutralize bases in a reaction that produces water and a salt.
- The net ionic equation for any neutralization is $H^+ + OH^- \rightarrow H_2O$.
- In a balanced neutralization equation, an equal number of moles of H^+ and OH^- must react.
- The concentration of an acid can be determined by titration.

◆ Learning Exercise 8.5A

Complete and balance each of the following reactions of acids:

1. ____ $Zn(s)$ + ____ $HCl(aq) \rightarrow$ ____ $ZnCl_2(aq)$ + _____

2. ____ $HCl(aq)$ + ____ $Li_2CO_3(s) \rightarrow$ ____ + _____

3. ____ $HCl(aq)$ + ____ $NaHCO_3(s) \rightarrow$ ____ $CO_2(g)$ ____ + $H_2O(l)$ + ____ $NaCl(aq)$

4. ____ $Al(s)$ + ____ $H_2SO_4(aq) \rightarrow$ ____ $Al_2(SO_4)_3(aq)$ + _____

Answers
1. $1 Zn(s) + 2 HCl(aq) \rightarrow 1 ZnCl_2(aq) + H_2(g)$
2. $2 HCl(aq) + 1 Li_2CO_3(s) \rightarrow 1 CO_2(g) + 1 H_2O(l) + 2 LiCl(aq)$
3. $1 HCl(aq) + 1 NaHCO_3(s) \rightarrow 1 CO_2(g) + 1 H_2O(l) + 1 NaCl(aq)$
4. $2 Al(s) + 3 H_2SO_4(aq) \rightarrow 1 Al_2(SO_4)_3(aq) + 3 H_2(g)$

◆ Learning Exercise 8.5B

Balance each of the following neutralization reactions:

1. ____ $NaOH(aq)$ + ____ $H_2SO_4(aq) \rightarrow$ ____ $Na_2SO_4(aq)$ + ____ $H_2O(l)$

2. ____ $Mg(OH)_2(aq)$ + ____ $HCl(aq) \rightarrow$ ____ $MgCl_2(aq)$ + ____ $H_2O(l)$

3. ____ $Al(OH)_3(aq)$ + ____ $HNO_3(aq) \rightarrow$ ____ $Al(NO_3)_3(aq)$ + ____ $H_2O(l)$

4. ____ $Ca(OH)_2(aq)$ + ____ $H_3PO_4(aq) \rightarrow$ ____ $Ca_3(PO_4)_2(s)$ + ____ $H_2O(l)$

Answers   1. $2\,NaOH(aq) + 1\,H_2SO_4(aq) \rightarrow 1\,Na_2SO_4(aq) + 2\,H_2O(l)$
2. $1\,Mg(OH)_2(aq) + 2\,HCl(aq) \rightarrow 1\,MgCl_2(aq) + 2\,H_2O(l)$
3. $1\,Al(OH)_3(aq) + 3\,HNO_3(aq) \rightarrow 1\,Al(NO_3)_3(aq) + 3\,H_2O(l)$
4. $3\,Ca(OH)_2(aq) + 2\,H_3PO_4(aq) \rightarrow 1\,Ca_3(PO_4)_2(s) + 6\,H_2O(l)$

◆ Learning Exercise 8.5C

Complete each of the following neutralization reactions and then balance:

a. $KOH(aq) + \underline{\hspace{1cm}} H_3PO_4(aq) \rightarrow \underline{\hspace{1cm}} + \underline{\hspace{1cm}} H_2O(l)$

b. $NaOH(aq) + \underline{\hspace{1cm}} \rightarrow \underline{\hspace{1cm}} Na_2SO_4(aq) + \underline{\hspace{1cm}}$

c. $\underline{\hspace{1cm}} + \underline{\hspace{1cm}} \rightarrow \underline{\hspace{1cm}} AlCl_3(aq) + \underline{\hspace{1cm}}$

d. $\underline{\hspace{1cm}} + \underline{\hspace{1cm}} \rightarrow \underline{\hspace{1cm}} Fe_2(SO_4)_3(s) + \underline{\hspace{1cm}}$

Answers   **a.** $3\,KOH(aq) + 1\,H_3PO_4(aq) \rightarrow 1\,K_3PO_4(aq) + 3\,H_2O(l)$
b. $2\,NaOH(aq) + 1\,H_2SO_4(aq) \rightarrow 1\,Na_2SO_4(aq) + 2\,H_2O(l)$
c. $Al(OH)_3(aq) + 3\,HCl(aq) \rightarrow 1\,AlCl_3(aq) + 3\,H_2O(l)$
d. $2\,Fe(OH)_3(aq) + 3\,H_2SO_4(aq) \rightarrow 1\,Fe_2(SO_4)_3(s) + 6\,H_2O(l)$

◆ Learning Exercise 8.5D

1. A 24.6-mL sample of HCl reacts with 33.0 mL of 0.222 M NaOH solution. What is the molarity of the HCl solution?

2. A 15.7-mL sample of H_2SO_4 reacts with 27.7 mL of 0.187 M KOH solution. What is the molarity of the H_2SO_4 solution?

Answers   1. 0.298 M HCl   2. 0.165 M H_2SO_4

8.6 Buffers

- A buffer solution resists a change in pH when small amounts of acid or base are added.
- A buffer contains either (1) a weak acid and its salt, or (2) a weak base and its salt. The weak acid picks up excess OH^-, and the anion of the salt picks up excess H_3O^+.

◆ **Learning Exercise 8.6**

State whether each of the following represents a buffer system or not.

a. HCl + NaCl _____

c. H_2CO_3 _____

b. K_2SO_4 _____

d. H_2CO_3 + $NaHCO_3$ _____

Answers
a. No. A strong acid is not a buffer.
b. No. A salt alone cannot act as a buffer.
c. No. A weak acid alone cannot act as a buffer.
d. Yes. A weak acid and its salt act as a buffer system.

Checklist for Chapter 8

You are ready to take the practice test for Chapter 8. Be sure that you have accomplished the following learning goals for this chapter. If you are not sure, review the section listed at the end of the goal. Then apply your new skills and understanding to the practice test.
After studying Chapter 8, I can successfully:

_____ Describe the properties of Arrhenius acids and bases and write their names (8.1).

_____ Describe the Brønsted–Lowry concept of acids and bases; write conjugate acid–base pairs for an acid–base reaction (8.1).

_____ Write equations for the ionization of strong and weak acids and bases (8.2).

_____ Use the ion product of water to calculate $[H_3O^+]$ and $[OH^-]$ (8.3).

_____ Calculate pH from the $[H_3O^+]$ of a solution (8.4).

_____ Calculate $[H_3O^+]$ from the pH of a solution (8.4).

_____ Write a balanced equation for the reactions of acids with metals, carbonates, and/or bases (8.5).

_____ Describe the role of buffers in maintaining the pH of a solution and calculate the pH of a buffer solution. (8.6).

Practice Test for Chapter 8

1. An acid is a compound which when placed in water yields this characteristic ion:
A. H_3O^+
B. OH^-
C. Na^+
D. Cl^-
E. CO_3^{2-}

2. $MgCl_2$ would be classified as a(n)
A. acid
B. base
C. salt
D. buffer
E. nonelectrolyte

3. $Mg(OH)_2$ would be classified as a
A. weak acid
B. strong base
C. salt
D. buffer
E. nonelectrolyte

4. In the K_w expression for pure H_2O, the $[H_3O^+]$ has the value (at 25 °C)
A. 1×10^{-7} M
B. 1×10^{-1} M
C. 1×10^{-14} M
D. 1×10^{-6} M
E. 1×10^{-12} M

5. Of the following pH values, which is the most acidic?
A. 8.0
B. 5.5
C. 1.5
D. 3.2
E. 9.0

6. Of the following pH values, which is the most basic pH?
- **A.** 10.0
- **B.** 4.0
- **C.** 2.2
- **D.** 11.5
- **E.** 9.0

For questions 7 through 9, consider a solution with $[H_3O^+] = 1 \times 10^{-11}$ M.

7. The pH of the solution is
- **A.** 1.0
- **B.** 2.0
- **C.** 3.0
- **D.** 11.0
- **E.** 14.0

8. The hydroxide ion concentration is
- **A.** 1×10^{-1} M
- **B.** 1×10^{-3} M
- **C.** 1×10^{-4} M
- **D.** 1×10^{-7} M
- **E.** 1×10^{-11} M

9. The solution is
- **A.** acidic
- **B.** basic
- **C.** neutral
- **D.** a buffer
- **E.** neutralized

For questions 10 through 12, consider a solution with a $[OH^-] = 1 \times 10^{-5}$ M.

10. The hydrogen ion concentration of the solution is
- **A.** 1×10^{-5} M
- **B.** 1×10^{-7} M
- **C.** 1×10^{-9} M
- **D.** 1×10^{-10} M
- **E.** 1×10^{-14} M

11. The pH of the solution is
- **A.** 2.0
- **B.** 5.0
- **C.** 9.0
- **D.** 11
- **E.** 14

12. The solution is
- **A.** acidic
- **B.** basic
- **C.** neutral
- **D.** a buffer
- **E.** neutralized

13. Acetic acid is a weak acid because
- **A.** it forms a dilute acid solution.
- **B.** it is isotonic.
- **C.** it is less than 50% ionized in water.
- **D.** it is a nonpolar molecule.
- **E.** it can form a buffer.

14. A weak base when added to water
- **A.** makes the solution slightly basic.
- **B.** does not affect the pH.
- **C.** dissociates completely.
- **D.** does not dissociate.
- **E.** makes the solution slightly acidic.

15. Which is an equation for neutralization?
- **A.** $CaCO_3(s) \rightarrow CaO(s) + CO_2(g)$
- **B.** $Na_2SO_4(s) \rightarrow 2Na^+(aq) + SO_4^{2-}(aq)$
- **C.** $H_2SO_4(aq) + 2NaOH(aq) \rightarrow Na_2SO_4(aq) + 2H_2O(l)$
- **D.** $Na_2O(s) + SO_3(g) \rightarrow Na_2SO_4(s)$
- **E.** $H_2CO_3(s) \rightarrow CO_2(g) + H_2O(l)$

16. What is the name given to components in the body that keep blood pH within its normal 7.35 to 7.45 range?
- **A.** nutrients
- **B.** buffers
- **C.** metabolites
- **D.** fluids
- **E.** neutralizers

17. What is true of a typical buffer system?
- **A.** It maintains a pH of 7.0.
- **B.** It contains a weak base.
- **C.** It contains a salt.
- **D.** It contains a strong acid and its salt.
- **E.** It maintains the pH of a solution.

18. Which of the following would act as a buffer system?
- **A.** HCl
- **B.** Na_2CO_3
- **C.** $NaOH + NaNO_3$
- **D.** NH_4OH
- **E.** $NaHCO_3 + H_2CO_3$

19. Which of the following pairs is a conjugate acid–base pair?

    A. HCl/HNO_3 **B.** HNO_3/NO_3^- **C.** NaOH/KOH

    D. HSO_4^-/HCO_3^- **E.** Cl^-/F^-

20. The conjugate base of HSO_4^- is

    A. SO_4^{2-} **B.** H_2SO_4 **C.** HS^-

    D. H_2S **E.** SO_3^{2-}

21. In which reaction does H_2O act as an acid?

    A. $H_3PO_4(aq) + H_2O(l) \rightarrow H_3O^+(aq) + H_2PO_4^-(aq)$

    B. $H_2SO_4(aq) + H_2O(l) \rightarrow H_3O^+(aq) + HSO_4^-(aq)$

    C. $H_2O(l) + HS^-(aq) \rightarrow H_3O^+(aq) + S^{2-}(aq)$

    D. $NaOH(aq) + HCl(aq) \rightarrow NaCl(aq) + H_2O(l)$

    E. $NH_3(aq) + H_2O(l) \rightarrow NH_4^+(aq) + OH^-(aq)$

22. If 23.7 mL of HCl reacts with 19.6 mL of 0.179 M NaOH, what is the molarity of the HCl solution?

    A. 6.76 M   **B.** 0.216 M   **C.** 0.148 M

    D. 0.163 M   **E.** 0.333 M

23. A solution has a pH of 4.0. It has an $[H_3O^+]$ of

    A. 1×10^{-4} M   **B.** 1×10^4 M   **C.** 1×10^{-10} M

    D. 4.0 M   **E.** 1×10^{-14} M

24. A solution has a pH of 8.0. It has an $[OH^-]$ of

    A. 1×10^{-8} M   **B.** 1×10^8 M   **C.** 1×10^{-6} M

    D. 6.0 M   **E.** 1×10^{-14} M

25. A solution has a pH of 5.7. It has an $[H_3O^+]$ of

    A. 5×10^{-9} M   **B.** 2×10^{-6} M   **C.** 5×10^{-7} M

    E. 2×10^{-5} M   **E.** 1×10^{-14} M

Answers to the Practice Test

1. A		**2.** C		**3.** B		**4.** A		**5.** C
6. D		**7.** D		**8.** B		**9.** B		**10.** C
11. C		**12.** B		**13.** C		**14.** A		**15.** C
16. B		**17.** E		**18.** E		**19.** B		**20.** A
21. E		**22.** C		**23.** A		**24.** C		**25.** B

Answers and Solutions to Selected Text Problems

8.1 According to the Arrhenius theory,

    a. acids taste sour.   **b.** acids neutralize bases.

    c. acids produce H_3O^+ ions in water.   **d.** potassium hydroxide is the name of a base.

8.3 The names of nonoxy acids begin with *hydro-*, followed by the name of the anion. The names of oxyacids use the element root with *-ic acid*. Acids with one oxygen less than the common *-ic acid* name are named as *-ous acids.*

    a. hydrochloric acid   **b.** calcium hydroxide   **c.** carbonic acid

    d. nitric acid   **e.** sulfurous acid   **f.** lithium hydroxide

8.5 **a.** $Mg(OH)_2$ **b.** HF   **c.** H_3PO_4

      d. LiOH   **e.** $Cu(OH)_2$ **f.** H_2SO_4

8.7 The acid donates a proton (H^+), while the base accepts a proton.

    a. acid (proton donor) HI   proton acceptor (base) H_2O
    b. acid (proton donor) H_2O proton acceptor (base) F^-

8.9 To form the conjugate base, remove a proton (H^+) from the acid.

    a. F^-, fluoride ion   **b.** OH^-, hydroxide ion
    c. HCO_3^-, bicarbonate ion or hydrogen carbonate   **d.** SO_4^{2-}, sulfate ion

8.11 To form the conjugate acid, add a proton (H^+) to the base.

    a. HCO_3^-, bicarbonate ion or hydrogen carbonate   **b.** H_3O^+, hydronium ion
    c. H_3PO_4, phosphoric acid   **d.** HBr, hydrobromic acid

8.13 The conjugate acid is a proton donor and the conjugate base is a proton acceptor.

    a. acid H_2CO_3; conjugate base HCO_3^-; base H_2O; conjugate acid H_3O^+
    b. acid NH_4^+; conjugate base NH_3; base H_2O; conjugate acid H_3O^+
    c. acid HCN; conjugate base CN^-; base NO_2^- conjugate acid HNO_2

8.15 Use Table 8.3 to answer.

    a. HBr   **b.** HSO_4^- **c.** H_2CO_3

8.17 Use Table 8.3 to answer.

    a. HSO_4^- **b.** HF   **c.** HCO_3^-

8.19 In pure water, a small fraction of the water molecules breaks apart to form H^+ and OH^-. The H^+ combines with H_2O to form H_3O^+. Every time a H^+ is formed a OH^- is also formed. Therefore, the concentration of the two must be equal in pure water.

8.21 In an acidic solution, $[H_3O^+]$ is greater than $[OH^-]$, which means that the $[H_3O^+]$ is greater than 1×10^{-7} M and the $[OH^-]$ is less than 1×10^{-7} M.

8.23 A neutral solution has $[OH^-] = [H_3O^+] = 1.0 \times 10^{-7}$ M. If $[OH^-]$ is greater than 1×10^{-7}, the solution is basic; if $[H_3O^+]$ is greater than 1×10^{-7} M, the solution is acidic.

    a. Acidic; $[H_3O^+]$ is greater than 1×10^{-7} M.
    b. Basic; $[H_3O^+]$ is less than 1×10^{-7} M.
    c. Basic; $[OH^-]$ is greater than 1×10^{-7} M.
    d. Acidic; $[OH^-]$ is less than 1×10^{-7} M.

8.25 The $[H_3O^+]$ multiplied by the $[OH^-]$ is equal to K_w, which is 1.0×10^{-14}. When $[H_3O^+]$ is known, the $[OH^-]$ can be calculated.

    Rearranging the K_w gives $[OH^-] = K_w/[H_3O^+]$.

    a. 1.0×10^{-9} M   **b.** 1.0×10^{-6} M
    c. 2.0×10^{-5} M   **d.** 4.0×10^{-13} M

8.27 The value of the $[H_3O^+]$ multiplied by the value of the $[OH^-]$ is always equal to K_w, which is 1×10^{-14}. When $[H_3O^+]$ is known, the $[OH^-]$ can be calculated.

    Rearranging the K_w gives $[OH^-] = K_w/[H_3O^+]$.

    a. 1.0×10^{-11} M   **b.** 2.0×10^{-9} M
    c. 5.6×10^{-3} M   **d.** 2.5×10^{-2} M

8.29 In neutral solutions, the $[H_3O^+]$ is equal to 1.0×10^{-7} M. The pH is the $-\log [H_3O^+]$ and the $-\log [1.0 \times 10^{-7}] = 7.00$. Note that the pH has two *decimal places* because the coefficient 1.0 has two significant figures.

8.31 An acidic solution has a pH less than 7. A neutral solution has a pH equal to 7. A basic solution has a pH greater than 7.

a. basic	**b.** acidic	**c.** basic
d. acidic	**e.** acidic	**f.** basic

8.33 The value of $[H_3O^+][OH^-]$ is equal to K_w, which is 1.0×10^{-14}. Rearranging the K_w gives $[H_3O^+] = K_w/[OH^-]$ and pH $= -\log [H_3O^+]$.

a. pH $= -\log [1 \times 10^{-4}] = 4.0$
b. pH $= -\log [3 \times 10^{-9}] = 8.5$
c. $[H_3O^+] = 1 \times 10^{-9}$; pH $= -\log [1 \times 10^{-9}] = 9.0$
d. pH $= -\log [4.0 \times 10^{-4}] = 3.40$
e. pH $= -\log [6.7 \times 10^{-8}] = 7.17$
f. $[H_3O^+] = 1.2 \times 10^{-11}$; pH $= -\log [1.2 \times 10^{-11}] = 10.92$

8.35 On a calculator, pH is calculated by entering -log, followed by the coefficient EE (EXP) key and the power of 10 followed by the change sign $(+/-)$ key. On some calculators the concentration is entered first (coefficient EXP – power) followed by log and $+/-$ key.

$[H_3O^+]$	$[OH^-]$	pH	Acidic, Basic, or Neutral?
1×10^{-8} M	1×10^{-6} M	8.0	Basic
1×10^{-3} M	1×10^{-11} M	3.0	Acidic
2×10^{-5} M	5×10^{-10} M	4.7	Acidic
1×10^{-12} M	1×10^{-2} M	12.0	Basic
2.4×10^{-5} M	4.2×10^{-10} M	4.62	Acidic

8.37 Acids react with active metals to form H_2 and a salt of the metal. The reaction of acids with carbonates yields CO_2, H_2O, and a salt of the metal. In a neutralization reaction, an acid and a base form a salt and H_2O.

a. $ZnCO_3(s) + 2\,HBr(aq) \rightarrow ZnBr_2(aq) + CO_2(g) + H_2O(l)$
b. $Zn(s) + 2\,HCl(aq) \rightarrow ZnCl_2(aq) + H_2(g)$
c. $HCl(aq) + NaHCO_3(s) \rightarrow NaCl(aq) + H_2O(l) + CO_2(g)$
d. $H_2SO_4(aq) + Mg(OH)_2(s) \rightarrow MgSO_4(aq) + 2\,H_2O(l)$

8.39 In balancing a neutralization equation, the number of H^+ and OH^- must be equalized by placing coefficients in front of the formulas for the acid and base.

a. $2\,HCl(aq) + Mg(OH)_2(s) \rightarrow 2\,H_2O(l) + MgCl_2(aq)$
b. $H_3PO_4(aq) + 3\,LiOH(aq) \rightarrow 3\,H_2O(l) + Li_3PO_4(aq)$

8.41 In balancing a neutralization equation, the number of H^+ and OH^- must be equalized by placing coefficients in front of the formulas for the acid and base.

    a. $H_2SO_4(aq) + 2\,NaOH(aq) \rightarrow Na_2SO_4(aq) + 2\,H_2O(l)$
    b. $3\,HCl(aq) + Fe(OH)_3(aq) \rightarrow FeCl_3(aq) + 3\,H_2O(l)$
    c. $H_2CO_3(aq) + Mg(OH)_2(s) \rightarrow MgCO_3(aq) + 2\,H_2O(l)$

8.43 In the equation, one mole of HCl reacts with one mole of NaOH.

$$28.6\ \text{mL} \times \frac{1\ \text{L}}{1000\ \text{mL}} \times \frac{0.145\ \text{mole NaOH}}{1\ \text{L}} = 0.00415\ \text{mole of NaOH}$$

$$0.00415\ \text{mole NaOH} \times \frac{1\ \text{mole HCl}}{1\ \text{mole NaOH}} = 0.00415\ \text{mole of HCl}$$

Molarity of HCl: $5.00\ \text{mL} \times \dfrac{1\ \text{L}}{1000\ \text{mL}} = 0.00500\ \text{L}$ $\dfrac{0.00415\ \text{mole HCl}}{0.00500\ \text{L}} = 0.829\ \text{M}$

8.45 In the equation, one mole of H_2SO_4 reacts with two moles of KOH.

$$38.2\ \text{mL} \times \frac{1\ \text{L}}{1000\ \text{mL}} \times \frac{0.163\ \text{mole KOH}}{1\ \text{L}} = 0.00623\ \text{mole of KOH}$$

Moles of H_2SO_4: $0.00623\ \text{mole KOH} \times \dfrac{1\ \text{mole } H_2SO_4}{2\ \text{moles KOH}} = 0.00312\ \text{mole of } H_2SO_4$

$25.0\ \text{mL} \times \dfrac{1\ \text{L}}{1000\ \text{mL}} = 0.0250\ \text{L}$ $\dfrac{0.00312\ \text{mole } H_2SO_4}{0.0250\ \text{L}} = 0.125\ \text{M } H_2SO_4$

8.47 A buffer system contains a weak acid and its salt, or a weak base and its salt.

    a. This is not a buffer system, because it only contains a strong acid.
    b. This is a buffer system; it contains the weak acid H_2CO_3 and its salt $NaHCO_3$.
    c. This is a buffer system; it contains HF, a weak acid, and its salt KF.
    d. This is not a buffer system because it contains the salts KCl and NaCl.

8.49   **a.** A buffer system keeps the pH of a solution constant.
    b. The salt of the acid in a buffer is needed to neutralize any acid added.
    c. When H^+ is added to the buffer, the F^-, which is the salt of the weak acid, reacts with the acid to neutralize it.
    d. When OH^- is added to the buffer solution, HF (weak acid) reacts to neutralize OH^-.

8.51 The name of an acid from a simple nonmetallic anion is formed by adding the prefix *hydro-* to the name of the anion and changing the anion ending to *-ic acid*. If the acid has a polyatomic anion, the name of the acid uses the name of the polyatomic anion and ends in *-ic acid* or *-ous acid*. There is no prefix *hydro*. Bases are named as ionic compounds containing hydroxide anions. A salt is named from the names of the positive and negative ions.

    a. base; lithium hydroxide   **b.** salt; calcium nitrate
    c. acid; hydrobromic acid   **d.** base; barium hydroxide
    e. acid; carbonic acid

8.53

Acid	Conjugate base
H_2O	OH^-
HCN	CN^-
HNO_2	NO_2^-
H_3PO_4	$H_2PO_4^-$

8.55 **a.** This diagram represents a weak acid; only a few HX molecules are broken up into H_3O^+ and X^- ions.

b. This diagram represents a strong acid; all the HX molecules are broken up into H_3O^+ and X^- ions.

c. This diagram represents a weak acid; only a few HX molecules are broken up into H_3O^+ and X^- ions.

8.57 **a.** During hyperventilation, a person will lose CO_2 and the blood pH will rise.

b. Breathing into a paper bag will increase the CO_2 concentration and lower the blood pH.

8.59 The name of an acid from a simple nonmetallic anion is formed by adding the prefix *hydro-* to the name of the anion and changing the anion ending to *-ic acid*. If the acid has polyatomic anion, the name of the acid uses the name of the polyatomic anion and ends in *-ic acid* or *-ous acid*. There is no prefix *hydro*. Bases are named as ionic compounds containing hydroxide anions.

    a. sulfuric acid   **b.** potassium hydroxide   **c.** calcium hydroxide
    d. hydrochloric acid   **e.** nitrous acid

8.61 **a.** acidic; pH 5.2 < 7.0   **b.** basic; pH 7.5 > 7.0   **c.** acidic; pH 3.8 < 7.0
    d. acidic; pH 2.5 < 7.0   **e.** basic; pH 12.0 > 7.0

8.63 Both strong and weak acids dissolve in water to give H_3O^+. They both neutralize bases, turn litmus red and phenolphthalein clear. Both taste sour and are electrolytes in solution. However, weak acids are only slightly dissociated in solution and are weak electrolytes. Strong acids, which are nearly completely dissociated in solution, are strong electrolytes.

8.65 **a.** $Mg(OH)_2$ is a strong base because all the base that dissolves is dissociated in aqueous solution.

b. $Mg(OH)_2(aq) + 2\,HCl(aq) \rightarrow 2\,H_2O(l) + MgCl_2(aq)$

8.67 If the $[OH^-]$ is given, the $[H_3O^+]$ can be found from $[H_3O^+][OH^-] = 1.0 \times 10^{-14}$.

    a. pH 7.70   **b.** pH 1.30
    c. $[H_3O^+] = 2.9 \times 10^{-11}$, pH 10.54   **d.** $[H_3O^+] = 2. \times 10^{-12}$, pH 11.7

8.69 **a.** $[H_3O^+] = 1.0 \times 10^{-7}\,M$ pH $= -\log[1.0 \times 10^{-7}] = 7.00$

    b. pH $= -\log[4.2 \times 10^{-3}] = 2.38$

    c. $[H_3O^+] = 0.0001\,M = 1 \times 10^{-4}$ pH $= -\log[1 \times 10^{-4}] = 4.0$

    d. $[H_3O^+] = 1.2 \times 10^{-6}\,M$ pH $= -\log[1.2 \times 10^{-6}] = 5.92$

8.71 If the pH is given, the $[H_3O^+]$ can be found by using the relationship $[H_3O^+] = 10^{-pH}$. The $[OH^-]$ can be found from $[H_3O^+][OH^-] = 1 \times 10^{-14}$.

    a. $[H_3O^+] = 1 \times 10^{-3}\,M$ $[OH^-] = 1 \times 10^{-11}\,M$
    b. $[H_3O^+] = 1 \times 10^{-6}\,M$ $[OH^-] = 1 \times 10^{-8}\,M$
    c. $[H_3O^+] = 1 \times 10^{-8}\,M$ $[OH^-] = 1 \times 10^{-6}\,M$
    d. $[H_3O^+] = 1 \times 10^{-11}\,M$ $[OH^-] = 1 \times 10^{-3}\,M$

8.73 **a.** Solution A, with a pH of 4.0 is more acidic.

    b. In A the $[H_3O^+] = 1 \times 10^{-4}$, in B the $[H_3O^+] = 1 \times 10^{-6}$.

    c. In A the $[OH^-] = 1 \times 10^{-10}$, in B the $[OH^-] = 1 \times 10^{-8}$.

8.75 In a buffer, the anion accepts H^+ and the cation provides H^+.

    a. $H_2PO_4^-(aq) + H_3O^+(aq) \rightarrow H_3PO_4(aq) + H_2O(l)$
    b. $H_3PO_4(aq) + OH^-(aq) \rightarrow H_2PO_4^-(aq) + H_2O(l)$

8.77 a. One mole HCl reacts with one mole of NaOH.

$$25.0 \text{ mL} \times \frac{1 \text{ L}}{1000 \text{ mL}} \times \frac{0.288 \text{ mole HCl}}{1 \text{ L}} = 0.00720 \text{ mole of HCl}$$

$$0.00720 \text{ mole HCl} \times \frac{1 \text{ mole NaOH}}{1 \text{ mole HCl}} = 0.00720 \text{ mole of NaOH}$$

Volume of NaOH: $0.00720 \text{ mole} \times \dfrac{1 \text{ L}}{0.150 \text{ mole}} \times \dfrac{1000 \text{ mL}}{1 \text{ L}} = 48.0 \text{ mL}$

b. One mole of H_2SO_4 reacts with two moles of NaOH.

$$10.0 \text{ mL} \times \frac{1 \text{ L}}{1000 \text{ mL}} \times \frac{0.560 \text{ mole}}{1 \text{ L}} = 0.00560 \text{ mole of } H_2SO_4$$

$$0.00560 \text{ mole } H_2SO_4 \times \frac{2 \text{ moles NaOH}}{1 \text{ mole } H_2SO_4} = 0.0112 \text{ mole of NaOH}$$

Volume of NaOH: $0.0112 \text{ mole} \times \dfrac{1 \text{ L}}{0.150 \text{ mole}} \times \dfrac{1000 \text{ mL}}{1 \text{ L}} = 74.7 \text{ mL}$

8.79 One mole H_2SO_4 reacts with two moles NaOH.

$$45.6 \text{ mL} \times \frac{1 \text{ L}}{1000 \text{ mL}} \times \frac{0.205 \text{ mole NaOH}}{1 \text{ L}} \times \frac{1 \text{ mole } H_2SO_4}{2 \text{ moles NaOH}} = 0.00467 \text{ mole of } H_2SO_4$$

Moles of H_2SO_4: $0.00935 \text{ mole NaOH} \times \dfrac{1 \text{ mole } H_2SO_4}{2 \text{ moles NaOH}} = 0.00467 \text{ mole of } H_2SO_4$

$$20.0 \text{ mL} \times \frac{1 \text{ L}}{1000 \text{ mL}} = 0.0200 \text{ L} \qquad \frac{0.00467 \text{ mole } H_2SO_4}{0.0200 \text{ L}} = 0.234 \text{ M}$$

8.81 a. 1) HS^- 2) $H_2PO_4^-$ 3) CO_3^{2-}

   b. H_2S **c.** H_3PO_4

8.83 a. $ZnCO_3(s) + H_2SO_4(aq) \rightarrow ZnSO_4(aq) + CO_2(g) + H_2O(l)$
   b. $2 \text{ Al}(s) + 6 \text{ HCl}(aq) \rightarrow 2 \text{ AlCl}_3(aq) + 3 H_2(g)$
   c. $2 H_3PO_4(aq) + 3 \text{ Ca(OH)}_2(s) \rightarrow Ca_3(PO_4)_2(aq) + 6 H_2O(l)$
   d. $KHCO_3(s) + HNO_3(aq) \rightarrow KNO_3(aq) + CO_2(g) + H_2O(l)$

8.85 $2.5 \text{ g HCl} \times \dfrac{1 \text{ mole HCl}}{36.5 \text{ g HCl}} = 0.068 \text{ mole of HCl}$

$$\frac{0.068 \text{ mole HCl}}{0.425 \text{ L}} = 0.16 \text{ M HCl}$$

Since HCl is a strong acid, the $[H_3O^+]$ is also 0.16 M. pH $= -\log [1.6 \times 10^{-1}] = 0.80$

8.87 a. $3 \text{ NaOH}(aq) + H_3PO_4(aq) \rightarrow Na_3PO_4(aq) + 3 H_2O(l)$

   b. $16.4 \text{ mL} \times \dfrac{1 \text{ L}}{1000 \text{ mL}} \times \dfrac{0.204 \text{ mole NaOH}}{1 \text{ L}}$

$$\times \frac{1 \text{ mole } H_3PO_4}{3 \text{ moles NaOH}} \times \frac{1}{50.0 \text{ mL}} \times \frac{1000 \text{ mL}}{1 \text{ L}} = 0.0223 \text{ M } H_3PO_4$$

8.89 a. The $[H_3O^+]$ can be found by using the relationship $[H_3O^+] = 10^{-\text{pH}}$

$$[H_3O^+] = 6 \times 10^{-5} \text{ M}; \quad [OH^-] = 2 \times 10^{-10} \text{ M}$$

   b. $[H_3O^+] = 3 \times 10^{-7} \text{ M}; \quad [OH^-] = 3 \times 10^{-8} \text{ M}$

Answers to Combining Ideas from Chapter 6 to 8

CI.13 **a.** CH_4

$$
\begin{array}{c}
\quad\;\; H \\
\quad\;\; | \\
H\!-\!C\!-\!H \\
\quad\;\; | \\
\quad\;\; H
\end{array}
$$

b. $7.0 \times 10^6 \,\cancel{gal} \times \dfrac{4 \,\cancel{qt}}{1 \,\cancel{gal}} \times \dfrac{1000 \,\cancel{mL}}{1.06 \,\cancel{qt}} \times \dfrac{0.45 \,\cancel{g}}{1 \,\cancel{mL}} \times \dfrac{1 \,kg}{1000 \,\cancel{g}} = 1.2 \times 10^7 \,kg$ of LNG

c. $7.0 \times 10^6 \,\cancel{gal} \times \dfrac{4 \,\cancel{qt}}{1 \,\cancel{gal}} \times \dfrac{1000 \,\cancel{mL}}{1.06 \,\cancel{qt}} \times \dfrac{0.45 \,\cancel{g}}{1 \,\cancel{mL}} \times \dfrac{1 \,\cancel{mole\ CH_4}}{16.0 \,\cancel{g}} \times \dfrac{22.4 \,L}{1 \,\cancel{mole\ CH_4}}$

$= 1.7 \times 10^{10} \,L$ of CH_4

d. $CH_4(g) + 2\,O_2(g) \rightarrow CO_2(g) + 2\,H_2O(g)$

e. $7.0 \times 10^6 \,\cancel{gal} + \dfrac{4 \,\cancel{qt}}{1 \,\cancel{gal}} \times \dfrac{1000 \,\cancel{mL}}{1.06 \,\cancel{qt}} \times \dfrac{0.45 \,\cancel{g}}{1 \,\cancel{mL}} \times \dfrac{1 \,\cancel{mole\ CH_4}}{16.0 \,\cancel{g}}$

$\times \dfrac{2 \,\cancel{moles\ O_2}}{1 \,\cancel{mole\ CH_4}} \times \dfrac{32.0 \,\cancel{g\ O_2}}{1 \,\cancel{mole\ O_2}} \times \dfrac{1 \,kg}{1000 \,\cancel{g}} = 4.8 \times 10^7 \,kg$ of O_2

f. $7.0 \times 10^6 \,\cancel{gal} \times \dfrac{4 \,\cancel{qt}}{1 \,\cancel{gal}} \times \dfrac{1000 \,\cancel{mL}}{1.06 \,\cancel{qt}} \times \dfrac{0.45 \,\cancel{g}}{1 \,\cancel{mL}} \times \dfrac{1 \,\cancel{mole\ CH_4}}{16.0 \,\cancel{g}}$

$\times \dfrac{211 \,kcal}{1 \,\cancel{mole\ CH_4}} = 1.6 \times 10^{11} \,kcal$

CI.15 **a.** A salt formed from a metal and nonmetal polyatomic ion is an ionic compound.

b. $1.42 \,\cancel{gal} \times \dfrac{4 \,\cancel{qt}}{1 \,\cancel{gal}} \times \dfrac{1000 \,mL}{1.06 \,\cancel{qt}} = 5360 \,mL$

$\dfrac{282 \,g}{5360 \,mL} \times 100 = 5.26\%$ (m/v)

c. $2\,NaOH(aq) + Cl_2(g) \rightarrow NaCl(aq) + NaClO(aq) + H_2O(l)$

d. $\dfrac{282 \,\cancel{g\ NaOCl}}{1 \,bottle} \times \dfrac{1 \,\cancel{mole\ NaOCl}}{74.5 \,\cancel{g\ NaOCl}} \times \dfrac{1 \,\cancel{mole\ Cl_2}}{1 \,\cancel{mole\ NaOCl}}$

$\times \dfrac{22.4 \,L}{1 \,\cancel{mole\ Cl_2}} = 84.8 \,L$ of Cl_2 per bottle

e. $[H_3O^+] = 10^{-pH} = 10^{-10.3} = 5 \times 10^{-11} \,M$

$[OH^-] = 1 \times 10^{-14}/5 \times 10^{-11} = 2 \times 10^{-4} \,M$

CI.17 **a.** $2\,M(s) + 6\,HCl(aq) \rightarrow 2\,MCl_3(aq) + 3\,H_2(g)$

b. $34.8 \,\cancel{mL} \times \dfrac{0.520 \,\cancel{mole\ HCl}}{1000 \,\cancel{mL}} \times \dfrac{3 \,\cancel{moles\ H_2}}{6 \,\cancel{moles\ HCl}} \times \dfrac{22.4 \,\cancel{L}}{1 \,\cancel{mole\ H_2}} \times \dfrac{1000 \,mL}{1 \,\cancel{L}}$

$= 203 \,mL$ of H_2

c. $34.8 \,\cancel{mL} \times \dfrac{0.520 \,\cancel{mole\ HCl}}{1000 \,\cancel{mL}} \times \dfrac{2 \,moles\ M}{6 \,\cancel{moles\ HCl}} = 6.03 \times 10^{-3} \,mole$ of M

d. $\dfrac{0.420 \,g\ M}{0.00603 \,mole\ M} = 69.7 \,g/mole$ of M

e. Gallium (Ga) has an atomic mass of 69.7.

f. $2\,Ga(s) + 6\,HCl(aq) \rightarrow 2\,GaCl_3(aq) + 3\,H_2(g)$

Nuclear Radiation

Study Goals

- Identify the types of radiation as alpha particles, beta particles, or gamma radiation.
- Describe the methods required for proper shielding for each type of radiation.
- Write an equation for an atom that undergoes radioactive decay.
- Identify some radioisotopes used in nuclear medicine.
- Calculate the amount of radioisotope that remains after a given number of half-lives.
- Describe nuclear fission and fusion.

Think About It

1. What is nuclear radiation?

2. Why do you receive more radiation if you live in the mountains or travel on an airplane?

3. In nuclear medicine, iodine-125 is used for detecting a tumor in the thyroid. What does the number 125 used in the name indicate?

4. What type of radiation is emitted by smoke detectors?

5. How are living cells damaged by radiation?

6. How does nuclear fission differ from nuclear fusion?

7. Why is there a concern about radon in our homes?

Key Terms

Match each of the following key terms with the correct definition.

- **a.** radioactive nucleus
- **b.** half-life
- **c.** curie
- **d.** nuclear fission
- **e.** alpha particle

1. _____ A particle identical to a helium nucleus produced in a radioactive nucleus

2. _____ The time required for one-half of a radioactive sample to undergo radioactive decay

3. _____ A unit of radiation that measures the activity of a sample

4. _____ A process in which large nuclei split into smaller nuclei with the release of energy

5. _____ A nucleus that spontaneously emits radiation

Answers   **1.** e   **2.** b   **3.** c   **4.** d   **5.** a

9.1 Natural Radioactivity

- Radioactive isotopes have unstable nuclei that break down (decay) spontaneously emitting alpha (α), beta (β), and gamma (γ) radiation.
- An alpha particle is the same as a helium nucleus; it contains two protons and two neutrons. A beta particle is a high-energy electron and a gamma ray is high-energy radiation.
- Because radiation can damage the cells in the body, proper protection must be used: shielding, time limitation, and distance.

Study Note

It is important to learn the symbols used to describe the different types of radiation:

p or ^1_1H	^1_0n	β or $^0_{-1}e$	α or ^4_2He	β^+ or $^0_{+1}e$	$^0_0\gamma$ or γ
proton	*neutron*	*electron*	*alpha particle*	*positron*	*gamma*

◆ Learning Exercise 9.1A

Match the description in column B with the terms in column A:

A		**B**
1. _____ $^{18}_8O$		**a.** beta particle
2. _____ γ		**b.** alpha particle
3. _____ ^4_2He		**c.** positron
4. _____ β		**d.** atom of oxygen
5. _____ $^0_{+1}e$		**e.** gamma radiation

Answers   **1.** d   **2.** e   **3.** b   **4.** a   **5.** c

◆ Learning Exercise 9.1B

Essay: Discuss some things you can do to minimize the amount of radiation received if you work with a radioactive substance. Describe how each method helps to limit the amount of radiation you would receive.

Answer
Three ways to minimize exposure to radiation are: (1) use shielding, (2) keep time short in the radioactive area, and (3) keep as much distance as possible from the radioactive materials. Shielding such as clothing and gloves stops alpha and beta particles from reaching your skin, while lead or concrete will absorb gamma rays. Limiting the time spent near radioactive samples reduces exposure time. Increasing the distance from a radioactive source reduces the intensity of radiation.

◆ Learning Exercise 9.1C

What type(s) of radiation (alpha, beta, and/or gamma) would each of the following shielding materials protect you from?

a. clothing _____   **d.** concrete _____

b. skin _____   **e.** lead wall _____

c. paper _____

Answers **a.** alpha, beta   **b.** alpha   **c.** alpha
d. alpha, beta, gamma **e.** alpha, beta, gamma

9.2 Nuclear Reactions

- A balanced nuclear equation is used to represent the changes that take place in the nuclei of the reactants and products.
- The new isotopes and the type of radiation emitted can be determined from the symbols that show the mass numbers and atomic numbers of the isotopes in the nuclear reaction.

Radioactive nucleus → new nucleus + radiation
   Total of the mass numbers are equal

$$^{11}_{6}C \rightarrow ^{7}_{4}Be + ^{4}_{2}He$$

   Total of the atomic numbers are equal

- A new isotope is produced when a nonradioactive isotope is bombarded by a small particle such as a proton, an alpha, or a beta particle.

$$^{10}_{5}B \quad + \quad ^{4}_{2}He \quad \rightarrow \quad ^{13}_{7}N \quad + \quad ^{1}_{0}n$$

stable	*bombarding*	*new*	*emitted*
nucleus	*particle (α)*	*nucleus*	*neutron*

Study Note

When balancing nuclear equations for radioactive decay, be sure that
1. The mass number of the reactant equals the sum of the mass numbers of the products.
2. The atomic number of the reactant equals the sum of the atomic numbers of the products.

◆ Learning Exercise 9.2A

Write a nuclear symbol that completes each of the following nuclear equations:

a. $^{66}_{29}Cu \rightarrow ^{66}_{30}Zn + ?$ **a.** _____

b. $^{127}_{53}I \rightarrow ^{1}_{0}n + ?$ **b.** _____

c. $^{238}_{92}U \rightarrow ^{4}_{2}He + ?$ **c.** _____

d. $^{24}_{11}Na \rightarrow ^{0}_{-1}e + ?$ **d.** _____

e. $? \rightarrow ^{30}_{14}Si + ^{0}_{-1}e$ **e.** _____

Answers **a.** $^{0}_{-1}e$ **b.** $^{126}_{53}I$ **c.** $^{234}_{90}Th$ **d.** $^{24}_{12}Mg$ **e.** $^{30}_{13}Al$

Study Note

In balancing nuclear transmutation equations (bombardment by small particles), be sure that
   1. The sum of the mass numbers of the reactants equals the sum of the mass numbers of the products.
   2. The sum of the atomic numbers of the reactants equals the sum of the atomic numbers of the products.

◆ **Learning Exercise 9.2B**

Complete the following equations for bombardment reactions:

a. $^{40}_{20}Ca + ? \rightarrow ^{40}_{19}K + ^{1}_{1}H$ a. _____

b. $^{27}_{13}Al + ^{1}_{0}n \rightarrow ^{24}_{11}Na + ?$ b. _____

c. $^{10}_{5}B + ^{1}_{0}n \rightarrow ^{4}_{2}He + ?$ c. _____

d. $^{23}_{11}Na + ? \rightarrow ^{23}_{12}Mg + ^{1}_{0}n$ d. _____

e. $^{197}_{79}Au + ^{1}_{1}H \rightarrow ? + ^{1}_{0}n$ e. _____

Answers   a. $^{1}_{0}n$ b. $^{4}_{2}He$ c. $^{7}_{3}Li$ d. $^{1}_{1}H$ e. $^{197}_{80}Hg$

9.3 Radiation Measurement

- A Geiger counter is used to detect radiation. When radiation passes through the gas in the counter tube, some atoms of gas are ionized producing an electrical current.
- The activity of a radioactive sample measures the number of nuclear transformations per second. The curie (Ci) is equal to 3.7×10^{10} disintegrations in 1 second.
- The radiation dose absorbed by a gram of material such as body tissue is measured in units of rad and gray.
- The biological damage of different types of radiation on the body is measured in radiation units of rem and sievert.

◆ **Learning Exercise 9.3**

Match each type of measurement unit with the radiation process measured.

     a. curie   **b.** rad   **c.** gray   **d.** rem

1. _____ The amount of radiation absorbed by one gram of material

2. _____ An activity of 3.7×10^{10} disintegrations per second

3. _____ The biological damage caused by different kinds of radiation

4. _____ A unit of absorbed dose equal to 100 rads

Answers   **1.** b   **2.** a   **3.** d   **4.** c

9.4 Half-Life of a Radioisotope

- The half-life of a radioactive sample is the time required for one-half of the sample to decay (emit radiation).
- Most radioisotopes used in medicine, such as Tc-99m and I-131, have short half-lives. By comparison, many naturally occurring radioisotopes, such as C-14, Ra-226, and U-238, have long half-lives. For example, potassium-42 has a half-life of 12 hours, whereas potassium-40 takes 1.3×10^{9} years for one-half of a radioactive sample to decay.

◆ Learning Exercise 9.4

a. Suppose you have an 80-mg sample of iodine-125. If iodine-125 has a half-life of 60 days, how many mg are radioactive

1. after one half-life?

2. after two half-lives?

3. after 240 days?

b. ^{99m}Tc has a half-life of 6 hours. If a technician picked up a 16-mg sample at 8 A.M., how much of the radioactive sample remained at 8 P.M. that same day?

c. Phosphorus-32 has a half-life of 14 days. How much of a 240-μg sample will be radioactive after 56 days?

d. Iodine-131 has a half-life of 8 days. How many days will it take for an 80-mg sample of I-131 to decay to 5 mg?

e. Suppose a group of archaeologists digs up some pieces of a wooden boat at an ancient site. When a sample of the wood is analyzed for C-14, scientists determine that 12.5% or 1/8 of the original amount of C-14 remains. If the half-life of carbon-14 is 5730 years, how long ago was the boat made?

Answers   **a.** (1) 40 mg (2) 20 mg (3) 5 mg   **b.** 4.0 mg
c. 15 μg **d.** 32 days   **e.** 17 200 years ago

9.5 Medical Applications Using Radioactivity

- In nuclear medicine, radioactive isotopes are given orally or intravenously that go to specific sites in the body.
- For diagnostic work, radioisotopes are used that emit gamma rays and produce nonradioactive products.
- By detecting the radiation emitted by medical radioisotopes, evaluations can be made about the location and extent of an injury, disease, or tumor, blood flow, or the level of function of a particular organ.

◆ Learning Exercise 9.5

Write the nuclear symbol for each of the following radioactive isotopes:

a. _____ Iodine-131 used to study thyroid gland activity

b. _____ Phosphorus-32 used to locate brain tumors

c. _____ Sodium-24 used to determine blood flow and to locate a blood clot or embolism

d. _____ Nitrogen-13 used in positron emission tomography

Answers   **a.** $^{131}_{53}$I   **b.** $^{32}_{15}$P   **c.** $^{24}_{11}$Na   **d.** $^{13}_{7}$N

9.6 Nuclear Fission and Fusion

- In fission, a large nucleus breaks apart into smaller pieces releasing one or more types of radiation, and a great amount of energy.
- In fusion, small nuclei combine to form a larger nucleus, which releases great amounts of energy.

◆ Learning Exercise 9.6

Essay: Discuss the nuclear processes of fission and fusion for the production of energy.

Answer   *Nuclear fission* is a splitting of the atom into two or more nuclei accompanied by the release of large amounts of energy and radiation. In the process of *nuclear fusion,* two or more nuclei combine to form a heavier nucleus and release a large amount of energy. However, fusion requires a considerable amount of energy to initiate the process.

Checklist for Chapter 9

You are ready to take the practice test for Chapter 9. Be sure that you have accomplished the following learning goals for this chapter. If you are not sure, review the section listed at the end of the goal. Then apply your new skills and understanding to the practice test.

After studying Chapter 9, I can successfully:

_____ Describe alpha, beta, and gamma radiation (9.1).

_____ Write a nuclear equation showing mass number and atomic number for radioactive decay (9.2).

_____ Write a nuclear equation for the formation of a radioactive isotope (9.2).

_____ Describe the detection and measurement of radiation (9.3).

_____ Given a half-life, calculate the amount of radioisotope remaining after one or more half-lives (9.4).

_____ Describe the use of radioisotopes in medicine (9.5).

_____ Describe the processes of nuclear fission and fusion (9.6).

Practice Test for Chapter 9

1. The correctly written symbol for an atom of sulfur would be
   A. $^{30}_{16}Su$ **B.** $^{14}_{30}Si$ **C.** $^{30}_{16}S$ **D.** $^{30}_{16}Si$ **E.** $^{16}_{30}S$

2. Alpha particles are composed of
   A. protons.   **B.** neutrons.   **C.** electrons.
   D. protons and electrons.   **E.** protons and neutrons.

3. Gamma radiation is a type of radiation that
   A. originates in the electron shells.
   B. is most dangerous.
   C. is least dangerous.
   D. is the heaviest.
   E. goes the shortest distance.

4. The charge on an alpha particle is
 A. -1 **B.** $+1$ **C.** -2 **D.** $+2$ **E.** $+4$

5. Beta particles formed in a radioactive nucleus are
 A. protons.   **B.** neutrons.   **C.** electrons.
 D. protons and electrons.   **E.** protons and neutrons.

For questions 6 through 10, select from the following:

 A. $^{0}_{-1}X$ **B.** $^{4}_{2}X$ **C.** $^{1}_{1}X$ **D.** $^{1}_{0}X$ **E.** $^{0}_{0}X$

6. An alpha particle

7. A beta particle

8. A gamma ray

9. A proton

10. A neutron

11. Shielding from gamma rays is provided by
 A. skin.   **B.** paper.   **C.** clothing.   **D.** lead.   **E.** air.

12. The skin will provide shielding from
 A. alpha particles. **B.** beta particles.   **C.** gamma rays.   **D.** ultraviolet rays. **E.** X rays.

13. The radioisotope iodine-131 is used as a radioactive tracer for studying thyroid gland activity. The symbol for iodine-131 is
 A. I   **B.** $_{131}I$ **C.** $^{131}_{53}I$ **D.** $^{53}_{131}I$ **E.** $^{78}_{53}I$

14. When an atom emits an alpha particle, its atomic mass will
 A. increase by 1. **B.** increase by 2. **C.** increase by 4. **D.** decrease by 4. **E.** not change.

15. When a nucleus emits a beta particle, the atomic number of the new nucleus
 A. increases by 1. **B.** increases by 2. **C.** decreases by 1. **D.** decreases by 2. **E.** will not change.

16. When a nucleus emits a gamma ray, the atomic number of the new nucleus
 A. increases by 1. **B.** increases by 2. **C.** decreases by 1. **D.** decreases by 2. **E.** will not change.

For questions 17 through 20, select the particle that completes each of the equations.
 A. neutron   **B.** alpha particle   **C.** beta particle   **D.** gamma ray

17. $^{126}_{50}Sn \rightarrow ^{126}_{51}Sb + \ ?$

18. $^{69}_{30}Zn \rightarrow ^{69}_{31}Ga + \ ?$

19. $^{99m}_{43}Tc \rightarrow ^{99}_{43}Tc + \ ?$

20. $^{149}_{62}Sm \rightarrow ^{145}_{60}Nd + \ ?$

21. What symbol completes the following reaction?
 $^{14}_{7}N + ^{1}_{0}n \rightarrow \ ? + ^{1}_{1}H$
 A. $^{15}_{8}O$ **B.** $^{15}_{6}C$ **C.** $^{14}_{8}O$ **D.** $^{14}_{6}C$ **E.** $^{15}_{7}N$

22. To complete this nuclear equation, you need to write
$$^{54}_{26}Fe + ? \rightarrow ^{57}_{28}Ni + ^{1}_{0}n$$
    A. an alpha particle   **B.** a beta particle   **C.** gamma
    D. neutron   **E.** proton

23. The name of the unit used to measure the number of disintegrations per second is
    A. curie   **B.** rad   **C.** rem   **D.** RBE   **E.** MRI

24. The rem is a unit used to measure
    A. activity of a radioactive sample.
    B. biological damage of different types of radiation.
    C. radiation absorbed.
    D. background radiation.

25. Radiation can cause
    A. nausea.   **B.** a lower white cell count.   **C.** fatigue.
    D. hair loss.   **E.** All of these

26. Radioisotopes used in medical diagnosis
    A. have short half-lives.   **B.** emit only gamma rays.
    C. locate in specific organs.   **D.** produce nonradioactive nuclei.
    E. All of these

27. The imaging technique that uses energy emitted by exciting the nuclei of hydrogen atoms is
    A. computerized tomography (CT).   **B.** positron emission tomography (PET).
    C. radioactive tracer.   **D.** magnetic resonance imaging (MRI).
    E. radiation.

28. The imaging technique that detects the absorption of X rays by body tissues is
    A. computerized tomography (CT).   **B.** positron emission tomography (PET).
    C. radioactive tracer.   **D.** magnetic resonance imaging (MRI).
    E. radiation.

29. The time required for a radioisotope to decay is measured by its
    A. half-life.   **B.** protons.   **C.** activity.   **D.** fusion.   **E.** radioisotope.

30. Oxygen-15 used in PET imaging has a half-life of 2 min. How many half-lives have occurred in the 10 minutes it takes to prepare the sample?
    A. 2   **B.** 3   **C.** 4   **D.** 5   **E.** 6

31. Iodine-131 has a half-life of 8 days. How long will a 160-mg sample take to decay to 10 mg?
    A. 8 days   **B.** 16 days   **C.** 32 days   **D.** 40 days   **E.** 48 days

32. Phosphorus-32 has a half-life of 14 days. After 28 days, how many mg of a 100-mg sample will still be radioactive?
    A. 75 mg   **B.** 50 mg   **C.** 40 mg   **D.** 25 mg   **E.** 12.5 mg

33. The "splitting" of a large nucleus to form smaller particles accompanied by a release of energy is called
    A. radioisotope.   **B.** fission.   **C.** fusion.   **D.** rem.   **E.** half-life.

34. The process of combining small nuclei to form larger nuclei is
    A. radioisotope.   **B.** fission.   **C.** fusion.   **D.** rem.   **E.** half-life.

35. The fusion reaction
    A. occurs in the sun.
    B. forms larger nuclei from smaller nuclei.
    C. requires extremely high temperatures.
    D. releases a large amount of energy.
    E. All of these

Answers to the Practice Test

1. C	**2.** E	**3.** B	**4.** D	**5.** C
6. B	**7.** A	**8.** E	**9.** C	**10.** D
11. D	**12.** A	**13.** C	**14.** D	**15.** A
16. E	**17.** C	**18.** C	**19.** D	**20.** B
21. D	**22.** A	**23.** A	**24.** B	**25.** E
26. E	**27.** D	**28.** A	**29.** A	**30.** D
31. C	**32.** D	**33.** B	**34.** C	**35.** E

Answers and Solutions to Selected Text Problems

9.1   **a.** An α-particle and a helium nucleus both contain 2 protons and 2 neutrons. However, an α-particle has no electrons and carries a 2+ charge. Alpha-particles are emitted from unstable nuclei during radioactive decay.

       b. α, ^4_2He

       c. An α-particle is emitted from an unstable nucleus during radioactive decay.

9.3   **a.** $^{39}_{19}\text{K}$, $^{40}_{19}\text{K}$, $^{41}_{19}\text{K}$

       b. Each isotope has 19 protons and 19 electrons, but they differ in the number of neutrons present. Potassium-39 has 20 neutrons, potassium-40 has 21 neutrons, and potassium-41 has 22 neutrons.

9.5

Medical Use	Nuclear Symbol	Mass Number	Number of Protons	Number of Neutrons
Heart imaging	$^{201}_{81}\text{Tl}$	201	81	120
Radiation therapy	$^{60}_{27}\text{Co}$	60	27	33
Abdominal scan	$^{67}_{31}\text{Ga}$	67	31	36
Hyperthyroidism	$^{131}_{53}\text{I}$	131	53	78
Leukemia treatment	$^{32}_{15}\text{P}$	32	15	17

9.7   **a.** α, ^4_2He **b.** ^1_0n **c.** β, $^0_{-1}\text{e}$ **d.** $^{15}_7\text{N}$ **e.** $^{125}_{53}\text{I}$

9.9   **a.** β (or $^0_{-1}\text{e}$)   **b.** α (or ^4_2He)   **c.** ^1_0n **d.** $^{24}_{11}\text{Na}$ **e.** $^{14}_6\text{C}$

9.11   **a.** β particles have less mass and move faster than α-particles, they penetrate further into tissue.

       b. Ionizing radiation breaks bonds and forms reactive species that cause undesirable reactions in the cells.

       c. X-ray technicians leave the room to increase their distance from the radiation source. They are also shielded by a thick wall or one that contains a lead lining.

       d. Wearing gloves shields the skin from α and β radiation.

9.13 The mass number of the radioactive atom is reduced by 4 when an alpha particle is emitted. The unknown product will have an atomic number that is 2 less than the atomic number of the radioactive atom.

    a. $^{208}_{84}\text{Po} \rightarrow {}^{204}_{82}\text{Pb} + {}^{4}_{2}\text{He}$ **b.** $^{232}_{92}\text{Th} \rightarrow {}^{228}_{88}\text{Ra} + {}^{4}_{2}\text{He}$

    c. $^{251}_{102}\text{No} \rightarrow {}^{247}_{100}\text{Fm} + {}^{4}_{2}\text{He}$ **d.** $^{220}_{86}\text{Rn} \rightarrow {}^{216}_{84}\text{Po} + {}^{4}_{2}\text{He}$

9.15 **a.** $^{25}_{11}\text{Na} \rightarrow {}^{25}_{12}\text{Mg} + {}^{0}_{-1}e$ **b.** $^{20}_{8}\text{O} \rightarrow {}^{20}_{9}\text{F} + {}^{0}_{-1}e$

    c. $^{92}_{38}\text{Sr} \rightarrow {}^{92}_{39}\text{Y} + {}^{0}_{-1}e$ **d.** $^{42}_{19}\text{K} \rightarrow {}^{42}_{20}\text{Ca} + {}^{0}_{-1}e$

9.17 **a.** $^{28}_{13}\text{Al} \rightarrow {}^{28}_{14}\text{Si} + {}^{0}_{-1}e$ beta decay   **b.** $^{87}_{36}\text{K} \rightarrow {}^{86}_{36}\text{Kr} + {}^{1}_{0}n$ neutron emission

    c. $^{66}_{29}\text{Cu} \rightarrow {}^{66}_{30}\text{Zn} + {}^{0}_{-1}e$ beta decay   **d.** $^{228}_{90}\text{Th} \rightarrow {}^{224}_{88}\text{Ra} + {}^{4}_{2}\text{He}$ alpha decay

9.19 **a.** $^{9}_{4}\text{Be} + {}^{1}_{0}n \rightarrow {}^{10}_{4}\text{Be}$ **b.** $^{32}_{16}\text{S} \rightarrow {}^{0}_{-1}e \rightarrow {}^{32}_{15}\text{P}$

    c. $^{27}_{13}\text{Al} + {}^{1}_{0}n \rightarrow {}^{24}_{11}\text{Na} + {}^{4}_{2}\text{He}$ **d.** $^{27}_{13}\text{Al} + {}^{4}_{2}\text{He} \rightarrow {}^{30}_{15}\text{P} + {}^{1}_{0}n$

9.21 **a.** When radiation enters the Geiger counter, it ionizes a gas in the detection tube. The ions created in the tube move toward an electrode of opposite charge (recall that opposite electrical charges attract one another). This flow of charge produces an electric current, which is detected by the instrument.

    b. The becquerel (Bq) is the SI unit for activity. The curie (Ci) is the original unit for activity of radioactive samples.

    c. The SI unit for absorbed dose is the gray (Gy). The rad (radiation absorbed dose) is a unit of radiation absorbed per gram of sample. It is the older unit.

    d. A kilogray is 1000 gray (Gy), which is equivalent to 100 000 rads.

9.23 $70.0 \text{ kg} \times \dfrac{4.20 \; \mu\text{Ci}}{1 \text{ kg}} = 294 \; \mu\text{Ci}$

9.25 While flying a plane, a pilot is exposed to higher levels of background radiation because there is less atmosphere to act as a shield against cosmic radiation.

9.27 Half-life is the time required for one-half of a radioactive sample to decay.

9.29 **a.** After one half-life, one-half of the sample would be radioactive: $80.0 \text{ mg} \times {}^{1}\!/_{2} = 40.0 \text{ mg}$

    b. After two half-lives, one-fourth of the sample would still be radioactive:
    $80.0 \text{ mg} \times {}^{1}\!/_{2} \times {}^{1}\!/_{2} = 80.0 \text{ mg} \times {}^{1}\!/_{4} = 20.0 \text{ mg}$

    c. $18 \text{ h} \times \dfrac{1 \text{ half-life}}{6.0 \text{ h}} = 3.0 \text{ half-lives}$
    $80.0 \text{ mg} \times {}^{1}\!/_{2} \times {}^{1}\!/_{2} \times {}^{1}\!/_{2} = 80.0 \text{ mg} \times {}^{1}\!/_{8} = 10.0 \text{ mg}$

    d. $24 \text{ h} \times \dfrac{1 \text{ half-life}}{6.0 \text{ h}} = 4.0 \text{ half-lives}$
    $80.0 \text{ mg} \times {}^{1}\!/_{2} \times {}^{1}\!/_{2} \times {}^{1}\!/_{2} \times {}^{1}\!/_{2} = 80.0 \text{ mg} \times {}^{1}\!/_{16} = 5.00 \text{ mg}$

9.31 The radiation level in a radioactive sample is cut in half with each passing half-life. To answer the question we must first determine the number of half-lives. ${}^{1}\!/_{4} = {}^{1}\!/_{2} \times {}^{1}\!/_{2} = 2$ half-lives

Because each half-life is 64 days, it will take 128 days for the radiation level of strontium-85 to fall to one-fourth of its original value. 2 half-lives $\times$ 64 days/half-life $= 128$ days

To determine the amount of time for the strontium-85 to drop to one-eighth its original activity, we calculate the number of half-lives. ${}^{1}\!/_{8} = {}^{1}\!/_{2} \times {}^{1}\!/_{2} \times {}^{1}\!/_{2} = 3$ half-lives

Because each half-life is 64 days, it will take 192 days for the radiation level of strontium-85 to fall to one-eighth of its original value. 3 half-lives $\times$ 64 days/half-life $= 192$ days

9.33 **a.** Because the elements calcium and phosphorus are part of bone, any calcium and/or phosphorus atom, regardless of isotope, will be carried to and become part of the bony structures in the body. Once there, the radiation emitted by the radioisotope can be used for diagnosis or treatment of bone diseases.

b. $^{89}_{38}\text{Sr} \rightarrow {}^{89}_{39}\text{Y} + {}^{0}_{-1}\beta$

Strontium (Sr) acts much like calcium (Ca) because both are Group 2A (2) elements. The body will accumulate radioactive strontium in bones in the same way that it incorporates calcium. Once the strontium isotope is absorbed by the bone, the beta radiation will destroy cancer cells.

9.35 $4.0 \, \cancel{\text{mL}} \times \dfrac{45 \, \mu\text{Ci}}{1 \, \cancel{\text{mL}}} = 180 \, \mu\text{Ci}$

9.37 Nuclear fission is the splitting of a large atom into smaller fragments with a simultaneous release of large amounts of energy.

9.39 $^{235}_{92}\text{U} + {}^{1}_{0}\text{n} \rightarrow {}^{131}_{50}\text{Sn} + {}^{103}_{42}\text{Mo} + 2{}^{1}_{0}\text{n}$

9.41 **a.** fission   **b.** fusion   **c.** fission   **d.** fusion

9.43 **a.** $^{11}_{6}\text{C}$

b.
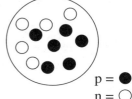

9.45

9.47 Half of a radioactive sample decays with each half-life:

$^{1}/_{2}$ lives   (1)   (2)   (3)

$6.4 \, \mu\text{Ci} \rightarrow 3.2 \, \mu\text{Ci} \rightarrow 1.6 \, \mu\text{Ci} \rightarrow 0.80 \, \mu\text{Ci}$

Therefore, the activity of carbon-14 drops to $0.80 \, \mu\text{Ci}$ in three half-lives or 3×5730 years, which makes the age of the painting 17 200 years.

9.51 **a.** Alpha (α) and beta (β) radiation consist of particles emitted from an unstable nucleus, while gamma (γ) rays are radiation emitted as pure energy.

b. Alpha radiation is abbreviated as α and $^{4}_{2}\text{He}$. Beta radiation is abbreviated as β and $^{0}_{-1}\text{e}$. Gamma radiation is abbreviated as γ and $^{0}_{0}\gamma$.

9.53 **a.** gamma radiation   **b.** positron emission   **c.** alpha decay

9.55 **a.** $^{225}_{90}\text{Th} \rightarrow {}^{221}_{88}\text{Ra} + {}^{4}_{2}\text{He}$ **b.** $^{210}_{83}\text{Bi} \rightarrow {}^{206}_{81}\text{Ti} + {}^{4}_{2}\text{He}$

c. $^{137}_{55}\text{Cs} \rightarrow {}^{137}_{56}\text{Ba} + {}^{0}_{-1}e$ **d.** $^{126}_{50}\text{Sn} \rightarrow {}^{126}_{51}\text{Sb} + {}^{0}_{-1}e$

9.57 **a.** $^{14}_{7}\text{N} \rightarrow {}^{4}_{2}\text{He} \rightarrow {}^{17}_{8}\text{O} + {}^{1}_{1}\text{H}$ **b.** $^{27}_{13}\text{Al} + {}^{4}_{2}\text{He} \rightarrow {}^{30}_{14}\text{Si} + {}^{1}_{1}\text{H}$

c. $^{235}_{92}\text{U} + {}^{1}_{0}\text{n} \rightarrow {}^{90}_{38}\text{Sr} + 3{}^{1}_{0}\text{n} + {}^{143}_{54}\text{Xe}$

9.59 **a.** $^{16}_{8}O + ^{16}_{8}O \rightarrow ^{4}_{2}He + ^{28}_{14}Si$

 b. $^{249}_{98}Cf + ^{18}_{8}O \rightarrow ^{263}_{106}Sg + 4^{1}_{0}n$

 c. $^{222}_{86}Rn \rightarrow ^{4}_{2}He + ^{218}_{84}Po$

 Then the polonium-218 decays: $^{218}_{84}Po \rightarrow ^{4}_{2}He + ^{214}_{82}Pb$

9.61 **a.** $^{26}_{14}Si \rightarrow ^{26}_{13}Al + ^{0}_{+1}e$ **b.** $^{54}_{27}Co \rightarrow ^{54}_{26}Fe + ^{0}_{+1}e$

 c. $^{77}_{37}Rb \rightarrow ^{77}_{36}Kr + ^{0}_{+1}e$ **d.** $^{93}_{45}Rh \rightarrow ^{93}_{44}Ru + ^{0}_{+1}e$

9.63 Half of a radioactive sample decays with each half-life:

 Half-lives   (1)   (2)

           1.2 g $\rightarrow$ 0.60 g $\rightarrow$ 0.30 g

 Therefore, the amount of phosphorus-32 will drop to 0.30 g in two half-lives, which is 28 days. 28 days/2 half-lives = 14 days/half-life. One half-life is 14 days.

9.65 **a.** $^{131}_{53}I \rightarrow ^{0}_{-1}e + ^{131}_{54}Xe$

 b. First we must determine the number of half-lives.

 $$40 \text{ days} \times \frac{1 \text{ half-life}}{8.0 \text{ days}} = 5.0 \text{ half-lives}$$

 Now we can calculate the number of grams of iodine-131 remaining:

 $$12.0 \text{ g} \times \frac{1}{2} \times \frac{1}{2} \times \frac{1}{2} \times \frac{1}{2} \times \frac{1}{2} = 12.0 \text{ g} \times \frac{1}{32} = 0.375 \text{ g}$$

 c. One-half of a radioactive sample decays with each half-life:

 Half-lives (1)   (2)   (3)   (4)

       48 g $\rightarrow$ 24 g $\rightarrow$ 12 g $\rightarrow$ 6.0 g $\rightarrow$ 3.0 g

 When 3.0 g remain, four half-lives must have passed. Because each half-life is 8.0 days, we can calculate the number of days that the sample required to decay to 3.0 g.

 $$4 \text{ half-lives} \times \frac{8.0 \text{ days}}{1 \text{ half-life}} = 32 \text{ days}$$

9.67 First, calculate the number of half-lives that have passed since the nurse was exposed:

 $$36 \text{ hrs} \times \frac{1 \text{ half-life}}{12 \text{ hrs}} = 3.0 \text{ half-lives}$$

 Because the activity of a radioactive sample is cut in half with each half-life, the activity must have been double its present value prior to each half-life. For 3.0 half-lives, we need to double the value 3 times. 2.0 μCi $\times$ (2 $\times$ 2 $\times$ 2) = 16 μCi

9.69 First, calculate the number of half-lives:

 $$24 \text{ hrs} \times \frac{1 \text{ half-life}}{6.0 \text{ hrs}} = 4.0 \text{ half-lives}$$

 And now calculate the amount of technetium-99m that remains after 4 half-lives have passed:

 $$120 \text{ mg} \times (\tfrac{1}{2} \times \tfrac{1}{2} \times \tfrac{1}{2} \times \tfrac{1}{2}) = 120 \text{ mg} \times \frac{1}{16} = 7.5 \text{ mg}$$

9.71 Irradiating foods kills bacteria that are responsible for food spoilage. As a result, shelf life of the food is extended.

9.73 Nuclear fission is the splitting of a large atom into smaller fragments with a simultaneous release of large amounts of energy. Nuclear fusion occurs when two (or more) nuclei combine (fuse) to form a larger species, with a simultaneous release of large amounts of energy.

9.75 Fusion reactions naturally occur in the stars, such as our sun.

9.77 **a.** $^{238}_{92}U \rightarrow ^{234}_{90}Th + ^{4}_{2}He$

b. $^{234}_{90}Th \rightarrow ^{234}_{91}Pa + ^{0}_{-1}e$

c. $^{226}_{88}Ra \rightarrow ^{222}_{86}Rn + ^{4}_{2}He$

9.79 **a.** $^{23m}_{12}Mg \rightarrow ^{23}_{12}Mg + ^{0}_{0}\gamma$

b. $^{61}_{30}Zn \rightarrow ^{61}_{29}Cu + ^{0}_{+1}e$

c. $^{249}_{98}Cf + ^{12}_{6}C \rightarrow ^{257}_{104}Rf + 4^{0}_{1}n$

$^{23m}_{12}Mg \rightarrow ^{23}_{12}Mg + ^{0}_{0}\gamma \quad ^{61}_{30}Zn \rightarrow ^{61}_{29}Cu + ^{0}_{-1}e \quad ^{249}_{98}Cf + ^{12}_{6}C \rightarrow ^{257}_{104}Rf + 4^{0}_{1}n$

9.81 $^{1}/_{2}$ life $= 4.5$ days $\qquad ^{47}Ca$ 4.0 μCi after 18 days

$$18 \ \cancel{days} \times \frac{1 \ \text{half-life}}{4.5 \ \cancel{days}} = 4 \ \text{half-lives}$$

$$\begin{array}{cccc} (1) & (2) & (3) & (4) \end{array}$$
$$64 \ \mu Ci \rightarrow 32 \ \mu Ci \rightarrow 16 \ \mu Ci \rightarrow 8.0 \ \mu Ci \rightarrow 4.0 \ \mu Ci$$

Introduction to Organic Chemistry: Alkanes

Study Goals

- Identify the number of bonds for carbon and other atoms in organic compounds.
- Describe the tetrahedral shape of carbon with single bonds in organic compounds.
- Draw expanded and condensed structural formulas for alkanes.
- Write the IUPAC names for alkanes.
- Describe the physical properties of alkanes.
- Write equations for the combustion of alkanes.
- Describe the properties that are characteristic of organic compounds.
- Identify the functional groups in organic compounds.

Think About It

1. What is the meaning of the term "organic"?

2. What two elements are found in all organic compounds?

3. In a salad dressing, why is there a layer of vegetable oil floating on the vinegar and water layer?

Key Terms

1. Match the statements shown below with the following key terms.

   a. alkene   **b.** isomers   **c.** hydrocarbon   **d.** alcohol
   e. functional group   **f.** alkane   **g.** condensed structural formula
   h. main chain   **i.** combustion   **j.** cycloalkane

   1. _____ An atom or group of atoms that influences the chemical reactions of an organic compound

   2. _____ A class of organic compounds with one or more hydroxyl (—OH) groups

   3. _____ A type of hydrocarbon with one or more carbon–carbon double bonds

   4. _____ Organic compound consisting of only carbon and hydrogen atoms

   5. _____ Compounds having the same molecular formula but a different arrangement of atoms

   6. _____ A hydrocarbon that contains only carbon–carbon single bonds

   7. _____ An alkane that exists as a cyclic structure

   8. _____ The chemical reaction of an alkane and oxygen that yields CO_2, H_2O, and heat

   9. _____ The type of formula that shows the arrangement of the carbon atoms grouped with their attached H atoms

   10. _____ The longest continuous chain of carbon atoms in a structural formula

   Answers   **1.** e   **2.** d   **3.** a   **4.** c   **5.** b
                   6. f   **7.** j   **8.** i   **9.** g   **10.** h

10.1 Organic Compounds

- Organic compounds are compounds of carbon and hydrogen that have covalent bonds, low melting and boiling points, burn vigorously, are nonelectrolytes, and are usually more soluble in nonpolar solvents than in water.
- Each carbon in an alkane has four bonds arranged so that the bonded atoms are in the corners of a tetrahedron.

◆ Learning Exercise 10.1A

Identify the following as typical of organic (O) or inorganic (I) compounds.

1. _____ have covalent bonds

2. _____ have low boiling points

3. _____ burn in air

4. _____ are soluble in water

5. _____ have high melting points

6. _____ are soluble in nonpolar solvents

7. _____ have ionic bonds

8. _____ form long chains

9. _____ contain carbon

10. _____ are not very combustible

11. _____ have a formula of Na_2SO_4

12. _____ have a formula of $CH_3—CH_2—CH_3$

Answers   **1.** O   **2.** O   **3.** O   **4.** I   **5.** I   **6.** O
              7. I   **8.** O   **9.** O   **10.** I   **11.** I   **12.** O

◆ Learning Exercise 10.1B

1. What is the name of the three-dimensional structure of methane?
2. Draw the expanded structural formula (two-dimensional) for methane.

Answers   **1.** tetrahedron   **2.**
$$H—\overset{\displaystyle H}{\underset{\displaystyle H}{\overset{|}{\underset{|}{C}}}}—H$$

10.2 Alkanes

- In an IUPAC name, the stem indicates the number of carbon atoms, and the suffix describes the family of the compound. For example, in the name *propane,* the stem *prop* indicates a chain of three carbon atoms and the ending *ane* indicates single bonds (alkane). The names of the first six alkanes follow:

Name	Carbon atoms	Condensed structural formula
Methane	1	CH_4
Ethane	2	$CH_3—CH_3$
Propane	3	$CH_3—CH_2—CH_3$
Butane	4	$CH_3—CH_2—CH_2—CH_3$
Pentane	5	$CH_3—CH_2—CH_2—CH_2—CH_3$
Hexane	6	$CH_3—CH_2—CH_2—CH_2—CH_2—CH_3$

- An expanded structural formula shows a separate line to each bonded atom; a condensed structural formula depicts each carbon atom and its attached hydrogen atoms as a group. A molecular formula gives the total numbers of atoms.

Expanded Structural Formula	Condensed Structural Formula	Molecular Formula

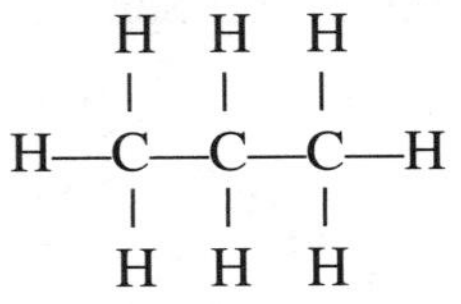 H H H
  | | |
H—C—C—C—H
  | | |
H H H
	$CH_3-CH_2-CH_3$	C_3H_8

◆ Learning Exercise 10.2A

Indicate if each of the following is a molecular formula (M), an expanded structural formula (E), or a condensed structural formula (C).

1. _____ CH_3-CH_3 **2.** _____ C_5H_{12} **3.** _____ $CH_3-CH_2-CH_3$

4. _____ 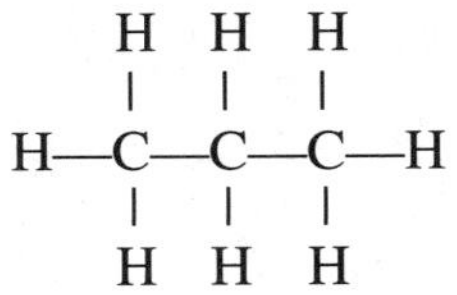 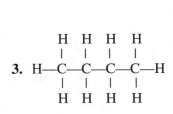 **5.** _____ $CH_3-\underset{\underset{CH_3}{|}}{CH}-CH_2-CH_3$ **6.** _____ C_8H_{18}

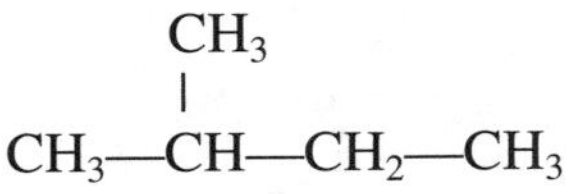

Answers **1.** C **2.** M **3.** C **4.** E **5.** C **6.** M

◆ Learning Exercise 10.2B

Write the condensed formulas for the following structural formulas.

1. 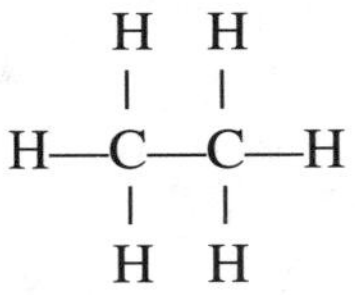 H H
  \| \|
H—C—C—H
  \| \|
H H

2. 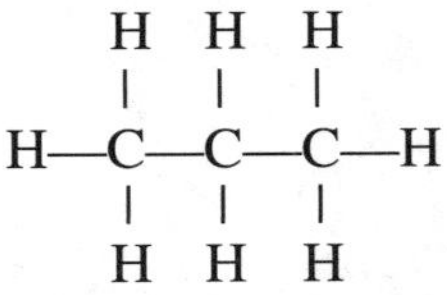 H H H
  \| \| \|
H—C—C—C—H
  \| \| \|
H H H

3. H H H H
  \| \| \| \|
H—C—C—C—C—H
  \| \| \| \|
H H H H

4.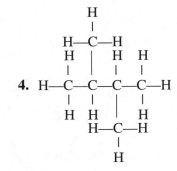

Answers **1.** CH_3-CH_3 **2.** $CH_3-CH_2-CH_3$

3. $CH_3-CH_2-CH_2-CH_3$ **4.** 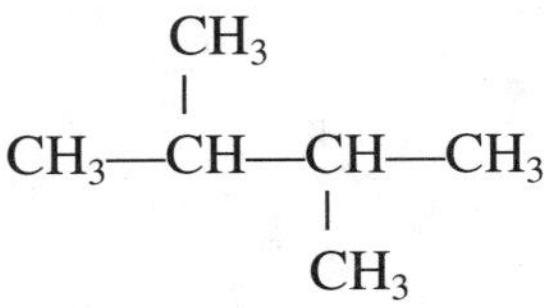 $CH_3-\underset{\underset{CH_3}{|}}{CH}-CH-CH_3$ with CH_3 above

◆ **Learning Exercise 10.2C**

Write the condensed structure and name for the straight-chain alkane of each of the following formulas.

1. C_2H_6 _____

2. C_3H_8 _____

3. C_4H_{10} _____

4. C_5H_{12} _____

5. C_6H_{14} _____

Answers
1. CH_3—CH_3, ethane
2. CH_3—CH_2—CH_3, propane
3. CH_3—CH_2—CH_2—CH_3, butane
4. CH_3—CH_2—CH_2—CH_2—CH_3, pentane
5. CH_3—CH_2—CH_2—CH_2—CH_2—CH_3, hexane

10.3 Alkanes with Substituents

- The IUPAC system is a set of rules used to name organic compounds in a systematic manner.
- Each substituent is numbered and listed alphabetically in front of the name of the longest chain.
- Carbon groups that are substituents are named as alkyl groups or alkyl substituents. An alkyl group is named by replacing the *ane* of the alkane name with *yl*. For example, CH_3— is named as methyl (from CH_4 methane), and CH_3—CH_2— is named as an ethyl group (from CH_3—CH_3 ethane).
- Structural isomers have the same molecular formula, but differ in the sequence of atoms in each of their structural formulas.
- A halogen atom is named as a substituent (fluoro, chloro, bromo, iodo) attached to the alkane chain.

Study Note

Example: Write the IUPAC name for the compound

$$CH_3$$
$$|$$
$$CH_3\text{—}CH_2\text{—}CH\text{—}CH_3$$

Solution: The four-carbon chain *butane* is numbered from the end nearest the side group, which places the *methyl* substituent on carbon 2: *2-methylbutane*.

◆ **Learning Exercise 10.3A**

Provide a correct IUPAC name for each of the following compounds.

$$CH_3$$
$$|$$
1. CH_3—CH—CH_3 _____

$$CH_3 \qquad\qquad CH_3$$
$$| \qquad\qquad\quad |$$
2. CH_3—CH—CH_2—CH—CH_2—CH_3 _____

$$\underset{\underset{\displaystyle CH_3}{|}}{3.\ CH_3-CH}-CH_2-CH_2-\underset{\underset{\displaystyle CH_3}{|}}{CH}-CH_2-CH_3 \rule{5cm}{0.4pt}$$

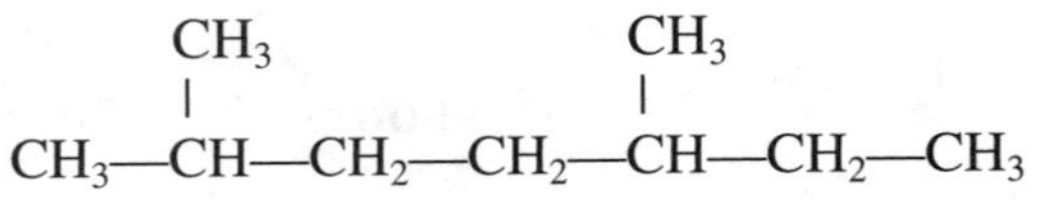

$$4.\ CH_3-\underset{\underset{\displaystyle CH_3}{|}}{\overset{\overset{\displaystyle CH_3}{|}}{C}}-CH_2-CH_3 \rule{5cm}{0.4pt}$$

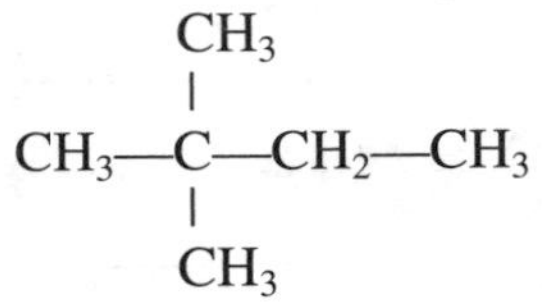

Answers   **1.** 2-methylpropane   **2.** 2,4-dimethylhexane
               3. 2,5-dimethylheptane   **4.** 2,2-dimethylbutane

◆ **Learning Exercise 10.3B**

Write the condensed formula for each of the following compounds:

1. hexane   **2.** methane

3. 2,4-dimethylpentane   **4.** propane

Answers   **1.** $CH_3-CH_2-CH_2-CH_2-CH_2-CH_3$ **2.** CH_4

    3. 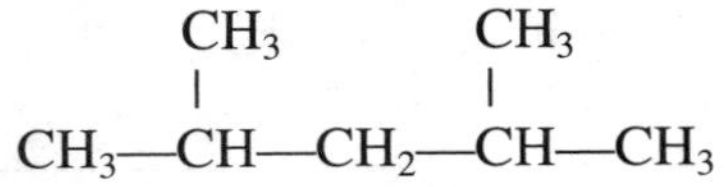 $\underset{\underset{\displaystyle CH_3}{|}}{CH_3-CH}-CH_2-\underset{\underset{\displaystyle CH_3}{|}}{CH}-CH_3$ **4.** $CH_3-CH_2-CH_3$

◆ **Learning Exercise 10.3C**

Write the condensed structural formula for each of the following alkanes.

1. pentane   **2.** 2-methylpentane

3. 4-ethyl-2-methylhexane   **4.** 2,2,4-trimethylhexane

Answers   **1.** $CH_3-CH_2-CH_2-CH_3-CH_3$ **2.** 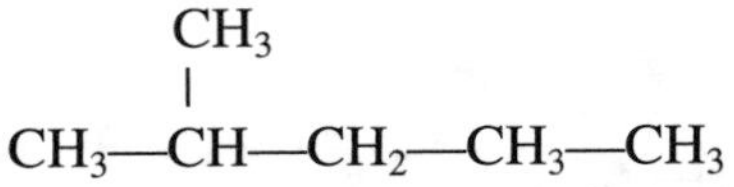 $\underset{\underset{\displaystyle CH_3}{|}}{CH_3-CH}-CH_2-CH_3-CH_3$

    3. $\underset{\underset{\displaystyle CH_3}{|}}{CH_3-CH}-CH_2-\underset{\underset{\displaystyle CH_2-CH_3}{|}}{CH}-CH_2-CH_3$ 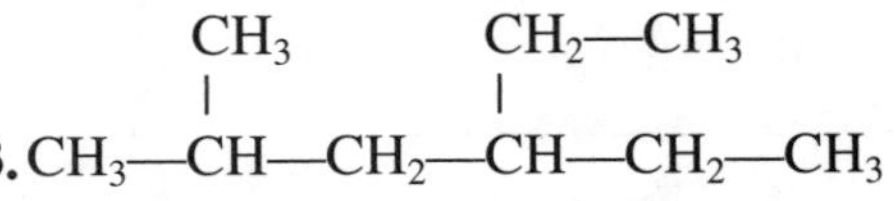   **4.** 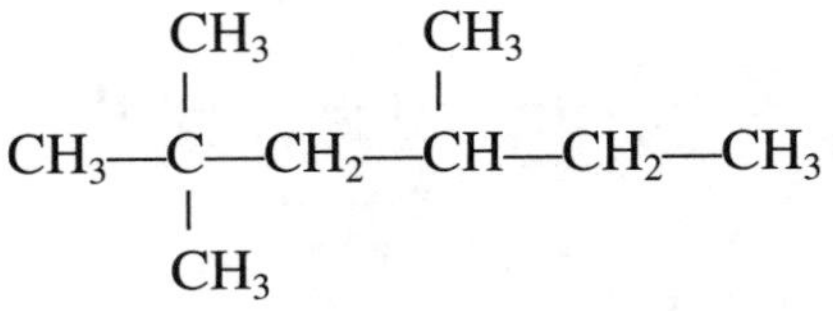 $CH_3-\underset{\underset{\displaystyle CH_3}{|}}{\overset{\overset{\displaystyle CH_3}{|}}{C}}-CH_2-\underset{\underset{\displaystyle CH_3}{|}}{CH}-CH_2-CH_3$

◆ Learning Exercise 10.3D

Write a correct IUPAC (or common name) for the following.

1. CH_3-CH_2-Br

2. 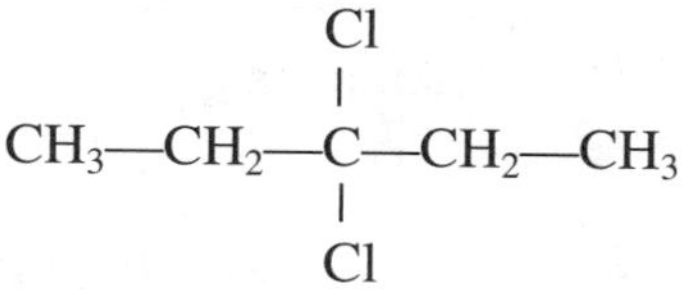
$$CH_3-CH_2-\underset{\underset{Cl}{|}}{\overset{\overset{Cl}{|}}{C}}-CH_2-CH_3$$

3.
$$CH_3-CH_2-\underset{\overset{|}{Cl}}{CH}-CH_2-\underset{\overset{|}{Br}}{CH}-CH_3$$

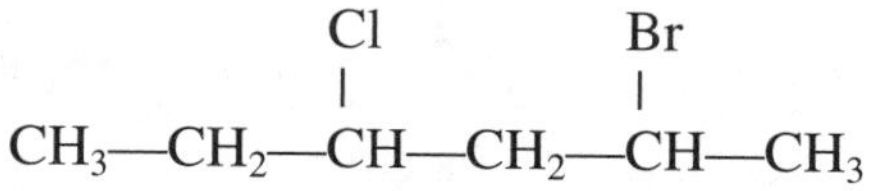

4. 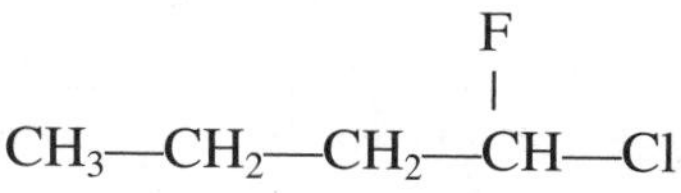
$$CH_3-CH_2-CH_2-\underset{\overset{|}{F}}{CH}-Cl$$

Answers
1. bromoethane (ethyl bromide)
3. 2-bromo-4-chlorohexane
2. 3,3-dichloropentane
4. 1-chloro-1-fluorobutane

◆ Learning Exercise 10.3E

Write the condensed formula for each of the following haloalkanes.

1. ethyl chloride

2. bromomethane

3. 3-bromo-1-chloropentane

4. 1,1-dichlorohexane

5. 2,2,3-trichlorobutane

6. 2,4-dibromo-2,4-dichloropentane

Answers
1. CH_3-CH_2-Cl

2. CH_3-Br

3. 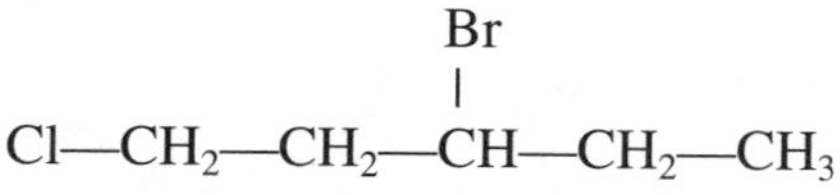 $Cl-CH_2-CH_2-\underset{\overset{|}{Br}}{CH}-CH_2-CH_3$

4. 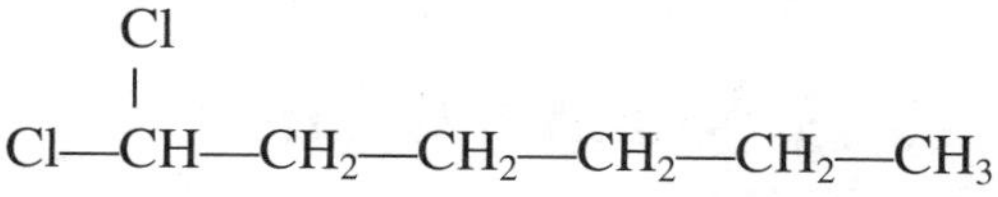 $Cl-\underset{\overset{|}{Cl}}{CH}-CH_2-CH_2-CH_2-CH_2-CH_3$

5. 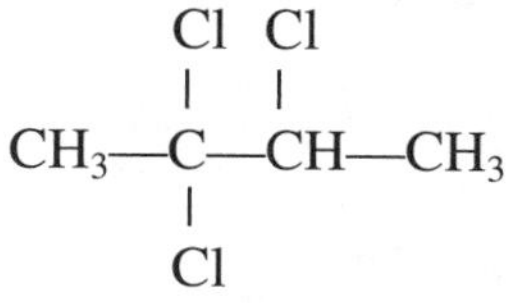 $CH_3-\underset{\underset{Cl}{|}}{\overset{\overset{Cl}{|}}{C}}-\underset{\overset{|}{Cl}}{CH}-CH_3$

6. 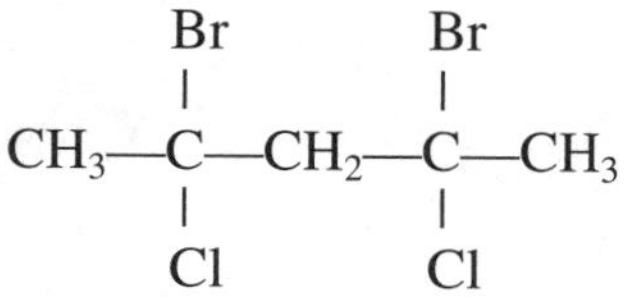 $CH_3-\underset{\underset{Cl}{|}}{\overset{\overset{Br}{|}}{C}}-CH_2-\underset{\underset{Cl}{|}}{\overset{\overset{Br}{|}}{C}}-CH_3$

◆ Learning Exercise 10.3F

Write the IUPAC name for each of the following cycloalkanes.

1.

2.

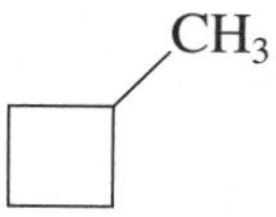

3.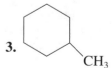

4.

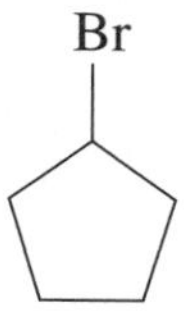

5.

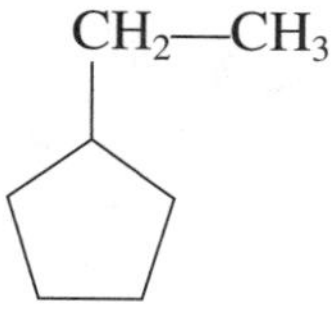

Answers
1. chlorocyclopropane
4. bromocyclopentane
2. methylcyclobutane
5. ethylcyclopentane
3. methylcyclohexane

169

10.4 Properties of Alkanes

- The alkanes are nonpolar, less dense than water, and mostly unreactive, except that they burn vigorously.
- Alkanes are found in natural gas, gasoline, and diesel fuels.
- In combustion, an alkane at a high temperature reacts rapidly with oxygen to produce carbon dioxide, water, and a great amount of heat.

Study Note

Example: Write the equation for the combustion of methane.

Solution: Write the molecular formulas for the reactants: methane (CH_4) and oxygen (O_2). Write the products CO_2 and H_2O and balance the equation.

$$CH_4(g) + O_2(g) \rightarrow CO_2(g) + H_2O(g) + \text{Heat}$$

$$CH_4(g) + 2\,O_2(g) \rightarrow CO_2(g) + 2\,H_2O(g) + \text{Heat (balanced)}$$

◆ Learning Exercise 10.4

Write a balanced equation for the complete combustion of the following:

1. propane   _____

2. hexane   _____

3. pentane   _____

4. cyclobutane   _____

Answers

1. $C_3H_8(g) + 5\,O_2(g) \rightarrow 3\,CO_2(g) + 4\,H_2O(g) + \text{Heat}$
2. $2\,C_6H_{14}(g) + 19\,O_2(g) \rightarrow 12\,CO_2(g) + 14\,H_2O(g) + \text{Heat}$
3. $C_5H_{12}(g) + 8\,O_2(g) \rightarrow 5\,CO_2(g) + 6\,H_2O(g) + \text{Heat}$
4. $C_4H_8(g) + 6\,O_2(g) \rightarrow 4\,CO_2(g) + 4\,H_2O(g) + \text{Heat}$

10.5 Functional Groups

- Organic compounds are classified by *functional groups,* which are atoms or groups of atoms where specific chemical reactions occur.
- Alkenes are hydrocarbons that contain one or more double bonds ($C=C$); alkynes contain a triple bond ($C\equiv C$).
- Alcohols contain a hydroxyl ($-OH$) group; ethers have an oxygen atom ($-O-$) between two alkyl groups.
- Aldehydes contain a carbonyl group ($C=O$) bonded to at least one H atom; ketones contain the carbonyl group bonded to two alkyl groups.
- Carboxylic acids have a carboxyl group ($-COO-$) attached to hydrogen ($-COOH$); esters contain the carboxyl groups attached to a carbon of an alkyl group.
- Amines are derived from ammonia (NH_3) in which alkyl groups replace one or more of the H atoms.
- Amides have the hydroxyl group of a carboxylic acid replaced by a nitrogen group.

◆ Learning Exercise 10.5A

Classify the organic compounds shown below according to their functional groups:

    a. alkane   **b.** alkene   **c.** alcohol   **d.** ether   **e.** aldehyde

1. _____ $CH_3—CH_2—CH=CH_2$ **2.** _____ $CH_3—CH_2—CH_3$

3. _____ $CH_3—CH_2—\overset{\overset{\displaystyle O}{||}}{C}—H$ **4.** _____ $CH_3—CH_2—CH_2—OH$

5. _____ $CH_3—CH_2—O—CH_2—CH_3$ **6.** _____ $CH_3—CH_2—CH_2—CH_3$

Answers   **1.** b   **2.** a   **3.** e   **4.** c   **5.** d   **6.** a

◆ Learning Exercise 10.5B

Classify the following compounds according to their functional groups:

    a. alcohol   **b.** aldehyde   **c.** ketone   **d.** ether   **e.** amine   **f.** amide

1. _____ $CH_3—CH_2—CH_2—\overset{\overset{\displaystyle O}{||}}{C}—H$ **2.** _____ $CH_3—CH_2—CH_2—NH_2$

3. _____ $CH_3—CH_2—\overset{\overset{\displaystyle O}{||}}{C}—CH_2—CH_3$ **4.** _____ $CH_3—CH_2—O—CH_3$

5. _____ $CH_3—\overset{\overset{\displaystyle O}{||}}{C}—NH_2$ **6.** _____ $CH_3—\overset{\overset{\displaystyle O}{||}}{C}—H$

7. _____ $CH_3—CH_2—\overset{\overset{\displaystyle NH_2}{|}}{CH}—CH_3$ **8.** _____ $CH_3—CH_2—\overset{\overset{\displaystyle OH}{|}}{CH}—CH_3$

Answers   **1.** b   **2.** e   **3.** c   **4.** d
                 5. f   **6.** b   **7.** e   **8.** a

Checklist for Chapter 10

You are ready to take the practice test for Chapter 10. Be sure that you have accomplished the following learning goals for this chapter. If you are not sure, review the section listed at the end of the goal. Then apply your new skills and understanding to the practice test.

After studying Chapter 10, I can successfully:

_____ Identify properties as characteristic of organic or inorganic compounds (10.1).

_____ Identify the number of bonds for carbon (10.1).

_____ Describe the tetrahedral shape of carbon in carbon compounds (10.1).

_____ Draw the expanded complete structural formula and the condensed structural formula for an alkane (10.2).

_____ Use the IUPAC system to write the names for alkanes (10.2).

_____ Use the IUPAC system to write the names for alkanes with substituents (10.3).

_____ Draw the structural formulas of alkanes from the name (10.3).

_____ Describe physical properties of alkanes (10.4).

_____ Identify the functional groups in organic compounds (10.5).

Practice Test for Chapter 10

For problems 1 through 8, indicate whether the following characteristics are typical of (O) organic compounds or (I) inorganic compounds.

1. _____ higher melting points   **5.** _____ ionic bonds

2. _____ fewer compounds   **6.** _____ combustible

3. _____ covalent bonds   **7.** _____ low boiling points

4. _____ soluble in water   **8.** _____ soluble in nonpolar solvents

Match the name of the hydrocarbon with each of the following structures 9 through 13.
     A. methane   **B.** ethane   **C.** propane   **D.** pentane   **E.** heptane

9. _____ $CH_3-CH_2-CH_3$

10. _____ $CH_3-CH_2-CH_2-CH_2-CH_2-CH_2-CH_3$

11. _____ CH_4

12. _____ $CH_3-CH_2-CH_2-CH_2-CH_3$

13. _____ CH_3-CH_3

Match the name of the hydrocarbon with each of the following structures 14 through 17.
     A. butane   **B.** methylcyclohexane   **C.** cyclopropane
     D. 3,5-dimethylhexane   **E.** 2,4-dimethylhexane

14. $CH_3-CH_2-CH_2-CH_3$ **15.**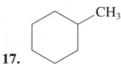

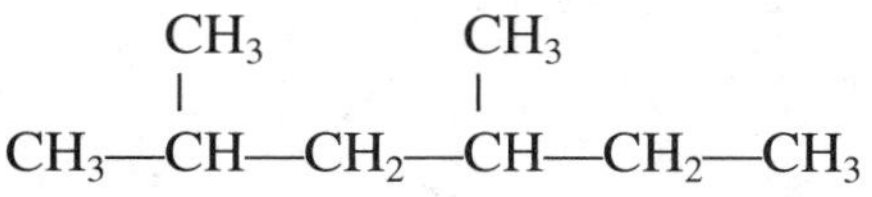

16.    **17.**

Match the name of the hydrocarbon (18–20) with each of the following structures.
     A. methylcyclopentane   **B.** cyclobutane
     C. cyclohexane   **D.** ethylcyclopentane

18.   **19.** 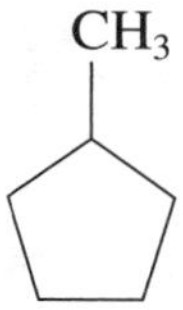   **20.**

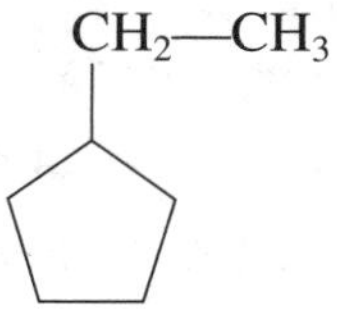

Match each of the following compounds (21–23) with the correct name.
- **A.** 2,4-dichloropentane
- **B.** chlorocyclopentane
- **C.** 1,2-dichloropentane
- **D.** 4,5-dichloropentane

21. 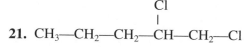 CH₃—CH₂—CH₂—CH—CH₂—Cl with Cl above CH

23. 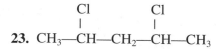 CH₃—CH—CH₂—CH—CH₃ with Cl above each CH

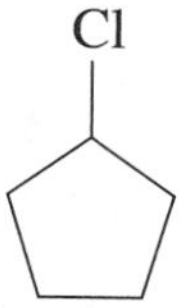

22.

24. The correctly balanced equation for the complete combustion of ethane is
- **A.** $C_2H_6 + O_2 \rightarrow 2\,CO + 3\,H_2O$.
- **B.** $C_2H_6 + O_2 \rightarrow CO_2 + H_2O$.
- **C.** $C_2H_6 + 2\,O_2 \rightarrow 2\,CO_2 + 3\,H_2O$.
- **D.** $2\,C_2H_6 + 7\,O_2 \rightarrow 4\,CO_2 + 6\,H_2O$.
- **E.** $2\,C_2H_6 + 4\,O_2 \rightarrow 4\,CO_2 + 6\,H_2O$.

Classify the compounds in problems 25 through 33 by the functional groups.
- **A.** alkane
- **B.** alkene
- **C.** alcohol
- **D.** aldehyde
- **E.** ketone
- **F.** ether
- **G.** amine
- **H.** amide

25. CH₃—CH₂—C(=O)—NH₂

30. CH₃—CH—CH₂—CH₃ with NH₂ above CH

26. CH₃—CH₂—C(=O)—CH₃

31. CH₃—C(=O)—H

27. CH₃—CH₂—CH₂—OH

32. CH₃—CH₂—CH=CH₂

28. CH₃—CH₂—CH—CH₃ with CH₃ above CH

33. CH₃—CH₂—CH—CH₃ with OH above CH

29. CH₃—CH₂—O—CH₃

Indicate whether the pairs of compounds in problems 34 through 38 are isomers (I), the same compound (S), or different compounds (D).

34. CH₃—CH₂—CH₂—CH₃ and CH₃—CH—CH₃ with CH₃ above CH

35. CH₃—CH₂—OH and CH₃—C(=O)—H

36. CH₃—CH₂—NH₂ and CH₃—N—CH₂—CH₃ with H above N

37. $\underset{\displaystyle CH_3-CH_2-\overset{\textstyle O}{\overset{\|}{C}}-OH}{}$ and $\underset{\displaystyle CH_3-\overset{\textstyle O}{\overset{\|}{C}}-O-CH_3}{}$

38. $CH_3-CH_2-CH_2-CH_3$ and $CH_3-C\equiv C-CH_3$

Answers to Practice Test

1. I	**2.** I	**3.** O	**4.** I	**5.** I
6. O	**7.** O	**8.** O	**9.** C	**10.** E
11. A	**12.** D	**13.** B	**14.** A	**15.** E
16. C	**17.** B	**18.** D	**19.** A	**20.** E
21. C	**22.** B	**23.** A	**24.** D	**25.** H
26. E	**27.** C	**28.** A	**29.** F	**30.** G
31. D	**32.** B	**33.** C	**34.** I	**35.** D
36. D	**37.** I	**38.** D		

Answers to Selected Problems

10.1   Organic compounds contain C and H and sometimes O, N, or a halogen atom. Inorganic compounds usually contain elements other than C and H.

    a. inorganic   **b.** organic   **c.** organic
    d. inorganic   **e.** inorganic   **f.** organic

10.3   **a.** Inorganic compounds are usually soluble in water.
    b. Organic compounds have lower boiling points than most inorganic compounds.
    c. Organic compounds often burn in air.
    d. Inorganic compounds are more likely to be solids at room temperature.

10.5   **a.** ethane   **b.** ethane   **c.** NaBr   **d.** NaBr

10.7   VSEPR theory predicts that the four bonds in CH_4 will be as far apart as possible, which means that the hydrogen atoms are at the corners of a tetrahedron.

10.9   **a.** Pentane is a carbon chain of five (5) carbon atoms.
    b. Ethane is a carbon chain of two (2) carbon atoms.
    c. Hexane is a carbon chain of six (6) carbon atoms.

10.11   **a.** CH_4 **b.** CH_3-CH_3 **c.** $CH_3-CH_2-CH_2-CH_2-CH_3$ **d.** △

10.13   Isomers have the same molecular formula, but different structural formulas. The same structure occurs when rotation about a single bond moves the attached groups into different positions, but does not change the arrangement of atoms.

    a. The same molecule.
    b. Same molecular formula, but different structures, which are isomers of C_5H_{12}.
    c. Same molecular formula, but different structures, which are isomers of C_6H_{14}.

10.15   **a.** 2-methylbutane   **b.** 2,2-dimethylpropane
    c. 2,3-dimethylpentane   **d.** 4-ethyl-2,2-dimethylhexane

10.17   **a.** A ring of four carbon atoms is cyclobutane.
    b. A ring of five carbon atoms with one methyl group is methylcyclopentane.
    c. A ring of six carbon atoms with an ethyl group is ethylcyclohexane.

10.19 Draw the main chain with the number of carbon atoms in the ending. For example, butane has a main chain of 4 carbon atoms, and hexane has a main chain of 6 carbon atoms. Attach substituents on the carbon atoms indicated. For example, in 3-methylpentane, a CH_3— group is bonded to carbon 3 of a five-carbon chain.

$$\begin{array}{c} CH_3 \\ | \\ \textbf{a.} \quad CH_3{-}CH{-}CH_2{-}CH_3 \end{array}$$

$$\begin{array}{c} CH_3 \\ | \\ CH_3{-}CH_2{-}C{-}CH_2{-}CH_3 \\ | \\ CH_3 \end{array}$$
b.

$$\textbf{c.} \quad \begin{array}{c} CH_3 \;\; CH_3 \qquad\; CH_3 \\ \;| \quad\;\; | \qquad\quad | \\ CH_3{-}CH{-}CH{-}CH_2{-}CH{-}CH_3 \end{array}$$
d.
$$\begin{array}{c} CH_3 \;\; CH_2{-}CH_3 \;\; CH_3 \\ | \qquad\;\; | \qquad\quad\; | \\ CH_3{-}CH{-}CH{-}CH_2{-}CH{-}CH_2{-}CH_2{-}CH_3 \end{array}$$

10.21 a.

b.

10.23 In the IUPAC system, the halogen substituent is named as a halo- on the longest carbon chain.

a. bromoethane b. 1-fluoropropane
c. 2-chloropropane d. trichloromethane

10.25 a.
$$\begin{array}{c} Cl \\ | \\ CH_3{-}CH{-}CH_3 \end{array}$$
b.
$$\begin{array}{c} Br \;\; Cl \\ | \quad\; | \\ CH_3{-}CH{-}CH{-}CH_3 \end{array}$$
c. CH_3Br d. CBr_4

10.27 Alkanes are nonpolar and less dense than water.

a. $CH_3{-}CH_2{-}CH_2{-}CH_2{-}CH_2{-}CH_2{-}CH_3$ b. liquid
c. insoluble in water d. float

10.29 In combustion, a hydrocarbon reacts with oxygen to yield CO_2 and H_2O.

a. $2\,C_2H_6 + 7\,O_2 \rightarrow 4\,CO_2 + 6\,H_2O$ b. $2\,C_3H_6 + 9\,O_2 \rightarrow 6\,CO_2 + 6\,H_2O$
c. $2\,C_8H_{18} + 25\,O_2 \rightarrow 16\,CO_2 + 18\,H_2O$ d. $C_6H_{12} + 9\,O_2 \rightarrow 6\,CO_2 + 6\,H_2O$

10.31 a. alcohol b. alkene c. aldehyde d. ester

10.33 a. Ethers have an —O— group. b. Alcohols have a —OH group.
c. Ketones have a C=O group between alkyl groups.
d. Carboxylic acids have a —COOH group. e. Amines contain a N atom.

10.35 a. aromatic, ether, alcohol, ketone b. aromatic, ether, alkene, ketone

10.37 aromatic, ether, amide

10.39 a. Organic compounds have covalent bonds; inorganic compounds have ionic as well as polar covalent and a few have nonpolar covalent bonds.
b. Most organic compounds are insoluble in water; inorganic compounds are soluble in water.
c. Most organic compounds have low melting points; inorganic compounds have high melting points.
d. Most organic compounds are flammable; inorganic compounds are not flammable.

10.41 a. butane; organic compounds have low melting points
b. butane; organic compounds burn vigorously in air
c. potassium chloride; inorganic compounds have high melting points
d. potassium chloride; inorganic compounds (ionic) produce ions in water
e. butane; organic compounds are more likely to be gases at room temperature

10.43 **a.** aldehyde, aromatic   **b.** alkene, aldehyde, aromatic   **c.** ketone

10.45 carboxylic acid, aromatic, amine, amide, ester

10.47 The alkyl name for a hydrocarbon substituent uses the name of the alkane, but changes the ending to −yl. A halogen atom is named as a substituent (fluoro, chloro, bromo, iodo).

    a. methyl   **b.** propyl   **c.** chloro

10.49 **a.** 2,2-dimethylbutane   **b.** chloroethane
    c. 2-bromo-4-ethylhexane   **d.** cyclohexane

10.51 Write the carbon atom in the main chain first. Attach the substituents listed in front of the alkane name or use the alkyl group indicated.

$$
\begin{array}{c}
\qquad\quad CH_2{-}CH_3 \\
\qquad\quad | \\
\textbf{a. } CH_3{-}CH_2{-}CH{-}CH_2{-}CH_2{-}CH_3
\end{array}
$$

$$
\begin{array}{c}
\quad CH_3\ \ CH_3 \\
\quad |\qquad | \\
\textbf{b. } CH_3{-}CH{-}CH{-}CH_2{-}CH_3
\end{array}
$$

$$
\begin{array}{c}
\qquad\qquad\qquad Cl \\
\qquad\qquad\qquad | \\
\textbf{c. } Cl{-}CH_2{-}CH_2{-}C{-}CH_2{-}CH_2{-}CH_2{-}CH_3 \\
\qquad\qquad\qquad | \\
\qquad\qquad\qquad CH_3
\end{array}
$$

10.53 **a.** $C_3H_8 + 5\,O_2 \rightarrow 3\,CO_2 + 4\,H_2O$ **b.** $C_5H_{12} + 8\,O_2 \rightarrow 5\,CO_2 + 6\,H_2O$
    c. $C_4H_8 + 6\,O_2 \rightarrow 4\,CO_2 + 4\,H_2O$ **d.** $2\,C_8H_{18} + 2\,5O_2 \rightarrow 16\,CO_2 + 18\,H_2O$
    e. $2\,C_3H_6 + 9\,O_2 \rightarrow 6\,CO_2 + 6\,H_2O$

10.55 **a.** amine   **b.** alcohol   **c.** ester   **d.** alkene

10.57 **a.** alcohol   **b.** alkene   **c.** aldehyde
    d. alkane   **e.** carboxylic acid   **f.** amine

10.59 **a.** $C_3H_8(g) + 5\,O_2(g) \rightarrow 3\,CO_2(g) + 4\,H_2O(g)$

    b. $5.0\ \cancel{\text{lb propane}} \times \dfrac{454\ \cancel{\text{g propane}}}{1\ \cancel{\text{lb propane}}} \times \dfrac{1\ \cancel{\text{mole propane}}}{44.0\ \cancel{\text{g propane}}}$

$$\times\ \dfrac{3\ \cancel{\text{moles CO}_2}}{1\ \cancel{\text{mole propane}}} \times \dfrac{44.0\ \cancel{\text{g CO}_2}}{1\ \cancel{\text{mole CO}_2}} \times \dfrac{1\ \text{kg CO}_2}{1000\ \cancel{\text{g CO}_2}} = 6.8\ \text{kg of CO}_2$$

10.61 **a.** $C_5H_{12}(g) + 8\,O_2(g) \rightarrow 5\,CO_2(g) + 6\,H_2O(g)$
    b. 72.0 g/mole

$$1\ \cancel{\text{gal}} \times \dfrac{3.78\ \cancel{L}}{1\ \cancel{\text{gal}}} \times \dfrac{1000\ \cancel{\text{mL}}}{1\ \cancel{L}} \times \dfrac{0.63\ \cancel{g}}{1\ \cancel{\text{mL}}} \times \dfrac{1\ \cancel{\text{mole C}_5\text{H}_{12}}}{72.0\ \cancel{g}} \times \dfrac{845\ \text{kcal}}{1\ \cancel{\text{mole}}}$$

    c. $= 2.8 \times 10^4\ \text{kcal}$

$$1\ \cancel{\text{gal}} \times \dfrac{3.78\ \cancel{L}}{1\ \cancel{\text{gal}}} \times \dfrac{1000\ \cancel{\text{mL}}}{1\ \cancel{L}} \times \dfrac{0.63\ \cancel{g}}{1\ \cancel{\text{mL}}} \times \dfrac{1\ \cancel{\text{mole C}_5\text{H}_{12}}}{72.0\ \cancel{g}} \times \dfrac{5\ \cancel{\text{moles CO}_2}}{1\ \cancel{\text{mole C}_5\text{H}_{12}}}$$

    d.

$$\times\ \dfrac{22.4\ \text{L}}{1\ \cancel{\text{mole}}} = 3700\ \text{L}$$

10.63

$$\begin{array}{ccc} & CH_3 \quad CH_3 \\ & | \qquad | \\ CH_3{-}CH{-}CH{-}CH_3 \end{array} \qquad \begin{array}{c} CH_3 \\ | \\ CH_3{-}C{-}CH_2{-}CH_3 \\ | \\ CH_3 \end{array}$$

10.65 **a.** $CH_3{-}CH_2{-}CH_3$

b. $C_3H_8 + 5 O_2 \rightarrow 3 CO_2 + 4 H_2O$

c. $12.0 \; \text{L } C_3H_8 \times \dfrac{1 \text{ mole } C_3H_8}{22.4 \text{ L } C_3H_8} \times \dfrac{5 \text{ moles } O_2}{1 \text{ mole } C_3H_8} \times \dfrac{32.0 \text{ g } O_2}{1 \text{ mole } O_2} = 85.7 \text{ g of } O_2$

d. $12.0 \; \text{L } C_3H_8 \times \dfrac{1 \text{ mole } C_3H_8}{22.4 \text{ L } C_3H_8} \times \dfrac{3 \text{ moles } CO_2}{1 \text{ mole } C_3H_8} \times \dfrac{44.0 \text{ g } CO_2}{1 \text{ mole } CO_2} = 70.7 \text{ g of } CO_2$

10.67 Condensed structural formula

$$\begin{array}{c} CH_3 \qquad\quad CH_3 \\ | \qquad\qquad\quad | \\ CH_3{-}C{-}CH_2{-}CH{-}CH_3 \\ | \\ CH_3 \end{array}$$

molecular formula: C_8H_{18}

The combustion reaction: $2 C_8H_{18}(g) + 25 O_2(g) \rightarrow 16 CO_2(g) + 18 H_2O(g)$

Study Goals

- Classify unsaturated compounds as alkenes and alkynes.
- Write IUPAC and common names for alkenes and alkynes.
- Write structural formulas and names for cis–trans isomers of alkenes.
- Write equations for hydration and hydrogenation of alkenes and alkynes.
- Describe the formation of a polymer from alkene monomers.
- Describe the bonding in benzene.
- Write structural formulas and give the names of aromatic compounds.

Think About It

1. The label on a bottle of vegetable oil says the oil is unsaturated. What does this mean?

2. What are polymers?

3. Margarine is partially hydrogenated. What does that mean?

Key Terms

Match the statements shown below with the following key terms.

    a. alkene   **b.** hydrogenation   **c.** alkyne   **d.** hydration   **e.** polymer

1. _____ A long-chain molecule formed by linking many small molecules

2. _____ The addition of H_2 to a carbon–carbon double bond

3. _____ An unsaturated hydrocarbon containing a carbon–carbon double bond

4. _____ The addition of H_2O to a carbon–carbon double bond

5. _____ A compound that contains a triple bond

Answers   **1.** e   **2.** b   **3.** a   **4.** d   **5.** c

11.1 Alkenes and Alkynes

- Alkenes are unsaturated hydrocarbons that contain one or more carbon–carbon double bonds.
- Alkynes are unsaturated hydrocarbons that contain a carbon–carbon triple bond.
- The IUPAC names of alkenes are derived by changing the *ane* ending of the parent alkane to *ene*. For example, the IUPAC name of $H_2C=CH_2$ is ethene. It has a common name of ethylene.
- In alkenes, the longest carbon chain containing the double bond is numbered from the end nearest the double bond.

$$CH_3-CH=CH_2$$
Propene (propylene)

$$CH_2=CH-CH_2-CH_3$$
1-Butene

$$CH_3-CH=\overset{\overset{\displaystyle CH_3}{|}}{C}-CH_3$$
2-Methyl-2-butene

- The alkynes are a family of unsaturated hydrocarbons that contain a triple bond. They use naming rules similar to the alkenes, but the parent chain ends with *yne*.

  HC≡CH   CH₃—C≡CH

  ethyne   propyne

◆ Learning Exercise 11.1A

Classify the following structural formulas as alkane, alkene, cycloalkene, or alkyne.

1. _____ CH₃—CH₂—CH₃

2. _____ CH₃—CH₂—CH=C—CH₂—CH₃ with CH₃ on the C

3. _____ CH₃—C≡C—CH₃

Answers   **1.** alkane   **2.** alkyne   **3.** alkene

◆ Learning Exercise 11.1B

Write the IUPAC (and common name, if any) for each of the following alkenes:

1. CH₃—CH=CH₂

2. CH₃—CH=CH—CH₃

3.

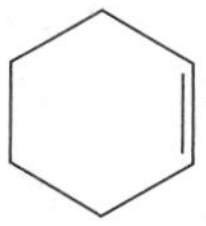

4. CH₂=CH—CH—CH₂—CH—CH₃ with Cl and CH₃ substituents

5. CH₃—CH=C—CH₂—CH₃ with CH₃ on the C

6. CH₃—CH₂—CH with CH₂ double bonded

Answers   **1.** propene (propylene)   **2.** 2-butene   **3.** cyclohexene
              4. 3-chloro-5-methyl-1-hexene   **5.** 3-methyl-2-pentene   **6.** 1-butene

◆ Learning Exercise 11.1C

Write the IUPAC and common name (if any) of each of the following alkynes.

1. HC≡CH

2. CH₃—C≡CH

3. CH₃—CH₂—C≡CH

4. CH₃—CH—C≡C—CH₃ with CH₃ substituent

Answers   **1.** ethyne (acetylene)   **2.** propyne
              3. 1-butyne   **4.** 4-methyl-2-pentyne

◆ **Learning Exercise 11.1D**

Draw the condensed structural formula for each of the following:

1. 2-pentyne

2. 2-chloro-2-butene

3. 3-bromo-2-methyl-2-pentene

4. cyclohexene

Answers

1. $CH_3—C≡C—CH_2—CH_3$

2. 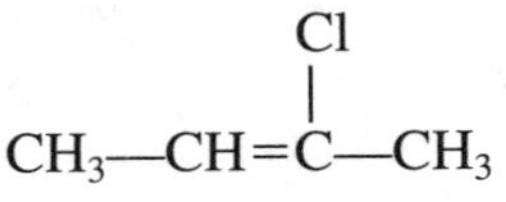 $CH_3—CH=\overset{\overset{\displaystyle Cl}{\displaystyle |}}{C}—CH_3$

3. $CH_3—C=\overset{\overset{\displaystyle Br}{\displaystyle |}}{\underset{\underset{\displaystyle CH_3}{\displaystyle |}}{C}}—CH_2—CH_3$

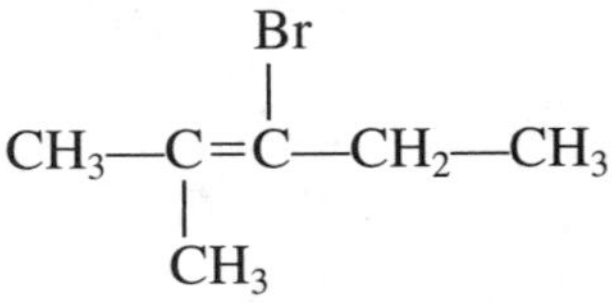

4.

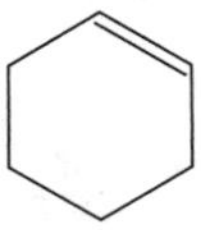

11.2 Cis–Trans Isomers

- Cis–trans isomers are possible for alkenes because there is no rotation around the rigid double bond.
- In the cis isomer, groups are attached on the same side of the double bond, whereas in the trans isomer, they are attached on the opposite sides of the double bond.

◆ **Learning Exercise 11.2A**

Write the cis–trans isomers of 2,3-dibromo-2-butene and name each.

Answers

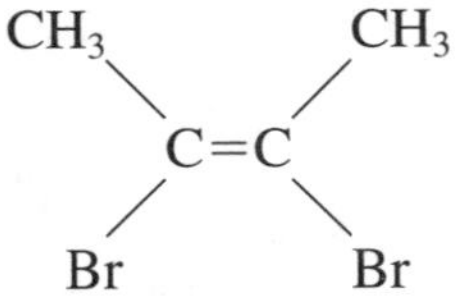

cis-2,3-dibromo-2-butene 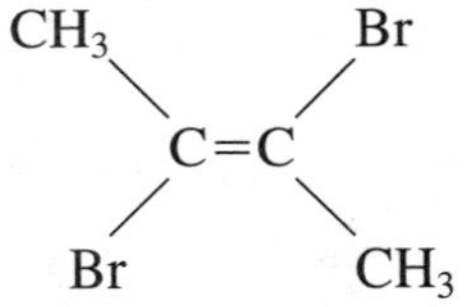   *trans*-2,3-dibromo-2-butene

In the cis isomer, the bromine atoms are attached on the same side of the double bond; in the trans isomer, they are on opposite sides.

◆ **Learning Exercise 11.2B**

Name the following alkenes using the cis–trans where isomers are possible.

1.

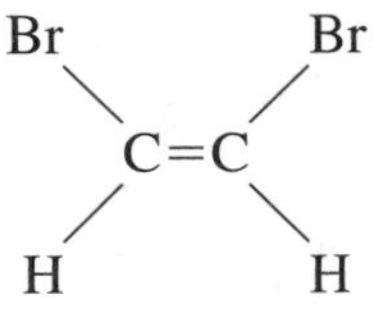

2.

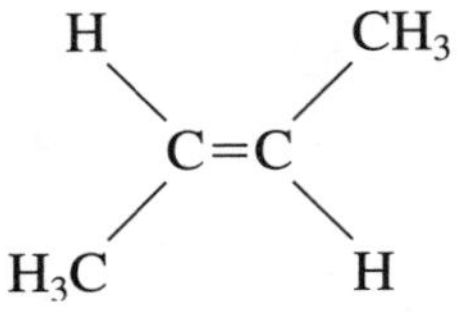

$$\underset{\substack{\text{Cl} \qquad\qquad \text{H} \\ | \qquad\qquad | \\ \text{3. } \text{H}_3\text{C}}}{}\;\text{C}=\text{C}\;\underset{\text{H}}{}$$

$$\underset{\substack{\text{H} \qquad\qquad \text{CH}_2\text{—CH}_3 \\ | \qquad\qquad | \\ \text{4. } \text{H}_3\text{C}}}{}\;\text{C}=\text{C}\;\underset{\text{H}}{}$$

Answers
1. *cis*-1,2-dibromoethene
3. 2-chloropropene (not a cis–trans isomer)
2. *trans*-2-butene
4. *trans*-2-pentene

11.3 Addition Reactions

- The addition of small molecules to the double bond is a characteristic reaction of alkenes.
- Hydrogenation adds hydrogen atoms to the double bond of an alkene or the triple bond of an alkyne to yield an alkane.

$$\text{CH}_2{=}\text{CH}_2 + \text{H}_2 \xrightarrow{\text{Pt}} \text{CH}_3{-}\text{CH}_3$$

$$\text{HC}{\equiv}\text{CH} + 2\text{H}_2 \xrightarrow{\text{Pt}} \text{CH}_3{-}\text{CH}_3$$

- Hydration adds water to a double bond. The H from water (HOH) bonds to the carbon in the double bond that has the greater number of hydrogen atoms.

$$\text{CH}_2{=}\text{CH}_2 + \text{HOH} \xrightarrow{\text{H}^+} \text{CH}_3{-}\text{CH}_2{-}\text{OH}$$

◆ Learning Exercise 11.3A

Write the products of the following addition reactions.

1. $\text{CH}_3{-}\text{CH}_2{-}\text{CH}{=}\text{CH}_2 + \text{H}_2 \xrightarrow{\text{Pt}}$

2. ⬠ $+ \text{H}_2 \xrightarrow{\text{Pt}}$

3. $\text{CH}_3{-}\text{CH}{=}\text{CH}{-}\text{CH}_2{-}\text{CH}_3 + \text{H}_2 \xrightarrow{\text{Pt}}$

4. $\text{CH}_3{-}\text{CH}{=}\text{CH}_2 + \text{H}_2 \xrightarrow{\text{Pt}}$

5. $\text{CH}_3{-}\text{CH}{=}\text{CH}{-}\text{CH}_3 + \text{H}_2 \xrightarrow{\text{Pt}}$

Answers

1. $\text{CH}_3{-}\text{CH}_2{-}\text{CH}_2{-}\text{CH}_3$
2. ⬠
3. $\text{CH}_3{-}\text{CH}_2{-}\text{CH}_2{-}\text{CH}_2{-}\text{CH}_3$
4. $\text{CH}_3{-}\text{CH}_2{-}\text{CH}_3$
5. $\text{CH}_3{-}\text{CH}_2{-}\text{CH}_2{-}\text{CH}_3$

◆ Learning Exercise 11.3B

1. $\text{CH}_3{-}\text{CH}{=}\text{CH}{-}\text{CH}_3 + \text{HOH} \xrightarrow{\text{H}^+}$

2. $\underset{\substack{\text{CH}_3 \\ | \\ \text{CH}_3{-}\text{C}{=}\text{CH}_2}}{}\; + \text{HOH} \xrightarrow{\text{H}^+}$

3. $CH_3-CH=CH_2 + HOH \xrightarrow{H^+}$

4. $CH_3-CH_2-CH=C(CH_3)-CH_3 + HOH \xrightarrow{H^+}$

5. 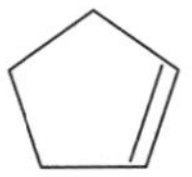 $+ H_2O \xrightarrow{H^+}$

Answers

1. $CH_3-CH_2-CH(OH)-CH_3$
2. $CH_3-C(CH_3)(OH)-CH_3$
3. $CH_3-CH(OH)-CH_3$
4. $CH_3-CH_2-CH_2-C(CH_3)(OH)-CH_3$
5. 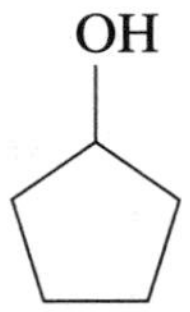

11.4 Polymers of Alkenes

- *Polymers* are large molecules prepared from the bonding of many small units called *monomers.*
- Many synthetic polymers are made from small alkene monomers.

◆ Learning Exercise 11.4A

Write the formula of the alkene monomer that would be used for each of the following polymers.

1. $-CH_2-CH_2-CH_2-CH_2-CH_2-CH_2-$
2. $-CH_2-CH(CH_3)-CH_2-CH(CH_3)-CH_2-CH(CH_3)-$
3. $-CF_2-CF_2-CF_2-CF_2-CF_2-CF_2-$

Answers

1. $H_2C=CH_2$
2. $H_2C=CH(CH_3)$
3. $F_2C=CF_2$

◆ **Learning Exercise 11.4B**

Write three sections of the polymer that would result when 1,1-difluoroethene is the monomer unit.

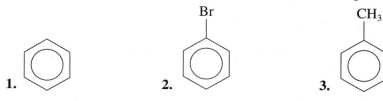

Answer

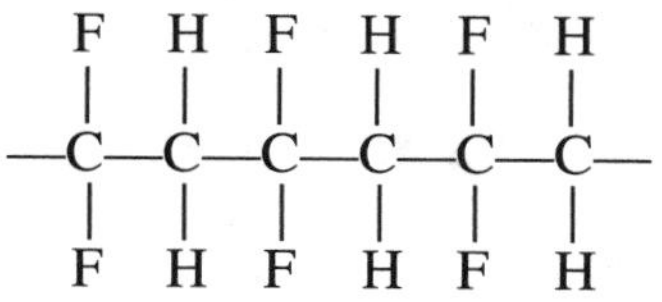

11.5 Aromatic Compounds

- Most aromatic compounds contain benzene, a cyclic structure containing six CH units. The structure of benzene is represented as a hexagon with a circle in the center.
- The names of many aromatic compounds use the parent name benzene, although many common names were retained as IUPAC names, such as toluene, phenol, and aniline.

◆ **Learning Exercise 11.5**

Write the IUPAC (or common name) for each of the following.

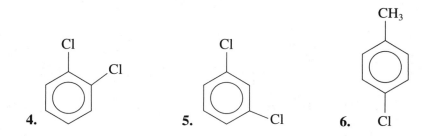

Answers   **1.** benzene   **2.** bromobenzene   **3.** methylbenzene; toluene
                4. 1,2-dichlorobenzene   **5.** 1,3-dichlorobenzene   **6.** 4-chlorotoluene

Checklist for Chapter 11

You are ready to take the practice test for Chapter 11. Be sure that you have accomplished the following learning goals for this chapter. If you are not sure, review the section listed at the end of the goal. Then apply your new skills and understanding to the practice test.
After studying Chapter 11, I can successfully:

_____ Identify the structural features of alkenes and alkynes, and name alkenes and alkynes using IUPAC rules (11.1)

_____ Identify alkenes that exist as cis–trans isomers; write their structural formulas and names (11.2)

_____ Write the structural formulas and names for the products of the addition of hydrogen and water to alkenes (11.3)

_____ Describe the process of forming polymers from alkene monomers (11.4)

_____ Write the names and structures for compounds that contain a benzene ring (11.5).

Practice Test for Chapter 11

Questions 1 through 4 refer to $H_2C=CH-CH_3$ and

(A)

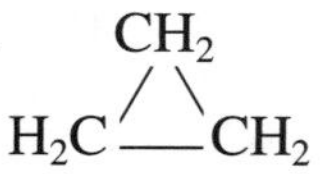

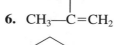

(B)

1. These compounds are
   A. aromatic.　　**B.** alkanes.　　**C.** isomers.　　**D.** alkenes.　　**E.** cycloalkanes.

2. Compound (A) is a(n)
   A. alkane.　　**B.** alkene.　　**C.** cycloalkane.　　**D.** alkyne.　　**E.** aromatic.

3. Compound (B) is named
   A. propane.　　**B.** propylene.　　**C.** cyclobutane.　　**D.** cyclopropane.　　**E.** cyclopropene.

4. Compound (A) is named
   A. propane.　　**B.** propene.　　**C.** 2-propene.　　**D.** propyne.　　**E.** 1-butene.

In questions 5 through 8, match the name of the alkene with the structural formula.

   A. cyclopentene　　**B.** methylpropene　　**C.** cyclohexene　　**D.** ethene

5. $CH_2=CH_2$

6.
$$CH_3-\overset{\overset{\displaystyle CH_3}{|}}{C}=CH_2$$

7.

8.

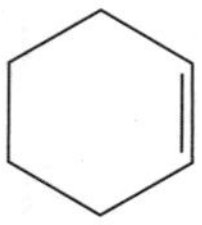

9. The *cis* isomer of 2-butene is
   A. $CH_2=CH-CH_2-CH_3$

   B. $CH_3-CH=CH-CH_3$

   C.
$$\underset{\displaystyle H}{\overset{\displaystyle CH_3}{\diagdown}}C=C\underset{\displaystyle CH_3}{\overset{\displaystyle H}{\diagup}}$$

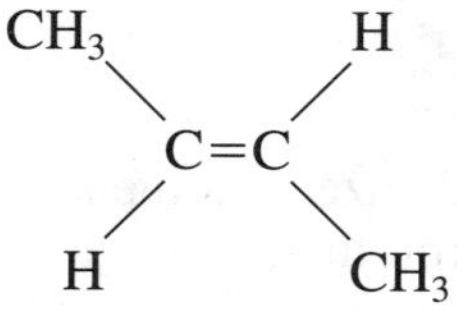

   D.
$$\underset{\displaystyle H}{\overset{\displaystyle CH_3}{\diagdown}}C=C\underset{\displaystyle H}{\overset{\displaystyle CH_3}{\diagup}}$$

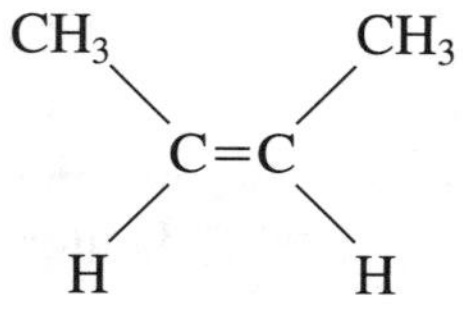

   E.
$$\overset{\overset{\displaystyle CH_3\ \ CH_3}{|\ \ \ \ |}}{CH=CH}$$

10. The name of this compound is
$$\underset{\displaystyle H}{\overset{\displaystyle Cl}{\diagdown}}C=C\underset{\displaystyle Cl}{\overset{\displaystyle H}{\diagup}}$$

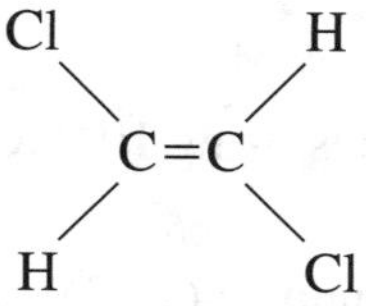

A. dichloroethene. **B.** *cis*-1,2-dichloroethene. **C.** *trans*-1,2-dichloroethene.
D. *cis*-chloroethene. **E.** *trans*-chloroethene.

11. Hydrogenation of CH_3—CH=CH_2 gives
A. $3CO_2 + 6H_2$. **B.** CH_3—CH_2—CH_3. **C.** CH_2=CH—CH_3.
D. no reaction. **E.** CH_3—CH_2—CH_2—CH_3.

12. Hydration of 2-butene gives
A. CH_3—CH_2—CH_2—CH_3 **B.** CH_3—CH_2—CH_2—CH_2—OH

C. CH_3—$\overset{\displaystyle OH}{\underset{|}{CH}}$—$CH_2$—$CH_3$ **D.** 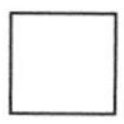 **E.**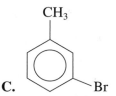

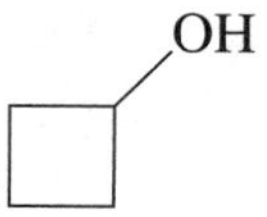

13. What is the common name of methylbenzene?
A. aniline **B.** phenol **C.** toluene
D. xylene **E.** toluidine

14. What is the IUPAC name of CH_3—CH_2—C≡CH?
A. methylacetylene **B.** propyne **C.** propylene
D. 4-butyne **E.** 1-butyne

15. The reaction CH_3—CH=CH_2 + H_2O → CH_3—$\overset{\displaystyle OH}{\underset{|}{CH}}$—$CH_3$ is called a
A. hydrogenation. **B.** halogenation. **C.** hydrohalogenation of an alkene.
D. hydration of an alkene. **E.** combustion.

For questions 16 through 19 identify the family for each compound as
A. alkane. **B.** alkene. **C.** alkyne. **D.** cycloalkene.

16. CH_3—CH=CH_2 **17.**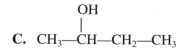

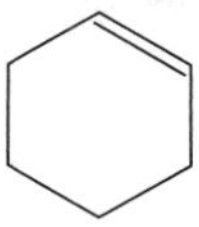

18. CH_3—CH_2—$\overset{\displaystyle CH_3}{\underset{|}{CH}}$—$CH_2$—$CH_3$

19. CH_3—CH_2—C≡CH

For problems 20 through 24, match the name of each of the following aromatic compounds with the correct structure.

A. **B.** **C.**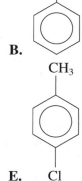

D. 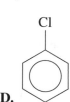 **E.**

20. _____ chlorobenzene

21. _____ benzene

22. _____ toluene

23. _____ 4-chlorotoluene

24. _____ 3-bromotoluene

Answers to the Practice Test

1. C	**2.** B	**3.** D	**4.** B	**5.** D
6. B	**7.** A	**8.** C	**9.** D	**10.** C
11. B	**12.** A	**13.** C	**14.** E	**15.** D
16. B	**17.** D	**18.** A	**19.** C	**20.** D
21. A	**22.** B	**23.** E	**24.** C	

Answers and Solutions to Selected Text Problems

11.1 **a.** An alkene has a double bond.
   b. An alkyne has a triple bond.
   c. A cycloalkene has a double bond in a ring.
   d. An alkene has a double bond.

11.3 Propene contains three carbon atoms, six hydrogen atoms, and a carbon–carbon double bond. Propyne contains three carbon atoms, four hydrogen atoms, and a carbon–carbon triple bond.

11.5 **a.** The two-carbon compound with a double bond is ethene.
   b. methylpropene
   c. 2-pentyne
   d. This four-carbon cyclic structure with a double bond is cyclobutene.

11.7 **a.** Propene is the three-carbon alkene. $H_2C=CH-CH_3$
   b. 1-pentene is the five-carbon compound with a double bond between carbon 1 and carbon 2.

   $H_2C=CH-CH_2-CH_2-CH_3$

   c. 2-methyl-1-butene has a four-carbon chain with a double bond between carbon 1 and carbon 2 and a methyl attached to carbon 2.

   $$\begin{array}{c} CH_3 \\ | \\ H_2C=C-CH_2-CH_3 \end{array}$$

   d. Cyclohexene is a six-carbon cyclic compound with a double bond.

   e. 1-butyne is a four-carbon chain with a triple bond between carbon 1 and 2.

   $CH_3-CH_2-C\equiv CH$

   f. 1-bromo-3-butyne is a six-carbon chain with a bromine atom on carbon 1 and a triple bond between carbon 3 and 4. $Br-CH_2-CH_2-C\equiv C-CH_2-CH_3$

11.9 **a.** *cis*-2-butene. This is a four-carbon compound with a double bond between carbon 2 and carbon 3. Both methyl groups are on the same side of the double bond; it is cis.
   b. *trans*-3-octene. This compound has eight carbons with a double bond between carbon 3 and carbon 4. The alkyl groups are on opposite sides of the double bond; it is trans.
   c. *cis*-3-heptene. This is a seven-carbon compound with a double bond between carbon 3 and carbon 4. Both alkyl groups are on the same side of the double bond; it is cis.

186

11.11 **a.** *trans*-2-butene has a four-carbon chain with a double bond between carbon 2 and carbon 3. The trans isomer has the two methyl groups on opposite sides of the double bond.

$$CH_3 \diagdown \qquad \diagup H$$
$$C=C$$
$$H \diagup \qquad \diagdown CH_3$$

b. *cis*-2-pentene has a five-carbon chain with a double bond between carbon 2 and carbon 3. The cis isomer has the alkyl groups on the same side of the double bond.

$$CH_3 \diagdown \qquad \diagup CH_2—CH_3$$
$$C=C$$
$$H \diagup \qquad \diagdown H$$

c. *trans*-3-heptene has a seven-carbon chain with a double bond between carbon 3 and carbon 4. The trans isomer has the alkyl groups on opposite sides of the double bond.

$$CH_3—CH_2 \diagdown \qquad \diagup H$$
$$C=C$$
$$H \diagup \qquad \diagdown CH_2—CH_2—CH_3$$

11.13 **a.** $CH_3—CH_2—CH_2—CH_2—CH_3$ pentane

b. $CH_3—\overset{\overset{\displaystyle OH}{|}}{CH}—CH_2—CH_3$ 2-butanol

c. When water is added to an alkene, the product is an alcohol. The name is cyclobutanol.

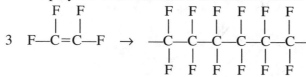

d. When H_2 is added to a cycloalkene the product is a cycloalkane. Cyclopentene would form cyclopentane.

$$\pentagon \quad + H_2 \overset{Pt}{\rightarrow} \quad \pentagon$$

cyclopentene   cyclopentane

e. When H_2 is added to an alkyne the product is an alkane. Pentyne would give:
$CH_3—CH_2—CH_2—CH_2—CH_3$ pentane

11.15 A polymer is a long-chain molecule consisting of many repeating smaller units. These smaller units are called monomers.

11.17 Teflon is a polymer of the monomer tetrafluoroethene.

$$3 \quad F—\overset{\overset{\displaystyle F}{|}}{C}=\overset{\overset{\displaystyle F}{|}}{C}—F \quad \rightarrow \quad —\overset{\overset{\displaystyle F}{|}}{\underset{\underset{\displaystyle F}{|}}{C}}—\overset{\overset{\displaystyle F}{|}}{\underset{\underset{\displaystyle F}{|}}{C}}—\overset{\overset{\displaystyle F}{|}}{\underset{\underset{\displaystyle F}{|}}{C}}—\overset{\overset{\displaystyle F}{|}}{\underset{\underset{\displaystyle F}{|}}{C}}—\overset{\overset{\displaystyle F}{|}}{\underset{\underset{\displaystyle F}{|}}{C}}—\overset{\overset{\displaystyle F}{|}}{\underset{\underset{\displaystyle F}{|}}{C}}—$$

11.19 Cyclohexane, C_6H_{12}, is a cycloalkane with 6 carbon atoms and 12 hydrogen atoms. The carbon atoms are connected in a ring by single bonds. Benzene, C_6H_6, is an aromatic compound with 6 carbon atoms and 6 hydrogen atoms. The carbon atoms are connected in a ring where the electrons are equally shared among the six carbon atoms.

11.21 **a.** 2-chlorotoluene
       b. ethylbenzene
       c. 1,3,5-trichlorobenzene
       d. 3-bromo-5-chlorotoluene

11.23 **a.**

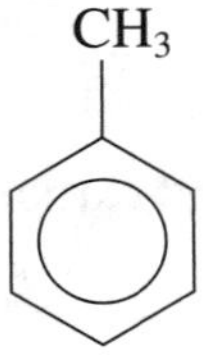

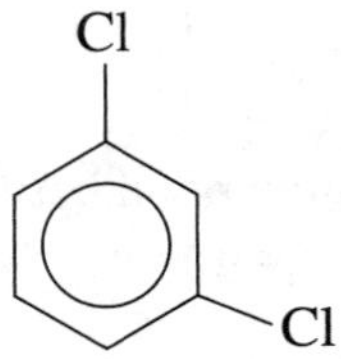

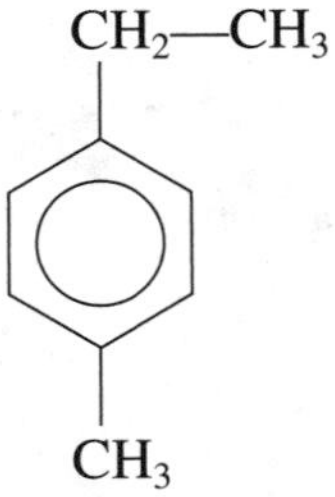

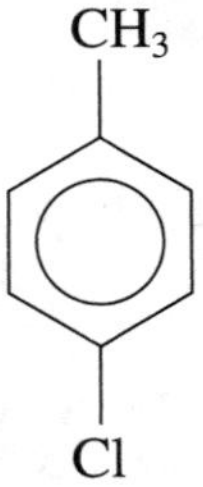

11.25

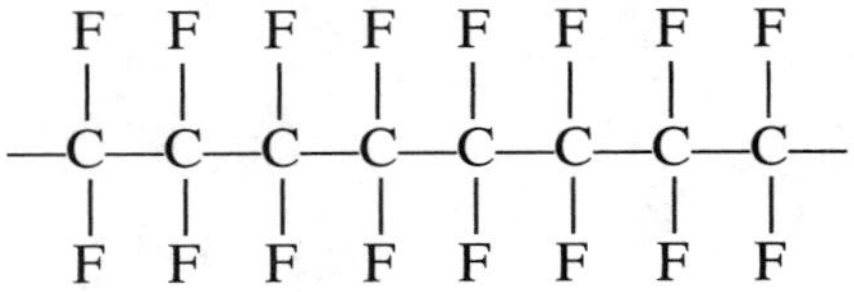

11.27 **a.**

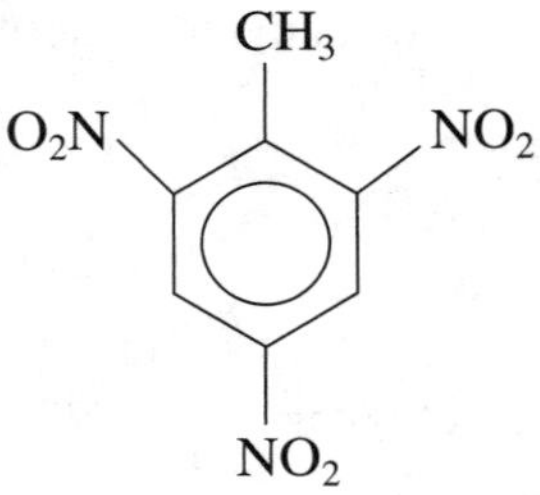

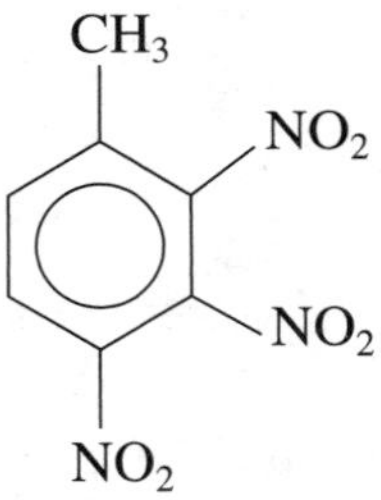

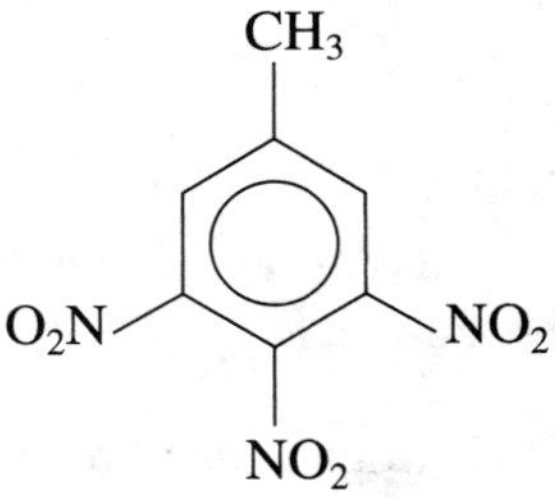

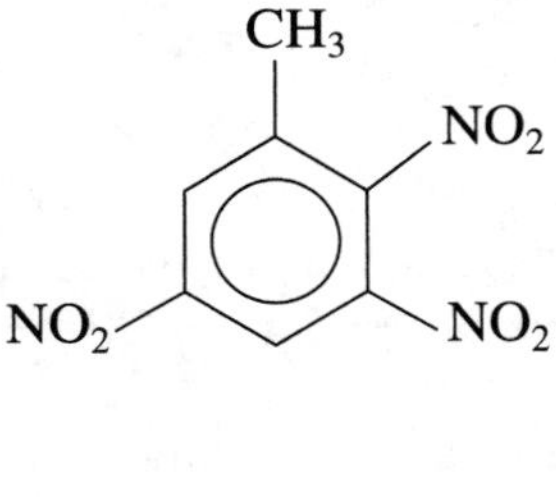

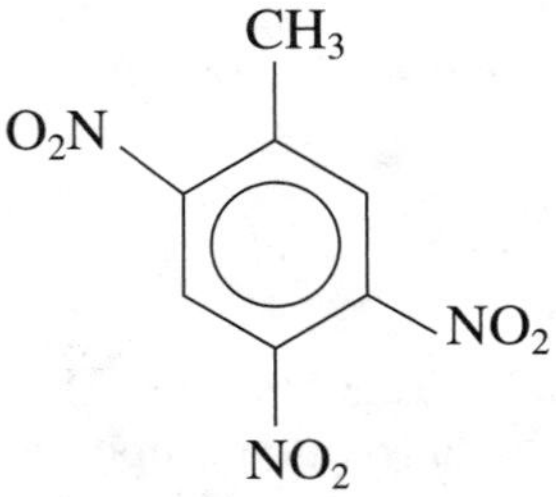

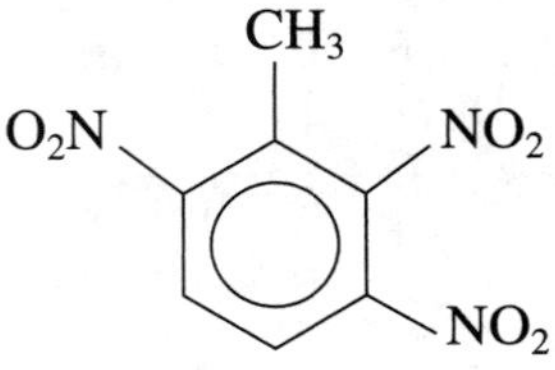

11.29 Propane is the three-carbon alkane with eight hydrogen atoms. All the carbon–carbon bonds in propane are single bonds. Cyclopropane is the three-carbon cycloalkane. All the carbon–carbon bonds in cyclopropane are single bonds. Propene is the three-carbon compound that has a carbon–carbon double bond. Cyclopropane and propene each has six hydrogen atoms. Propyne is the three-carbon compound with a carbon–carbon triple bond and four hydrogen atoms.

11.31 **a.** This compound contains a five-carbon chain with a double bond between carbon 1 and carbon 2 and a methyl group attached to carbon 2. The IUPAC name is 2-methyl-1-pentene.
b. This compound contains a five-carbon chain with a double bond between carbons 1 and 2 and a chlorine atom on carbon 5. The IUPAC name is 5-chloro-1-pentene.
c. This compound is a five-carbon cycloalkene. The IUPAC name is cyclopentene.

11.33 **a.** These structures represent a pair of isomers. In one isomer, the chlorine is attached to one of the carbons in the double bond; in the other isomer, the carbon bonded to the chlorine is not part of the double bond.
b. These structures are cis–trans isomers. In the cis isomer, the two methyl groups are on the same side of the double bond. In the trans isomer, the methyl groups are on opposite sides of the double bond.
c. These structures are identical and not isomers. Both have five-carbon chains with a double bond between carbon 1 and carbon 2.
d. These structures represent a pair of isomers. Both have the molecular formula C_7H_{16}. One isomer has a six-carbon chain with a methyl group attached; the other has a five-carbon chain with two methyl groups attached.

cis-2-pentene; both alkyl groups are on the same side of the

11.35 **a.** double bond

trans-2-pentene; both alkyl groups are on opposite sides of the double bond

b. *cis*-3-hexene; both alkyl groups are on the same side of the double bond

trans-3-hexene; both alkyl groups are on opposite sides of the double bond

11.37 The hydrogenation of unsaturated compounds adds hydrogen to form the saturated compounds (alkanes) with the same number of carbons and substituents.

a. butane   **b.** 3-methylpentane
c. cyclohexane   **d.** pentane

11.39 **a.** CH_3—CH_2—CH_2—CH_3

b. *cis*-2-butene; both alkyl groups are on the same side of the double bond

trans-2-butene; both alkyl groups are on opposite sides of the double bond

189

c.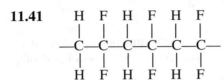

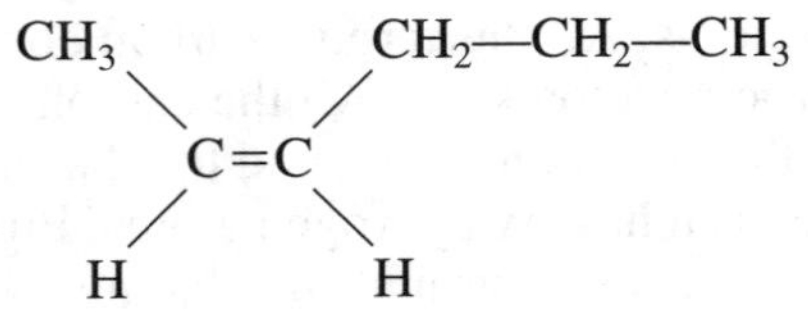

cis-2-hexene; both alkyl groups are on the same side of the double bond

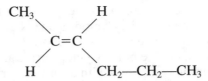

trans-2-hexene; both alkyl groups are on opposite sides of the double bond

11.41

$$\begin{array}{ccccccc}
 & H & F & H & F & H & F \\
 & | & | & | & | & | & | \\
-C & -C & -C & -C & -C & -C- \\
 & | & | & | & | & | & | \\
 & H & F & H & F & H & F
\end{array}$$

11.43 **a.** toluene
b. 2-chlorotoluene
c. 4-ethyltoluene
d. 1,3-diethylbenzene

11.45 2-butene CH_3—$CH=CH$—CH_3 C_4H_8

$$30.0 \text{ g } C_4H_8 \times \frac{1 \text{ mole } C_4H_8}{56.0 \text{ g } C_4H_8} \times \frac{1 \text{ mole } H_2}{1 \text{ mole } C_4H_8} \times \frac{2.0 \text{ g } H_2}{1 \text{ mole } H_2} = 1.07 \text{ g of } H_2$$

11.47 Bombykol $C_{16}H_{30}O = 238$ g/mole

$$50 \text{ ng} \times \frac{1 \text{ g}}{10^9 \text{ ng}} \times \frac{1 \text{ mole}}{238 \text{ g}} \times \frac{6.02 \times 10^{23} \text{ molecules}}{1 \text{ mole}} = 1 \times 10^{14} \text{ molecules}$$

Organic Compounds with Oxygen and Sulfur

Study Goals

- Classify alcohols as primary, secondary, or tertiary.
- Name and write the condensed structural formulas for alcohols, phenols, and thiols.
- Name and write the condensed structural formulas for ethers.
- Describe the solubility in water and boiling points of alcohols, phenols, and ethers.
- Write equations for combustion, dehydration, and oxidation of alcohols.
- Name and write the condensed structural formulas for aldehydes and ketones.
- Identify the chiral carbon atoms in organic molecules.
- Write equations for the oxidation of aldehydes.

Think About It

1. What are the functional groups of alcohols, phenols, ethers, and thiols?

2. Phenol is sometimes used in mouthwashes. Why does it form a solution with water?

3. What reaction of ethanol takes place when you make a fondue dish or a flambé dessert?

4. What are the functional groups of an aldehyde and ketone?

5. How is an alcohol changed to an aldehyde or ketone?

Key Terms

Match the following terms with the statements shown below.

a. primary alcohol	**b.** thiol	**c.** ether	**d.** phenol
e. tertiary alcohol	**f.** chiral carbon	**g.** aldehyde	**h.** ketone

1. _____ An organic compound with one alkyl group bonded to the carbon with the —OH group

2. _____ An organic compound that contains an —SH group

3. _____ An organic compound that contains an oxygen atom —O— attached to two alkyl groups

4. _____ An organic compound with three alkyl groups bonded to the carbon with the —OH group

5. _____ An organic compound that contains a benzene ring bonded to a hydroxyl group

6. _____ An organic compound with a carbonyl group attached to two alkyl groups

7. _____ A carbon that is bonded to four different groups

8. _____ An organic compound that contains a carbonyl group and a hydrogen atom at the end of the carbon chain

Answers   **1.** a   **2.** b   **3.** c   **4.** e   **5.** d   **6.** h   **7.** f   **8.** g

12.1 Alcohols, Thiols, and Ethers

- Alcohols are classified according to the number of alkyl groups attached to the carbon bonded to the —OH group.
- In a primary alcohol, there is one alkyl group attached to the carbon atom bonded to the —OH. In a secondary alcohol, there are two alkyl groups, and in a tertiary alcohol there are three alkyl groups attached to the carbon atom with the —OH functional group.
- An alcohol contains the hydroxyl group —OH attached to a carbon chain.
- In the IUPAC system alcohols are named by replacing the *ane* of the alkane name with *ol*. The location of the —OH group is given by numbering the carbon chain. Simple alcohols are generally named by their common names with the alkyl name preceding the term alcohol. For example, CH_3—OH is methyl alcohol, and CH_3—CH_2—OH is ethyl alcohol.

CH_3—OH	CH_3—CH_2—OH	CH_3—CH_2—CH_2—OH
methanol	ethanol	1-propanol
(methyl alcohol)	(ethyl alcohol)	(propyl alcohol)

- When a hydroxyl group is attached to a benzene ring, the compound is a phenol.
- Thiols are similar to alcohols, except they have an —SH functional group in place of the —OH group.
- In ethers, an oxygen atom is connected by single bonds to two alkyl or aromatic groups.
- In the common names of ethers, the alkyl groups are listed alphabetically followed by the name *ether*.

Study Note

Example: Identify the following as primary, secondary, or tertiary alcohols.
Solution: Determine the number of alkyl groups attached to the hydroxyl carbon atom.

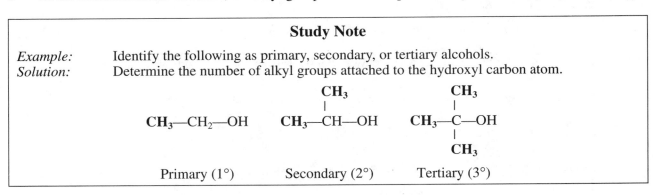

◆ Learning Exercise 12.1A

Give the correct IUPAC and common name (if any) for each of the following compounds.

1. CH_3—CH_2—OH

2. CH_3—CH_2—CH_2—OH

3. CH_3—$\overset{\underset{|}{OH}}{CH}$—$CH_2$—$CH_2$—$CH_3$

4. CH_3—CH_2—$\overset{\underset{|}{CH_3}}{CH}$—$\overset{\underset{|}{OH}}{CH}$—$CH_3$

5.

6.

Answers
1. ethanol (ethyl alcohol)
2. 1-propanol (propyl alcohol)
3. 2-pentanol
4. 3-methyl-2-pentanol
5. cyclopentanol
6. phenol

◆ **Learning Exercise 12.1B**

Write the correct condensed structural formula for each of the following compounds.
1. 2-butanol   **2.** 2-chloro-1-propanol

3. 2,4-dimethyl-1-pentanol   **4.** cyclohexanol

5. 3-methylcyclopentanol   **6.** 2-chlorophenol

Answers

1.
$$CH_3\!-\!\underset{\underset{OH}{|}}{CH}\!-\!CH_2\!-\!CH_3$$

2.
$$CH_3\!-\!\underset{\underset{Cl}{|}}{CH}\!-\!CH_2\!-\!OH$$

3.
$$CH_3\!-\!\underset{\underset{CH_3}{|}}{CH}\!-\!CH_2\!-\!\underset{\underset{CH_3}{|}}{CH}\!-\!CH_2\!-\!OH$$

4. 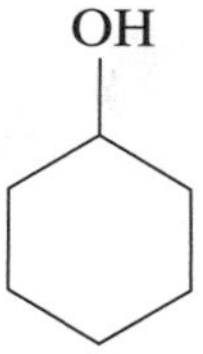 (cyclohexane ring with OH)

5. 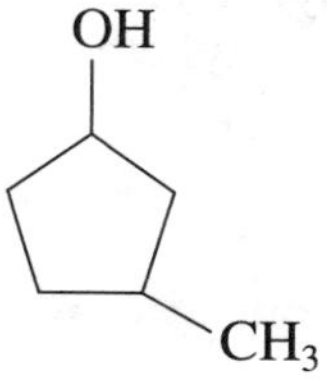 (cyclopentane ring with OH and CH$_3$)

6. 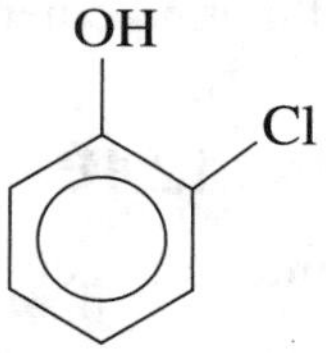 (benzene ring with OH and Cl)

◆ **Learning Exercise 12.1C**

Classify each of the following alcohols as primary (1°), secondary (2°), or tertiary (3°).

1. _____ $CH_3\!-\!CH_2\!-\!OH$

2. _____
$$CH_3\!-\!CH_2\!-\!\underset{\overset{OH}{|}}{CH}\!-\!CH_3$$

3. _____
$$CH_3\!-\!\underset{\overset{\overset{OH}{|}}{\underset{CH_3}{|}}}{C}\!-\!CH_2\!-\!CH_3$$

4. _____
$$CH_3\!-\!\underset{\overset{\overset{OH}{|}}{\underset{CH_3}{|}}}{C}\!-\!CH_2\!-\!CH_2\!-\!CH_3$$

5. _____
$$CH_3\!-\!\underset{\overset{\overset{CH_3}{|}}{\underset{CH_3}{|}}}{C}\!-\!CH_2\!-\!OH$$

6. _____ (cyclopentane ring with OH)

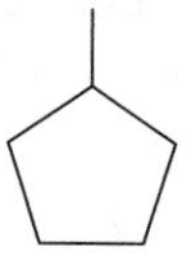

Answers   **1.** primary (1°)   **2.** secondary (2°)   **3.** tertiary (3°)
            4. tertiary (3°)   **5.** primary (1°)   **6.** secondary (2°)

Study Note

Example: Write the common name for $CH_3—CH_2—O—CH_3$.
Solution: The common name lists the alkyl groups alphabetically before the name *ether*.

   ethyl group   *methyl group*
   $CH_3—CH_2—O—CH_3$
   Common: ethyl methyl ether

◆ Learning Exercise 12.1D

Write the common name for each of the following ethers.

1. $CH_3—O—CH_3$

2. $CH_3—CH_2—O—CH_2—CH_3$

3. $CH_3—CH_2—CH_2—CH_2—O—CH_3$

4. $CH_3—O—CH_2—CH_3$

Answers   1. (di)methyl ether
             3. butyl methyl ether

             2. (di)ethyl ether
             4. ethyl methyl ether

◆ Learning Exercise 12.1E

Write the structural formula for each of the following ethers.

1. ethyl propyl ether

2. ethyl methyl ether

Answers   1. $CH_3—CH_2—O—CH_2—CH_2—CH_3$ 2. $CH_3—O—CH_2—CH_3$

12.2 Properties of Alcohols and Ethers

- The polar —OH group gives alcohols higher boiling points than alkanes and ethers of similar mass.
- Alcohols with one to four carbons are soluble in water because the —OH group forms hydrogen bonds with water molecules.
- Phenol is soluble in water and acts as a weak acid.
- Because ethers are less polar than alcohols, they have boiling points similar to alkanes. Ethers are soluble in water due to hydrogen bonding. Ethers are widely used as solvents but can be dangerous to use because their vapors are highly flammable.

◆ **Learning Exercise 12.2A**

Select the compound in each pair that is the more soluble in water.

1. CH_3—CH_3 or CH_3—CH_2—OH

2. CH_3—CH_2—CH_2—OH or CH_3—CH_2—CH_2—CH_2—CH_2—OH

3. CH_3—CH_2—CH_2—CH_3 or CH_3—CH_2—CH_2—CH_2—OH

4. Benzene or phenol

Answers   1. CH_3—CH_2—OH 2. CH_3—CH_2—CH_2—OH
           3. CH_3—CH_2—CH_2—CH_2—OH 4. phenol

◆ **Learning Exercise 12.2B**

Select the compound in each pair with the higher boiling point.

1. CH_3—CH_2—CH_3 or CH_3—CH_2—OH

2. CH_3—O—CH_2—CH_3 or CH_3—CH_2—CH_2—OH

3. CH_3—CH_2—CH_2—OH or CH_3—CH_2—CH_2—CH_3

Answers   1. CH_3—CH_2—OH 2. CH_3—CH_2—CH_2—OH
           3. CH_3—CH_2—CH_2—OH

12.3 Reactions of Alcohols and Thiols

• At high temperatures, an alcohol dehydrates in the presence of an acid to yield an alkene and water.

$$CH_3\text{—}CH_2\text{—}OH \xrightarrow[\text{Heat}]{H^+} H_2C{=}CH_2 + H_2O$$

• Using an oxidizing agent [O], primary alcohols oxidize to aldehydes, which usually oxidize further to carboxylic acids. Secondary alcohols are oxidized to ketones, but tertiary alcohols do not oxidize.

$$\underset{\text{1° alcohol}}{CH_3\text{—}CH_2\text{—}OH} \xrightarrow{[O]} \underset{\text{aldehyde}}{CH_3\overset{\overset{\displaystyle O}{\|}}{\text{—}C\text{—}}H} + H_2O$$

$$\underset{\text{2° alcohol}}{CH_3\overset{\overset{\displaystyle OH}{|}}{\text{—}CH\text{—}}CH_3} \xrightarrow{[O]} \underset{\text{ketone}}{CH_3\overset{\overset{\displaystyle O}{\|}}{\text{—}C\text{—}}CH_3} + H_2O$$

◆ **Learning Exercise 12.3A**

Write the condensed structural formulas of the products expected from dehydration of each of the following reactants:

1. CH_3—CH_2—CH_2—CH_2—OH $\xrightarrow{\text{H}^+,\ \text{heat}}$

2. $\xrightarrow{\text{H}^+,\ \text{heat}}$

3. CH_3—$\underset{\underset{\displaystyle OH}{|}}{CH}$—$CH_3$ $\xrightarrow{\text{H}^+,\ \text{heat}}$

4. CH_3—CH_2—$\underset{\underset{\displaystyle OH}{|}}{CH}$—$CH_2$—$CH_3$ $\xrightarrow{\text{H}^+,\ \text{heat}}$

Answers

1. CH_3—CH_2—CH=CH_2

2.

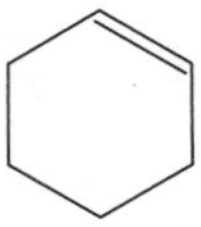

3. CH_3—CH=CH_2

4. CH_3—CH_2—CH=CH—CH_3

◆ **Learning Exercise 12.3B**

Write the condensed structural formulas of the products expected in the oxidation reaction of each of the following reactants.

1. CH_3—CH_2—CH_2—CH_2—OH $\xrightarrow{\text{[O]}}$

2. 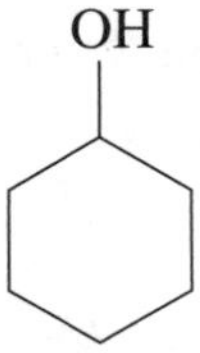 $\xrightarrow{\text{[O]}}$

3. CH_3—$\underset{\underset{\displaystyle OH}{|}}{CH}$—$CH_3$ $\xrightarrow{\text{[O]}}$

4. CH_3—CH_2—$\underset{\underset{\displaystyle OH}{|}}{CH}$—$CH_2$—$CH_3$ $\xrightarrow{\text{[O]}}$

Answers

1. CH_3—CH_2—CH_2—$\overset{\overset{\displaystyle O}{\|}}{C}$—H

2.

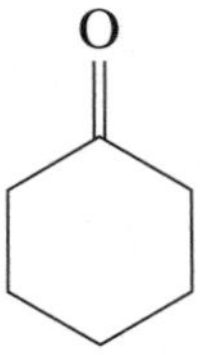

3. CH_3—$\overset{\overset{\displaystyle O}{\|}}{C}$—$CH_3$

4. CH_3—CH_2—$\overset{\overset{\displaystyle O}{\|}}{C}$—$CH_2$—$CH_3$

12.4 Aldehydes and Ketones

- In an aldehyde, the carbonyl group appears at the end of a carbon chain attached to at least one hydrogen atom.
- In a ketone, the carbonyl group occurs between carbon groups and has no hydrogens attached to it.
- In the IUPAC system, aldehydes and ketones are named by replacing the *e* in the longest chain containing the carbonyl group with *al* for aldehydes, and *one* for ketones. The location of the carbonyl group in a ketone is given if there are more than four carbon atoms in the chain.

$$
\begin{array}{cc}
\quad O & \quad O \\
\quad \| & \quad \| \\
CH_3-C-H & CH_3-C-CH_3 \\
\text{ethanal} & \text{propanone} \\
\text{(acetaldehyde)} & \text{(dimethyl ketone)}
\end{array}
$$

◆ Learning Exercise 12.4A

Classify each of the following compounds.

    a. alcohol   **b.** aldehyde   **c.** ketone   **d.** ether   **e.** thiol

$$
\begin{array}{l}
\qquad\qquad O \\
\qquad\qquad \| \\
\textbf{1.} \underline{\quad\quad} CH_3-CH_2-CH_2-C-H
\end{array}
\qquad
\textbf{2.} \underline{\quad\quad} CH_3-CH_2-CH_2-OH
$$

$$
\begin{array}{l}
\qquad\qquad O \\
\qquad\qquad \| \\
\textbf{3.} \underline{\quad\quad} CH_3-CH_2-C-CH_2-CH_3
\end{array}
\qquad
\textbf{4.} \underline{\quad\quad} CH_3-CH_2-O-CH_3
$$

$$
\begin{array}{l}
\qquad\qquad O \\
\qquad\qquad \| \\
\textbf{5.} \underline{\quad\quad} CH_3-C-CH_2-CH_3
\end{array}
\qquad
\begin{array}{l}
\qquad\qquad O \\
\qquad\qquad \| \\
\textbf{6.} \underline{\quad\quad} CH_3-C-H
\end{array}
$$

$$
\begin{array}{l}
\qquad\qquad SH \\
\qquad\qquad | \\
\textbf{7.} \underline{\quad\quad} CH_3-CH_2-CH-CH_3
\end{array}
\qquad
\begin{array}{l}
\qquad\qquad OH \\
\qquad\qquad | \\
\textbf{8.} \underline{\quad\quad} CH_3-CH_2-CH-CH_3
\end{array}
$$

Answers   **1.** b   **2.** a   **3.** c   **4.** d
                    5. c   **6.** b   **7.** e   **8.** a

◆ Learning Exercise 12.4B

Write the correct IUPAC (or common name) for the following aldehydes.

$$
\begin{array}{l}
\qquad O \\
\qquad \| \\
\textbf{1.}\ CH_3-C-H
\end{array}
\qquad\qquad
\begin{array}{l}
\qquad\qquad\qquad\qquad O \\
\qquad\qquad\qquad\qquad \| \\
\textbf{2.}\ CH_3-CH_2-CH_2-CH_2-C-H
\end{array}
$$

$$
\begin{array}{l}
\qquad\qquad CH_3 \qquad O \\
\qquad\qquad | \qquad\quad \| \\
\textbf{3.}\ CH_3-CH_2-CH-CH_2-CH_2-C-H
\end{array}
\qquad
\begin{array}{l}
\qquad\qquad\qquad\qquad O \\
\qquad\qquad\qquad\qquad \| \\
\textbf{4.}\ CH_3-CH_2-CH_2-C-H
\end{array}
$$

$$
\begin{array}{l}
\qquad O \\
\qquad \| \\
\textbf{5.}\ H-C-H
\end{array}
$$

Answers   **1.** ethanal; acetaldehyde   **2.** pentanal   **3.** 4-methylhexanal
             4. butanal; butyraldehyde   **5.** methanal; formaldehyde

◆ **Learning Exercise 12.4C**

Write the IUPAC (or common name) for the following ketones.

1. CH$_3$—C(=O)—CH$_3$

2. CH$_3$—CH$_2$—CH$_2$—C(=O)—CH$_3$

3. CH$_3$—CH$_2$—C(=O)—CH$_2$—CH$_3$

4. 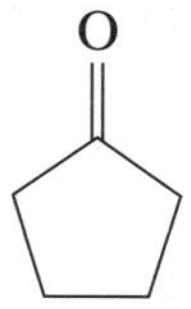 (cyclopentanone structure)

5. 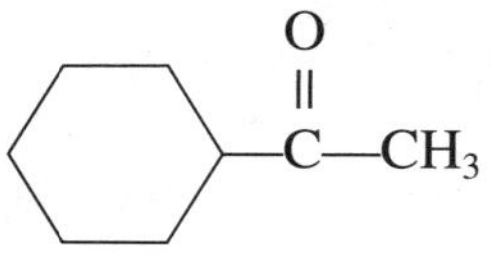 (cyclohexyl—C(=O)—CH$_3$ structure)

Answers
1. propanone; dimethyl ketone, acetone
2. 2-pentanone; methyl propyl ketone
3. 3-pentanone; diethyl ketone
4. cyclopentanone
5. cyclohexyl methyl ketone

◆ **Learning Exercise 12.4D**

Write the correct condensed formulas for the following.

1. ethanal

2. 2-methylbutanal

3. 2-chloropropanal

4. ethylmethylketone

5. 3-hexanone

6. benzaldehyde

Answers

1. CH$_3$—C(=O)—H

2. CH$_3$—CH$_2$—CH(CH$_3$)—C(=O)—H

3. CH$_3$—CH(Cl)—C(=O)—H

4. CH$_3$—CH$_2$—C(=O)—CH$_3$

5. CH$_3$—CH$_2$—C(=O)—CH$_2$—CH$_2$—CH$_3$

6. 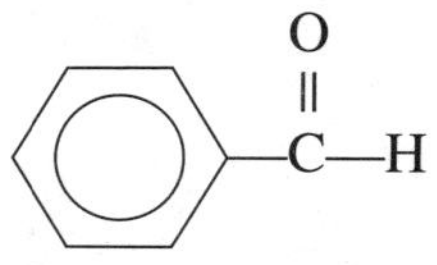 (benzene ring)—C(=O)—H

198

12.5 Properties of Aldehydes and Ketones

- The polarity of the carbonyl group makes aldehydes and ketones of one to four carbon atoms soluble in water.
- Using an oxidizing agent, aldehydes oxidize to carboxylic acids. Ketones do not oxidize further.

$$CH_3\!-\!\overset{\displaystyle O}{\overset{\|}{C}}\!-\!H \xrightarrow{[O]} CH_3\!-\!\overset{\displaystyle O}{\overset{\|}{C}}\!-\!OH$$

aldehyde   *carboxylic acid*

$$CH_3\!-\!\overset{\displaystyle O}{\overset{\|}{C}}\!-\!CH_3 \xrightarrow{[O]} \text{No reaction}$$

ketone

◆ Learning Exercise 12.5A

Indicate whether each of the following compounds is soluble (S) or not soluble (NS) in water.

. **1.** _____ 3-hexanone   **2.** _____ propanal   **3.** _____ acetaldehyde

4. _____ butanal   **5.** _____ cyclohexanone

Answers   **1.** NS   **2.** S   **3.** S   **4.** S   **5.** NS

◆ Learning Exercise 12.5B

Indicate the compound in each of the following pairs that will oxidize.

1. propanal or propanone   _____

2. butane or butanal   _____

3. ethane or acetaldehyde   _____

Answers   **1.** propanal   **2.** butanal   **3.** acetaldehyde

12.6 Chiral Molecules

- Chiral molecules have mirror images that cannot be superimposed.
- In a chiral molecule, there is one or more carbon atoms attached to four different atoms or groups.
- The mirror images of a chiral molecule represent two different molecules called enantiomers.
- In a Fischer projection (straight chain), the prefixes D- and L- are used to distinguish between the mirror images. In D-glyceraldehyde, the —OH is on the right of the chiral carbon; it is on the left in L-glyceraldehyde.

◆ Learning Exercise 12.6A

Indicate whether the following objects would be chiral or not.

1. _____ a piece of plain computer paper   **4.** _____ a volleyball net

2. _____ weight-lifting glove   **5.** _____ your left foot

3. _____ a baseball cap

Answers   **1.** not chiral   **2.** chiral   **3.** not chiral   **4.** not chiral   **5.** chiral

◆ Learning Exercise 12.6B

State whether each of the following molecules is chiral or not chiral.

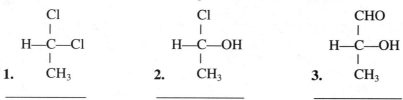

Answers   **1.** not chiral   **2.** chiral   **3.** chiral

◆ Learning Exercise 12.6C

Identify the following as a feature that is characteristic of a chiral compound or not.

   1. The central atom is attached to two identical groups   _____

   2. Contains a carbon attached to four different groups   _____

   3. Has identical mirror images   _____

Answers   **1.** not chiral   **2.** chiral   **3.** not chiral

◆ Learning Exercise 12.6D

Indicate whether each pair of Fischer projections represents enantiomers (E) or identical structures (I).

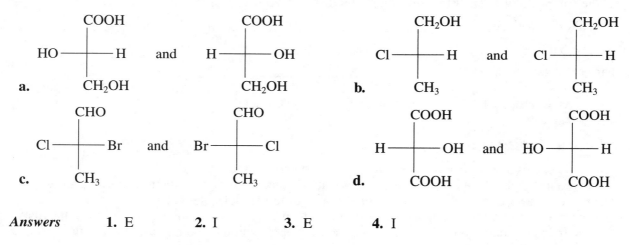

Answers   **1.** E   **2.** I   **3.** E   **4.** I

Check List for Chapter 12

You are ready to take the practice test for Chapter 12. Be sure that you have accomplished the following learning goals for this chapter. If you are not sure, review the section listed at the end of the goal. Then apply your new skills and understanding to the practice test.
After studying Chapter 12, I can successfully:

_____ Give the IUPAC or common name of an alcohol and classify an alcohol as primary, secondary, or tertiary (12.1).

_____ Describe the solubility of alcohols, phenols, and ethers in water (12.2).

_____ Write the products of alcohols that undergo dehydration or oxidation (12.3).

_____ Write the structural formulas for reactants and products of the oxidation of alcohols (12.3).

_____ Identify structural formulas as aldehydes and ketones (12.4).

_____ Give the IUPAC and common names of an aldehyde or ketone; draw the condensed structural formula from the name (12.4).

_____ Compare the physical properties of aldehydes and ketones with alcohols and alkanes (12.5).

_____ Identify a molecule as a chiral or not: write the D- and L-Fischer projections (12.6).

Practice Test for Chapter 12

Match the names of the following compounds with structures 1 to 5.

1. 1-propanol   **2.** cyclobutanol   **3.** 2-propanol   **4.** ethyl methyl ether   **5.** diethyl ether

1.
$$\begin{array}{c} OH \\ | \\ CH_3-CH-CH_3 \end{array}$$

2. $CH_3-CH_2-CH_2-OH$

3. $CH_3-O-CH_2-CH_3$

4.

5. $CH_3-CH_2-O-CH_2-CH_3$

6. The compound $CH_3-\overset{\overset{\displaystyle O}{\|}}{C}-CH_3$ is formed by the oxidation of

  A. 2-propanol.   **B.** propane.   **C.** 1-propanol.
  D. dimethyl ether.   **E.** methyl ethyl ketone.

7. Why are short-chain alcohols water soluble?

  A. They are nonpolar.   **B.** They can hydrogen bond.   **C.** They are organic.
  D. They are bases.   **E.** They are acids.

8. Phenol is

  A. the alcohol of benzene.   **B.** the aldehyde of benzene.   **C.** the phenyl group of benzene.
  D. the ketone of benzene.   **E.** cyclohexanol.

9. $CH_3-CH_2-OH \xrightarrow{[O]}$

  A. an alkane   **B.** an aldehyde   **C.** a ketone   **D.** an ether   **E.** a phenol

10. The dehydration of cyclohexanol gives

  A. cyclohexane.   **B.** cyclohexene.   **C.** cyclohexyne.   **D.** benzene.   **E.** phenol.

11. The formula of the thiol of ethane is

  A. CH_3-SH.   **B.** CH_3-CH_2-OH.   **C.** CH_3-CH_2-SH.
  D. $CH_3-CH_2-S-CH_3$.   **E.** CH_3-S-OH.

In questions 12 through 16, classify each alcohol as

  A. primary (1°).   **B.** secondary (2°).   **C.** tertiary (3°).

12. $CH_3-CH_2-CH_2-OH$

13. 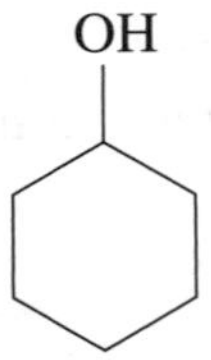

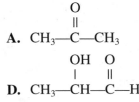

14. 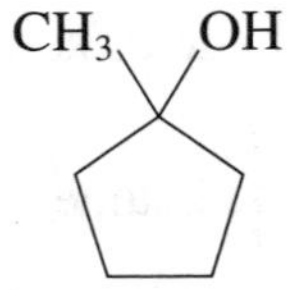

15. $CH_3-\overset{OH}{\underset{CH_3}{\overset{|}{\underset{|}{C}}}}-CH_2-CH_2-CH_3$

16. $CH_3-\overset{OH}{\overset{|}{C}H}-CH_2-CH_2-CH_2-CH_3$

Complete questions 17 through 20 by indicating the products (A–E) formed in each of the following reactions.

A. primary alcohol **B.** secondary alcohol **C.** aldehyde **D.** ketone **E.** carboxylic acid

17. _____ Oxidation of a primary alcohol

18. _____ Oxidation of a secondary alcohol

19. _____ Oxidation of an aldehyde

20. _____ Hydration of ethene

In 21–25, match the following compounds with the names given.

A. dimethyl ether **B.** acetaldehyde **C.** methanal **D.** dimethyl ketone **E.** propanal

21. _____ $H-\overset{O}{\overset{\|}{C}}-H$

22. _____ CH_3-O-CH_3

23. _____ $CH_3-\overset{O}{\overset{\|}{C}}-CH_3$

24. _____ $CH_3-\overset{O}{\overset{\|}{C}}-H$

25. _____ $CH_3-CH_2-\overset{O}{\overset{\|}{C}}-H$

26. The compound with the higher boiling point is

A. $CH_3-CH_2-CH_2-CH_3$

B. $CH_3-CH_2-CH_2-OH$

C. $CH_3-\overset{O}{\overset{\|}{C}}-CH_3$

D. $CH_3-CH_2-\overset{O}{\overset{\|}{C}}-H$

E. $CH_3-CH_2-O-CH_3$

27. Benedict's reagent will oxidize

A. $CH_3-\overset{O}{\overset{\|}{C}}-CH_3$

B. $CH_3-\overset{OH}{\overset{|}{C}H}-CH_2-OH$

C. $CH_3-\overset{O}{\overset{\|}{C}}-CH_2-OH$

D. $CH_3-\overset{OH}{\overset{|}{C}H}-\overset{O}{\overset{\|}{C}}-H$

E. $CH_3-\overset{OH}{\overset{|}{C}H}-\overset{O}{\overset{\|}{C}}-OH$

28. In the Tollens' test
    A. an aldehyde is oxidized and Ag^+ reduced.
    B. an aldehyde is reduced and Ag^+ oxidized.
    C. a ketone is oxidized and Ag^+ reduced.
    D. a ketone is reduced and Ag^+ oxidized.
    E. All of these

Identify each of the following structural formulas as

29. alcohol   **30.** ether

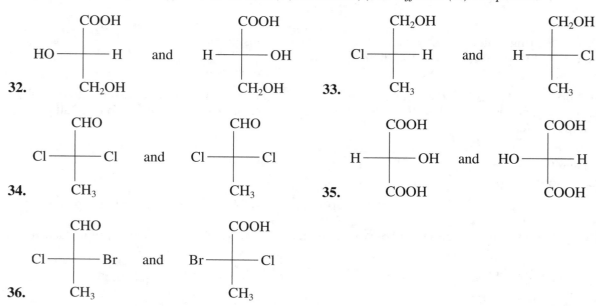

31. The structural formula for 4-bromo-3-methylcyclohexanone is

For questions 32–36, identify as enantiomers (E), identical (I), or different (D) compounds.

Answers to Practice Test

1. C		**2.** A		**3.** D		**4.** B		**5.** E
6. A		**7.** B		**8.** A		**9.** B		**10.** B
11. C		**12.** A		**13.** B		**14.** C		**15.** C
16. B		**17.** C, E		**18.** D		**19.** E		**20.** A
21. C		**22.** A		**23.** D		**24.** B		**25.** E
26. B		**27.** D		**28.** A		**29.** A		**30.** B
31. B		**32.** E		**33.** E		**34.** I		**35.** I
36. D								

Answers and Solutions to Selected Text Problems

12.1   **a.** This compound has a two-carbon chain (ethane). The final –e is dropped and –ol added to indicate an alcohol. The IUPAC name is ethanol.

b. This compound has a four-carbon chain with a hydroxyl attached to carbon 2. The IUPAC name is 2-butanol.

c. This compound has a five-carbon chain with a hydroxyl attached to carbon 2. The IUPAC name is 2-pentanol.

d. This compound is a six-carbon cycloalkane with a hydroxyl attached to carbon 1 and a methyl group attached to carbon 4. Since the hydroxyl is always attached to carbon 1, the number 1 is omitted in the name. The IUPAC name is 4-methylcyclohexanol.

e. This compound has a three-carbon chain with a thiol attached to carbon 1. The IUPAC name is 2-butanol.

12.3   **a.** 1-propanol has a three-carbon chain with a hydroxyl attached to carbon 1.
CH_3—CH_2—CH_2—OH

b. Methyl alcohol has a hydroxyl attached to a one-carbon alkane. CH_3—OH.

c. 3-pentanol has a five-carbon chain with a hydroxyl attached to carbon 3.

$$\begin{array}{c} \text{OH} \\ | \\ CH_3\text{—}CH_2\text{—}CH\text{—}CH_2\text{—}CH_3 \end{array}$$

d. 2-methyl-2-butanol has a four-carbon chain with a methyl and hydroxyl attached to carbon 2.

$$\begin{array}{c} \text{OH} \\ | \\ CH_3\text{—}C\text{—}CH_2\text{—}CH_3 \\ | \\ CH_3 \end{array}$$

12.5   A benzene ring with a hydroxyl group is called *phenol*. Substituents are numbered from the carbon bonded to the hydroxyl group as carbon 1.
a. phenol   **b.** 2-bromophenol   **c.** 3-bromophenol

12.7   In a primary (1°) alcohol, the carbon bonded to the hydroxyl group (—OH) is attached to one alkyl group (except for methanol); to two alkyl groups in a secondary alcohol (2°); and to three alkyl groups in a tertiary alcohol (3°).
a. 1°   **b.** 1°   **c.** 3°   **d.** 3°

12.9   **a.** ethyl methyl ether   **b.** dipropyl ether
c. cyclohexyl methyl ether   **d.** methyl propyl ether

12.11   **a.** Ethyl propyl ether has a two-carbon group and a three-carbon group attached to oxygen by single bonds. CH_3—CH_2—O—CH_2—CH_2—CH_3

b. Cyclopropyl ethyl ether has a two-carbon group and a three-carbon cycloalkyl group attached to oxygen by single bonds.

CH_3—CH_2—O—▷

c. CH_3—CH_2—O—CH_3

12.13 **a.** methanol; hydrogen bonding of alcohols gives higher boiling points than alkanes.
       b. 1-butanol; alcohols hydrogen bond, but ethers cannot.
       c. 1-butanol; hydrogen bonding of alcohols gives higher boiling points than alkanes.

12.15 **a.** yes; alcohols with 1–4 carbon atoms hydrogen bond with water
       b. yes; the water can hydrogen bond to the O in ether
       c. no; a carbon chain longer than 4 carbon atoms diminishes the effect of the —OH group
       d. no; alkanes are nonpolar and do not hydrogen bond

12.17 Dehydration is the removal of an —OH and a —H from adjacent carbon atoms.

       a. CH_3—CH_2—CH=CH_2 **b.**
       c. CH_3—CH_2—CH=CH—CH_3

12.19 **a.** A primary alcohol oxidizes to an aldehyde and then to a carboxylic acid.

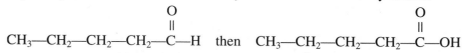

       b. A secondary alcohol oxidizes to a ketone.

       c. A secondary alcohol oxidizes to a ketone.

       d. A secondary alcohol oxidizes to a ketone.

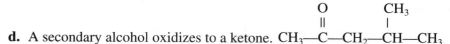

12.21 **a.** ketone   **b.** ketone   **c.** aldehyde

12.23 **a.** acetaldehyde   **b.** methyl propyl ketone   **c.** formaldehyde

12.25 **a.** propanal   **b.** 2-methyl-3-pentanone
       c. 3-methylcyclohexanone   **d.** benzaldehyde

12.27

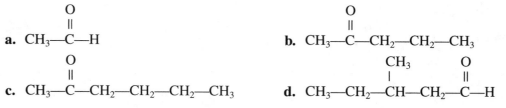

12.29 **a.** CH_3—CH_2—$\overset{\overset{\textstyle O}{\|}}{C}$—H has a polar carbonyl group.
       b. Pentanal has more carbons and thus a higher molar mass.
       c. 1-Butanol can hydrogen bond with other 1-butanol molecules.

12.31 **a.** CH_3—$\overset{\overset{\textstyle O}{\|}}{C}$—$\overset{\overset{\textstyle O}{\|}}{C}$—$CH_2$—$CH_3$: more hydrogen bonding
       b. acetone; lower number of carbon atoms

12.33 a. A primary alcohol will be oxidized to an aldehyde and then to a carboxylic acid.

$$CH_3-CH_2-CH_2-CH_2-\overset{\overset{\displaystyle O}{\|}}{C}-H \quad then \quad CH_3-CH_2-CH_2-CH_2-\overset{\overset{\displaystyle O}{\|}}{C}-OH$$

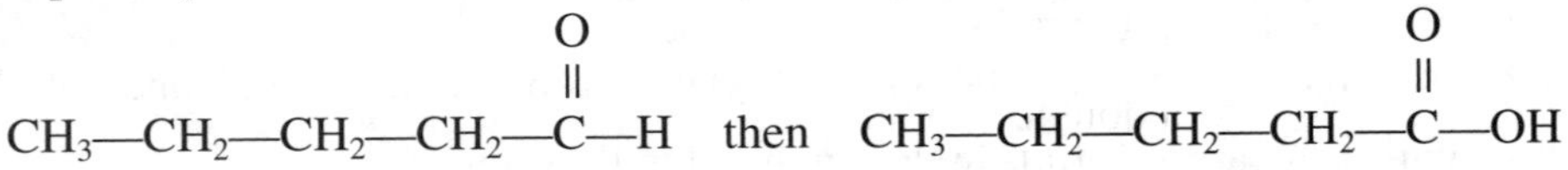

b. A secondary alcohol will be oxidized to a ketone. $CH_3-CH_2-\overset{\overset{\displaystyle O}{\|}}{C}-CH_3$

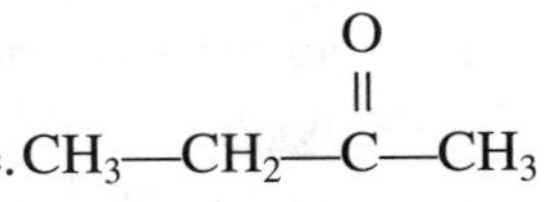

c. A secondary alcohol will be oxidized to a ketone.

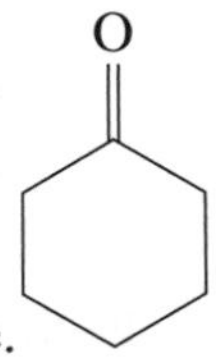

d. A secondary alcohol will be oxidized to a ketone. $CH_3-\overset{\overset{\displaystyle O}{\|}}{C}-CH_2-\overset{\overset{\displaystyle CH_3}{|}}{CH}-CH_3$

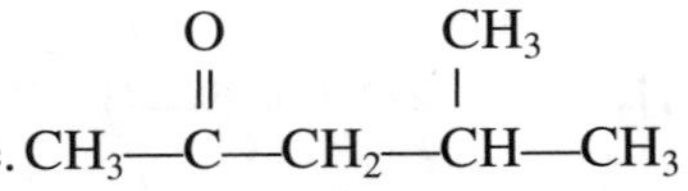

e. A primary alcohol will be oxidized to an aldehyde and then to a carboxylic acid.

$$CH_3-\overset{\overset{\displaystyle CH_3}{|}}{CH}-CH_2-\overset{\overset{\displaystyle O}{\|}}{C}-H \quad then \quad CH_3-\overset{\overset{\displaystyle CH_3}{|}}{CH}-CH_2-\overset{\overset{\displaystyle O}{\|}}{C}-OH$$

12.35 a. achiral

b. chiral $CH_3-\overset{\overset{\displaystyle Br}{|}}{CH}-CH_2-CH_3$ *chiral carbon*

c. chiral *chiral carbon* $CH_3-\overset{\overset{\displaystyle Br}{|}}{CH}-\overset{\overset{\displaystyle O}{\|}}{C}-H$

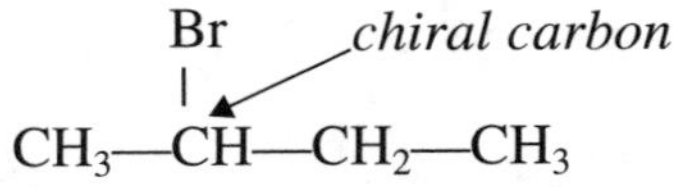

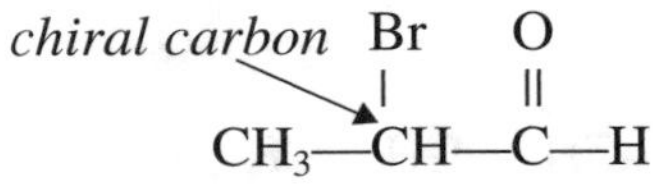

d. achiral

12.37 a. 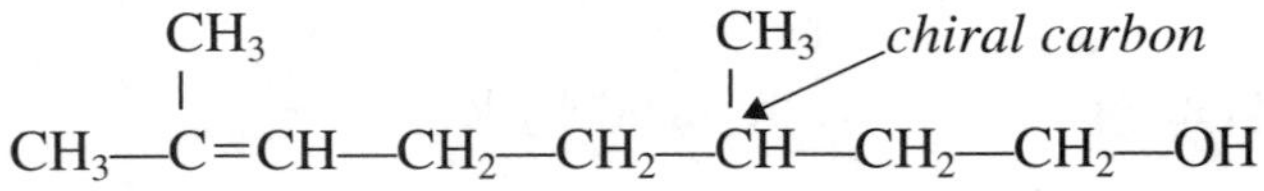 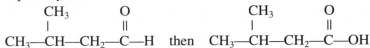 *chiral carbon*

$CH_3-\overset{\overset{\displaystyle CH_3}{|}}{C}=CH-CH_2-CH_2-\overset{\overset{\displaystyle CH_3}{|}}{CH}-CH_2-CH_2-OH$

b. 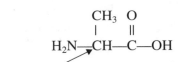 *chiral carbon*

$H_2N-\overset{\overset{\displaystyle CH_3}{|}}{CH}-\overset{\overset{\displaystyle O}{\|}}{C}-OH$

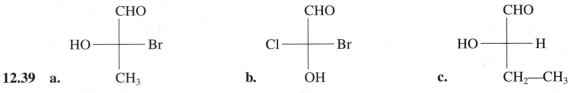

12.39 a. **b.** **c.**

12.41 a. identical **b.** enantiomers **c.** enantiomers **d.** enantiomers

12.43 a. aromatic, alcohol.

b. Toluene is the benzene ring with a methyl group.

12.45 **a.** phenol **b.** alcohol, cycloalkane

12.47 **a.** 3,7-dimethyl-6-octenal
  b. The *en* in octenal signifies a double bond. The 6 indicates the double bond is between carbon 6 and carbon 7 counting from the aldehyde as carbon 1.
  c. The *al* in octenal signifies that an aldehyde group is carbon 1.
  d. $C_{10}H_{18}O + 14O_2 \rightarrow 10CO_2 + 9H_2O$

12.49 **a.** 2° **b.** 1° **c.** 1° **d.** 2° **e.** 1° **f.** 3°

12.51 **a.** alcohol **b.** ether **c.** thiol **d.** alcohol
  e. ether **f.** alcohol **g.** phenol

12.53 **a.**

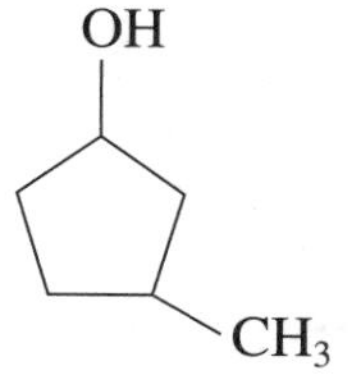

b.

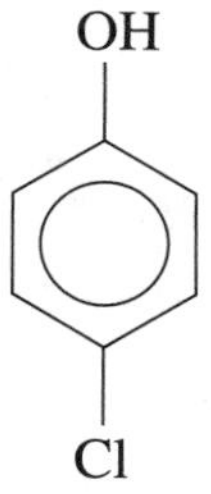

c. $H_3C-CH-CH-CH_2-CH_3$ (with CH_3 and OH substituents)

d.

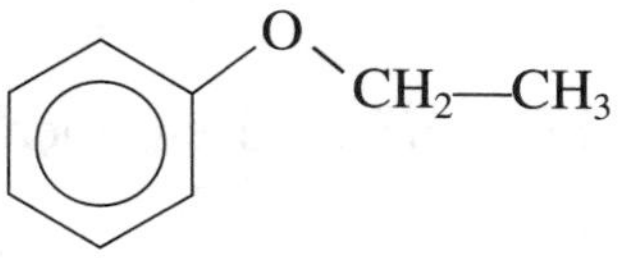

e. $CH_3-CH_2-\overset{\overset{O}{||}}{C}-CH_2-CH_3$

f.

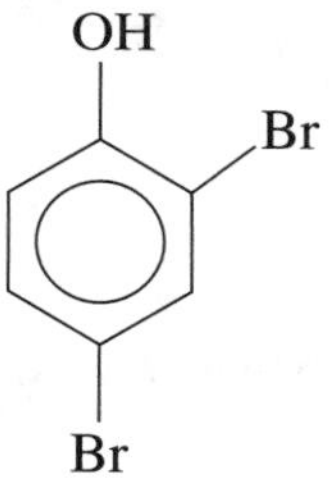

12.55 **a.** 1-propanol; hydrogen bonding **b.** 1-propanol; hydrogen bonding
  c. 1-butanol; larger molar mass

12.57 **a.** soluble; hydrogen bonding
  b. soluble; hydrogen bonding
  c. insoluble; carbon chains over four carbon atoms diminishes effect of polar —OH on hydrogen bonding

12.59 **a.** $CH_3-CH=CH_2$ **b.** $CH_3-CH_2-\overset{\overset{O}{||}}{C}-H$

c. $CH_3-CH_2-\overset{\overset{O}{||}}{C}-CH_3$ **d.** 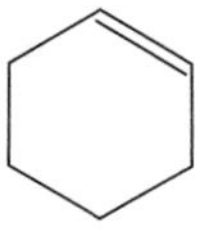 **e.**

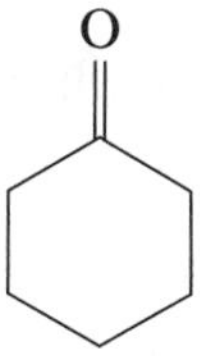

12.61 Testosterone contains cycloalkene, alcohol, and ketone functional groups.

12.63 **a.** 2,5-dichlorophenol is a benzene ring with a hydroxyl on carbon 1 and chlorine atoms on carbons 2 and 5.

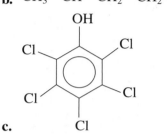

b. CH₃—CH—CH₂—CH₂—SH (with CH₃ and OH substituents)

c. (pentachlorophenol structure with OH and five Cl)

12.65 **a.** 2-bromo-4-chlorocyclopentanone   **b.** 3-bromo-4-chlorobenzaldehyde
 c. 3-chloropropanal   **d.** 2-chloro-3-pentanone
 e. 3-methylcyclohexanone

12.67 **a.** 3-methylcyclopentanone is a five-carbon cyclic structure with a methyl group located two carbons from the carbonyl group.

b. 4-chlorobenzaldehyde is a benzene with an aldehyde group and a chlorine on carbon 4.

c. 3–chloropropionaldehyde is a three-carbon aldehyde with a chlorine located two carbons from the carbonyl.

$$Cl—CH_2—CH_2—\overset{\displaystyle O}{\overset{\|}{C}}—H$$

d. Ethyl methyl ketone has a carbonyl bonded to a methyl and an ethyl group.

$$CH_3—\overset{\displaystyle O}{\overset{\|}{C}}—CH_2—CH_3$$

e. This is a six-carbon aldehyde with a methyl group on carbon 3.

$$CH_3-CH_2-CH_2-\overset{\overset{CH_3}{|}}{CH}-CH_2-\overset{\overset{O}{||}}{C}-H$$

f. This has a seven-carbon chain with a carbonyl group on carbon 2.

$$CH_3-\overset{\overset{O}{||}}{C}-CH_2-CH_2-CH_2-CH_2-CH_3$$

12.69 Compounds b, c, and d with oxygen atoms and less than five carbon atoms can hydrogen bond to be soluble.

12.71 **a.** CH_3-CH_2-OH; polar —OH group can hydrogen bond

b. $CH_3-CH_2-\overset{\overset{O}{||}}{C}-H$; polar carbonyl group can hydrogen bond

12.73 A chiral carbon is bonded to four different groups.

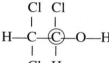

a. Cl H **b.** None. **c.** None.

d. $CH_3-\overset{\overset{NH_2}{|}}{CH}-\overset{\overset{O}{||}}{C}-H$ **e.** $CH_3-CH_2-\overset{\overset{Br}{|}}{CH}-CH_2-CH_2-CH_3$

12.75 **a.** identical **b.** enantiomers
c. enantiomers **d.** identical (two —OH groups)

12.77 Primary alcohols oxidize to aldehydes and then to carboxylic acids. Secondary alcohols oxidize to ketones.

a. $CH_3-CH_2-\overset{\overset{O}{||}}{C}-H$ that oxidizes further to $CH_3-CH_2-\overset{\overset{O}{||}}{C}-OH$

b. $CH_3-\overset{\overset{O}{||}}{C}-CH_2-CH_2-CH_3$

c. $CH_3-CH_2-CH_2-\overset{\overset{O}{||}}{C}-OH$

d.

12.79 $CH_3-CH_2-CH_2-CH_2-CH_2-OH$ 1-pentanol

$$\overset{\displaystyle OH}{\underset{\displaystyle |}{CH_3-CH-CH_2-CH_2-CH_3}}$$ 2-pentanol

$$\overset{\displaystyle OH}{\underset{\displaystyle |}{CH_3-CH_2-CH-CH_2-CH_3}}$$ 3-pentanol

$$\overset{\displaystyle CH_3}{\underset{\displaystyle |}{HO-CH_2-CH-CH_2-CH_3}}$$ 2-methyl-1-butanol

$$\overset{\displaystyle CH_3}{\underset{\displaystyle |}{HO-CH_2-CH_2-CH-CH_3}}$$ 3-methyl-1-butanol

$$\begin{array}{c} CH_3 \\ | \\ CH_3-C-CH_2-CH_3 \\ | \\ OH \end{array}$$ 2-methyl-2-butanol

$$\begin{array}{c} OH \quad CH_3 \\ | \qquad | \\ CH_3-CH-CH-CH_3 \end{array}$$ 3-methyl-2-butanol

$$\begin{array}{c} CH_3 \\ | \\ CH_3-C-CH_2-OH \\ | \\ CH_3 \end{array}$$ 2,2-dimethyl-1-propanol

12.81 $$\begin{array}{c} CH_3 \quad O \\ | \qquad || \\ CH_3-CH-C-H \end{array}$$

12.83   Compound A is $CH_3-CH_2-CH_2-OH$ 1-propanol

Compound B is an alkene $CH_3-CH=CH_2$ propene that forms when 1-propanol undergoes dehydration.

Compound C is the aldehyde that forms when a primary alcohol undergoes oxidation.

$$\overset{\displaystyle O}{\underset{\displaystyle ||}{CH_3-CH_2-C-H}}$$

Answers to Combining Ideas from Chapters 9 to 12

CI.19 **a.**

Isotope	Number of Protons	Number of Neutrons	Number of Electrons
^{27}Si	14	13	14
^{28}Si	14	14	14
^{29}Si	14	15	14
^{30}Si	14	16	14
^{31}Si	14	17	14

    b. 2, 8, 4

    c. Atomic mass calculated from the three stable isotopes is 28.09 amu.

    d. $^{27}_{14}Si \rightarrow {}^{27}_{13}Al + {}^{0}_{+1}e$

        $^{31}_{14}Si \rightarrow {}^{31}_{15}P + {}^{0}_{-1}e$

    e.

$$:\ddot{C}l:$$
$$|$$
$$:\ddot{C}l-Si-\ddot{C}l:\qquad \text{Tetrahedral}$$
$$|$$
$$:\ddot{C}l:$$

    f. $16\ \mu Ci \rightarrow 8.0\ \mu Ci \rightarrow 4.0\ \mu Ci \rightarrow 2.0\ \mu Ci$ is three half-lives.

$$3\ \text{half-lives} \times 2.6\ \frac{h}{\text{half-life}} = 7.8\ h$$

CI.21 **a.** $^{226}_{88}Ra \rightarrow {}^{222}_{86}Rn + {}^{4}_{2}He$

    b. $^{222}_{86}Rn \rightarrow {}^{218}_{84}Po + {}^{4}_{2}He$

    c. $15.2\ \text{days} \times \dfrac{1\ \text{half-life}}{3.8\ \text{days}} = 4\ \text{half-lives}$

        $24\ 000\ \text{atoms Rn-222} \times \dfrac{1}{2} \times \dfrac{1}{2} \times \dfrac{1}{2} \times \dfrac{1}{2} =$

        $24\ 000\ \text{atoms Rn-222} \times \dfrac{1}{16} = 1500\ \text{atoms}$

    d. $7.2 \times 10^4\ L \times \dfrac{2.5\ pCi}{1\ L} \times \dfrac{1\ Ci}{10^{12}\ pCi} \times \dfrac{3.7 \times 10^{10}\ \text{disintegrations}}{1\ Ci\ s}$

        $\times \dfrac{3600\ s}{1\ h} \times \dfrac{24\ h}{1\ day} \times \dfrac{1\ \alpha\ \text{particle}}{1\ \text{disintegration}} = 5.8 \times 10^8\ \alpha\ \text{particles/day}$

CI.23 **a.**

$$\overset{\displaystyle O}{\overset{\displaystyle \|}{CH_3-C-CH_3}}$$

    b. C_3H_6O 58.0 g/mole

    c.

$$\overset{\displaystyle OH}{\overset{\displaystyle |}{CH_3-CH-CH_3}}$$

Carboxylic Acids, Esters, Amines, and Amides

Study Goals

- Name and write structural formulas of carboxylic acids and esters.
- Describe the boiling points and solubility of carboxylic acids.
- Write equations for the ionization of carboxylic acids in water.
- Write equations for the esterification, hydrolysis, and saponification of esters.
- Name and write structural formulas of amines and amides.
- Describe the ionization of amines in water.
- Describe the boiling points of amines and amides compared to alkanes and alcohols.
- Describe the solubility of amines and amides in water.
- Write equations for the neutralization and amidation of amines.
- Describe acid and base hydrolysis of amides.

Think About It

1. Why do vinegar and citrus juices taste sour?

2. What type of compound gives flowers and fruits their pleasant aromas?

3. Fish smell "fishy," but lemon juice removes the "fishy" odor. Why?

4. What functional groups are often found in tranquilizers and hallucinogens?

Key Terms

Match the key term with the correct statement shown below.

a. carboxylic acid	**b.** saponification	**c.** esterification	**d.** hydrolysis
e. ester	**f.** amine	**g.** amide	**h.** alkaloid

1. _____ An organic compound containing the carboxyl group (—COOH)

2. _____ A reaction of a carboxylic acid and an alcohol in the presence of an acid catalyst

3. _____ A type of organic compound that produces pleasant aromas in flowers and fruits

4. _____ The hydrolysis of an ester with a strong base producing a salt of the carboxylic acid and an alcohol

5. _____ The splitting of a molecule such as an ester by the addition of water in the presence of an acid

6. _____ A nitrogen-containing compound that is active physiologically

7. _____ The hydrolysis of this compound produces a carboxylic acid and an amine

8. _____ An organic compound that contains an amino group

Answers	**1.** a	**2.** c	**3.** e	**4.** b
	5. d	**6.** h	**7.** g	**8.** f

13.1 Carboxylic Acids

- In the IUPAC system, a carboxylic acid is named by replacing the *ane* ending with *oic acid*. Simple acids usually are named by the common naming system using the prefixes: **form** (1C), **acet** (2C), **propion** (3C), **butyr** (4C), followed by *ic acid*.

$$\overset{\overset{\displaystyle O}{\|}}{H-C-OH}$$
methanoic acid
(formic acid)

$$\overset{\overset{\displaystyle O}{\|}}{CH_3-C-OH}$$
ethanoic acid
(acetic acid)

$$\overset{\overset{\displaystyle O}{\|}}{CH_3-CH_2-CH_2-C-OH}$$
butanoic acid
(butyric acid)

◆ Learning Exercise 13.1A

Give the IUPAC name for each of the following carboxylic acids.

1. $\overset{\overset{\displaystyle O}{\|}}{CH_3-C-OH}$

2. $CH_3-\overset{\overset{\displaystyle OH}{|}}{CH}-\overset{\overset{\displaystyle O}{\|}}{C}-OH$

3. $CH_3-\overset{\overset{\displaystyle CH_3}{|}}{CH}-CH_2-\overset{\overset{\displaystyle O}{\|}}{C}-OH$

4.
COOH
Cl

Answers
1. ethanoic acid
2. 2-hydroxypropanoic acid
3. 3-methylbutanoic acid
4. 4-chlorobenzoic acid

◆ Learning Exercise 13.1B

A. Write the formulas for the following carboxylic acids.

1. acetic acid

2. 2-ketobutanoic acid

3. benzoic acid

4. 3-hydroxypropanoic acid

5. formic acid

6. 3-methylpentanoic acid

Answers

1. $CH_3-\overset{\overset{\displaystyle O}{\|}}{C}-OH$

2. $CH_3-CH_2-\overset{\overset{\displaystyle O}{\|}}{C}-\overset{\overset{\displaystyle O}{\|}}{C}-OH$

3. (benzene ring with) $\overset{\overset{\displaystyle O}{\|}}{C}-OH$

4. $HO-CH_2-CH_2-\overset{\overset{\displaystyle O}{\|}}{C}-OH$

5. $H-\overset{\overset{\displaystyle O}{\|}}{C}-OH$

6. $CH_3-CH_2-\overset{\overset{\displaystyle CH_3}{|}}{CH}-CH_2-\overset{\overset{\displaystyle O}{\|}}{C}-OH$

13.2 Properties of Carboxylic Acids

- Carboxylic acids have higher boiling points than other polar compounds such as alcohols.
- Because they have two polar groups, two carboxylic acids form a dimer, which contains two sets of hydrogen bonds.
- Carboxylic acids with one to four carbon atoms are very soluble in water.
- As weak acids, carboxylic acids ionize slightly in water to form acidic solutions of H_3O^+ and a carboxylate ion.
- When bases neutralize carboxylic acids, the products are carboxylic acid salts and water.

◆ Learning Exercise 13.2A

Indicate whether each of the following carboxylic acids is soluble in water.

1. _____ hexanoic acid 3. _____ propanoic acid 5. _____ formic acid

2. _____ acetic acid 4. _____ benzoic acid 6. _____ octanoic acid

Answers   1. no   2. yes   3. yes   4. no   5. yes   6. no

◆ Learning Exercise 13.2B

Identify the compound in each pair that has the higher boiling point.

1. acetic acid or butyric acid

2. propanoic acid or 2-propanol

3. propanoic acid or propanone

4. acetic acid or acetaldehyde

Answers   1. butyric acid   2. propanoic acid
             3. propanoic acid   4. acetic acid

◆ **Learning Exercise 13.2C**

Write the products for the ionization of the following carboxylic acids in water.

1. $\underset{\substack{\| \\ O}}{CH_3-CH_2-C-OH} + H_2O \rightleftarrows$

2. benzoic acid + $H_2O \rightleftarrows$

Answers   1. $\underset{\substack{\| \\ O}}{CH_3-CH_2-C-O^-} + H_3O^+$ 2. 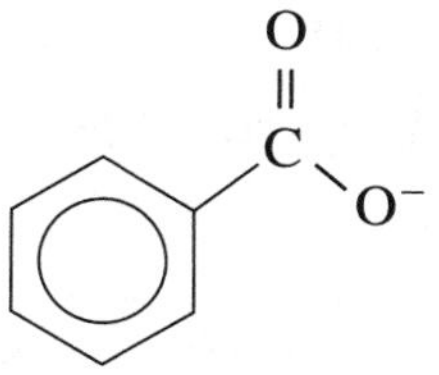 $+ H_3O^+$

◆ **Learning Exercise 13.2D**

Write the products and names for each of the following.

1. $\underset{\substack{\| \\ O}}{CH_3-CH_2-C-OH} + NaOH \rightarrow$

2. formic acid + KOH →

Answers   1. $\underset{\substack{\| \\ O}}{CH_3-CH_2-C-O^-Na^+} + H_2O$ 2. $\underset{\substack{\| \\ O}}{H-C-O^-K^+} + H_2O$
          sodium propanoate   potassium methanoate
          (sodium propionate)   (potassium formate)

13.3 Esters

● In the presence of a strong acid, carboxylic acids react with alcohols to produce esters and water.
● The names of esters consist of two words, one from the alcohol and the other from the carboxylic acid with the *ic* ending replaced by *ate*.

$\underset{\substack{\| \\ O}}{CH_3-C-O-CH_3}$ methyl ethanoate (IUPAC) or methyl acetate (common)

- In hydrolysis, esters are split apart by a reaction with water. When the catalyst is an acid, the products are a carboxylic acid and an alcohol.

$$CH_3-\overset{\overset{\displaystyle O}{\|}}{C}-O-CH_3 + H_2O \overset{H^+}{\rightarrow} CH_3-\overset{\overset{\displaystyle O}{\|}}{C}-OH + HO-CH_3$$

methyl acetate   *acetic acid*   *methyl alcohol*

- In saponification, long-chain fatty acids from fats react with strong bases to produce salts of the fatty acids, which are soaps.
- Saponification is the hydrolysis of an ester in the presence of a base, which produces a carboxylate salt and an alcohol.

$$CH_3-\overset{\overset{\displaystyle O}{\|}}{C}-O-CH_3 + NaOH \rightarrow CH_3-\overset{\overset{\displaystyle O}{\|}}{C}-O^-Na^+ + HO-CH_3$$

methyl acetate   *sodium acetate*   *methyl alcohol*

◆ **Learning Exercise 13.3A**

Write the products of the following reactions.

1. $CH_3-\overset{\overset{\displaystyle O}{\|}}{C}-OH + CH_3-OH \overset{H^+}{\rightarrow}$

2. $H-\overset{\overset{\displaystyle O}{\|}}{C}-OH + CH_3-CH_2-OH \overset{H^+}{\rightarrow}$

3. 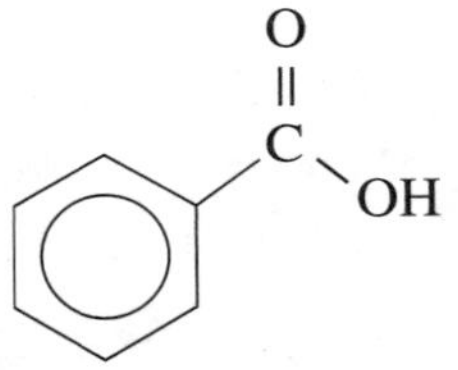
$+ HO-CH_3 \overset{H^+}{\rightarrow}$

4. propanoic acid and ethanol $\overset{H^+}{\rightarrow}$

Answers

1. $CH_3-\overset{\overset{\displaystyle O}{\|}}{C}-O-CH_3 + H_2O$ 2. $H-\overset{\overset{\displaystyle O}{\|}}{C}-O-CH_2-CH_3 + H_2O$

3. 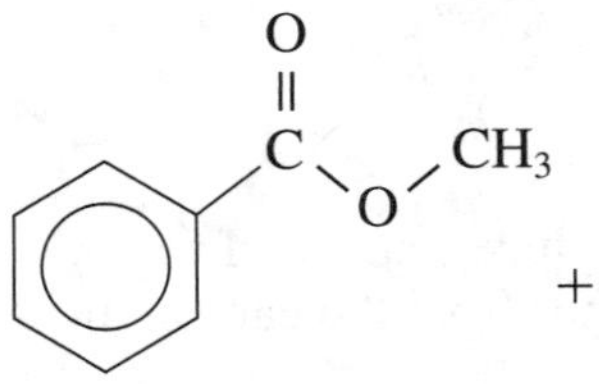
$+ H_2O$

4. $CH_3-CH_2-\overset{\overset{\displaystyle O}{\|}}{C}-O-CH_2-CH_3 + H_2O$

◆ **Learning Exercise 13.3B**

Name each of the following esters.

O
||
1. CH_3—C—O—CH_2—CH_3

O
||
2. CH_3—CH_2—CH_2—C—O—CH_3

O   OH
|| |
3. CH_3—CH_2—C—O—CH_2—CH—CH_3

O
||
C—O—CH_3
4.

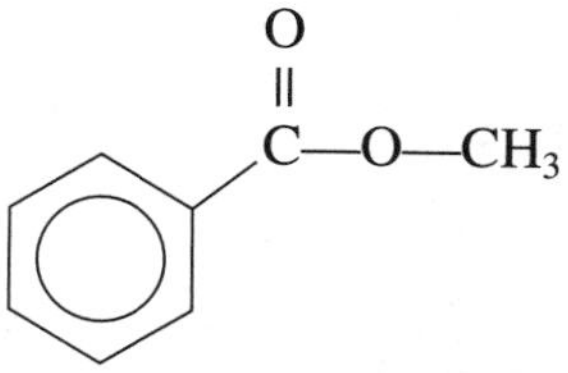

Answers　　**1.** ethyl ethanoate (ethyl acetate)
2. methyl butanoate (methyl butyrate)
3. 2-hydroxypropyl propanoate (2-hydroxypropyl propionate)
4. methyl benzoate

◆ **Learning Exercise 13.3C**

Write structural formulas for each of the following esters.

1. propyl acetate

2. ethyl butyrate

3. ethyl propanoate

4. ethyl benzoate

Answers

O
||
1. CH_3—C—O—CH_2—CH_2—CH_3

O
||
2. CH_3—CH_2—CH_2—C—O—CH_2—CH_3

O
||
C—O—CH_2—CH_3
4.

O
||
3. CH_3—CH_2—C—O—CH_2—CH_3

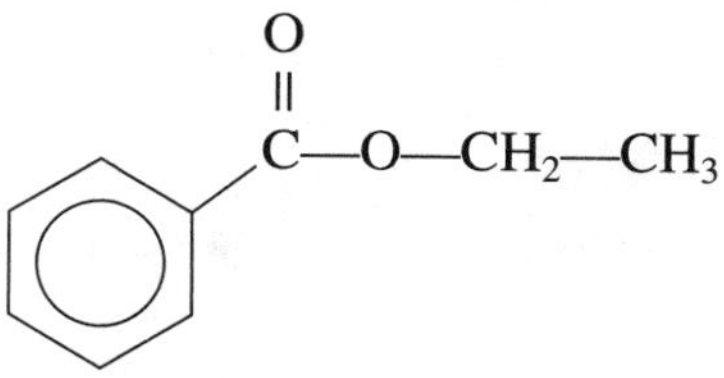

◆ Learning Exercise 13.3D

Write the products of hydrolysis or saponification for the following esters:

1. $CH_3-CH_2-CH_2-\overset{\overset{\displaystyle O}{\|}}{C}-O-CH_3 + H_2O \xrightarrow{H^+}$

2. $CH_3-\overset{\overset{\displaystyle O}{\|}}{C}-O-CH_3 + NaOH \rightarrow$

3. $CH_3-CH_2-\overset{\overset{\displaystyle O}{\|}}{C}-O-CH_2-CH_3 + KOH \rightarrow$

4. $CH_3-CH_2-\overset{\overset{\displaystyle O}{\|}}{C}-O-CH_3 + H_2O \xrightarrow{H^+}$

Answers

1. $CH_3-CH_2-CH_2-\overset{\overset{\displaystyle O}{\|}}{C}-OH + HO-CH_3$

2. $CH_3-\overset{\overset{\displaystyle O}{\|}}{C}-O^-Na^+ + CH_3-OH$

3. $CH_3-CH_2-\overset{\overset{\displaystyle O}{\|}}{C}-O^-K^+ + CH_3-CH_2-OH$

4. $CH_3-CH_2-\overset{\overset{\displaystyle O}{\|}}{C}-OH + HO-CH_3$

13.4 Amines

- Amines are derivatives of ammonia (NH_3), in which alkyl or aromatic groups replace one or more hydrogen atoms.
- Amines are classified as primary, secondary, or tertiary when the nitrogen atom is bonded to one, two, or three alkyl or aromatic groups.

CH_3-NH_2 $CH_3-\overset{\overset{\displaystyle CH_3}{|}}{N}-H$ $CH_3-\overset{\overset{\displaystyle CH_3}{|}}{N}-CH_3$
Primary (1°)   Secondary (2°)   Tertiary (3°)

- The N—H bonds in primary and secondary amines form hydrogen bonds.
- Amines have higher boiling points than hydrocarbons, but lower than alcohols of similar mass because the N atom is not as electronegative as the O atoms in alcohols.
- Hydrogen bonding allows amines with up to six carbon atoms to be soluble in water.
- In water, amines act as weak bases by accepting protons from water to produce ammonium and hydroxide ions.

$CH_3-NH_2 + H_2O \rightleftarrows CH_3-NH_3^+ + OH^-$
methylamine   *methylammonium hydroxide*

- Strong acids neutralize amines to yield ammonium salts.

$CH_3-NH_2 + HCl \rightarrow CH_3-NH_3^+ + Cl^-$
methylamine   *methylammonium chloride*

◆ **Learning Exercise 13.4A**

Classify each of the following as a primary (1°), secondary (2°), or tertiary (3°) amine.

1. _____ $\overset{\overset{\displaystyle H}{|}}{CH_3-N-CH_2-CH_3}$

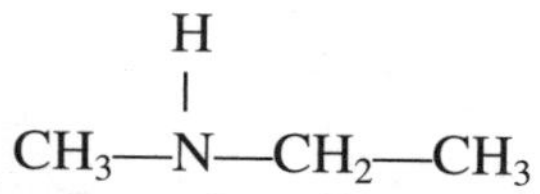

4. _____ $CH_3-CH_2-CH_2-CH_2-\overset{\overset{\displaystyle H}{|}}{N}-CH_2-CH_3$

2. _____ (cyclohexane with NH_2)

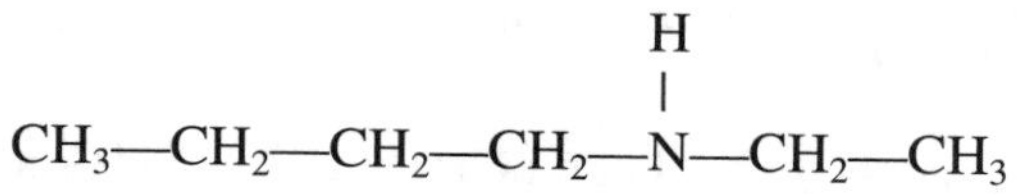

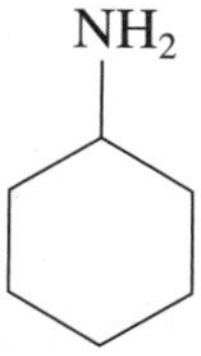

5. _____ (benzene ring with NH_2 and CH_3)

3. _____ $CH_3-CH_2-\overset{\overset{\displaystyle CH_3}{|}}{N}-CH_3$

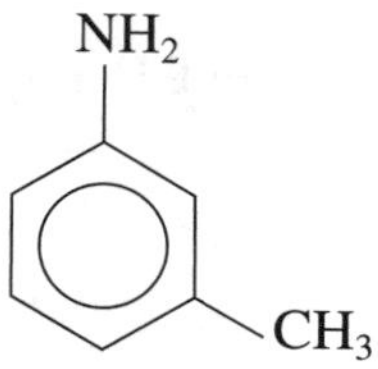

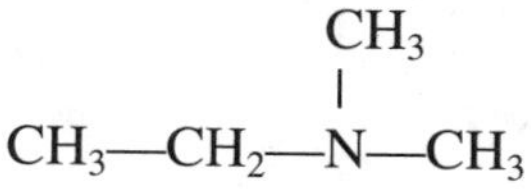

Answers   **1.** 2°   **2.** 1°   **3.** 3°   **4.** 2°   **5.** 1°

◆ **Learning Exercise 13.4B**

Name each of the amines in problem 13.4A.

1. _____   4. _____

2. _____   5. _____

3. _____

Answers   **1.** ethylmethylamine   **2.** cyclohexanamine   **3.** ethyldimethylamine
4. butylethylamine   **5.** 3-methylaniline

◆ **Learning Exercise 13.4C**

Write the structural formulas of the following amines.

1. isopropylamine

2. butylethylmethylamine

3. 3-bromoaniline

4. *N*-methylaniline

Answers

1. 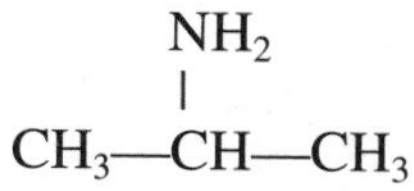 $\overset{\overset{\displaystyle NH_2}{|}}{CH_3-CH-CH_3}$

2. $CH_3-CH_2-\overset{\overset{\displaystyle CH_3}{|}}{N}-CH_2-CH_2-CH_2-CH_3$

3. (benzene ring with NH_2 and Br)

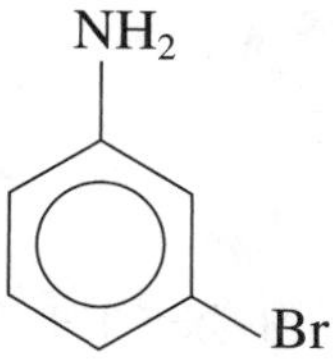

4. 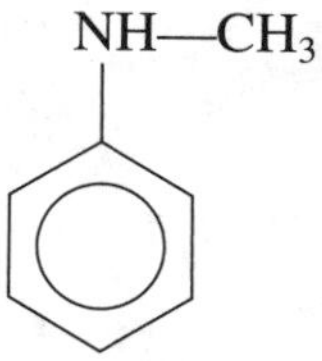 (benzene ring with $NH-CH_3$)

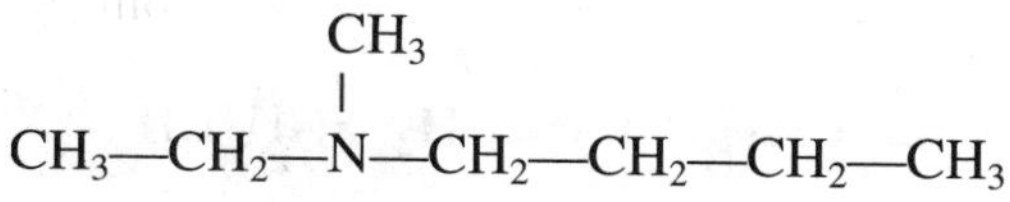

◆ **Learning Exercise 13.4D**

Select the compound in each pair that has the higher boiling point.

1. CH_3—NH_2 and CH_3—OH _____

2. CH_3—CH_2—CH_2 and CH_3—CH_2—NH_2 _____

3. CH_3—CH_2—NH_2 and CH_3—CH_2—CH_2—NH_2 _____

4. $\underset{\overset{|}{H}}{CH_3}$—N—$CH_3$ and CH_3—CH_2—NH_2 _____

Answers   **1.** CH_3—OH **2.** CH_3—CH_2—NH_2

              3. CH_3—CH_2—CH_2—NH_2 **4.** CH_3—CH_2—NH_2

◆ **Learning Exercise 13.4E**

Write the products of the following reactions.

1. CH_3—CH_2—NH_2 + H_2O $\rightleftarrows$

2. CH_3—CH_2—CH_2—NH_2 + HCl $\rightarrow$

3. CH_3CH_2—NH—CH_3 + HCl $\rightarrow$

4.

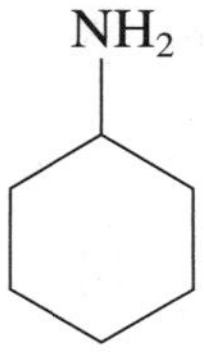

    + HBr $\rightarrow$

5.

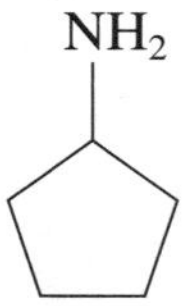

    + H_2O $\rightleftarrows$

6. CH_3—CH_2—$NH_3^+Cl^-$ + NaOH $\rightarrow$

Answers

1. CH_3—CH_2—$NH_3^+OH^-$ 2. CH_3—CH_2—CH_2—$NH_3^+Cl^-$ 3. CH_3—CH_2—$\overset{+}{N}H_2$—CH_3Cl^-

4.

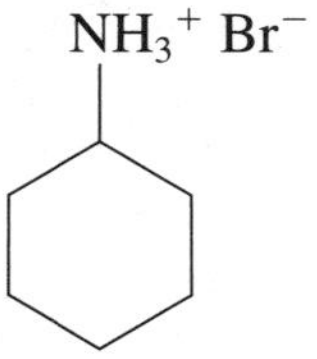

5.

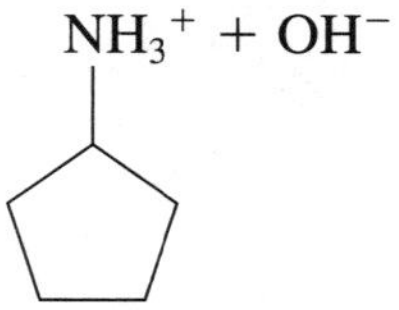

6. CH_3—CH_2—NH_2 + NaCl + H_2O

13.5 Amides

- Amides are derivatives of carboxylic acids in which an amine group replaces the —OH group in the carboxylic acid.
- Amides are named by replacing the *ic acid* or *oic acid* ending by *amide*.

$$CH_3-\overset{\overset{\displaystyle O}{\|}}{C}-NH_2 \qquad \text{ethanamide (acetamide)}$$

- Amides undergo acid and base hydrolysis to produce the carboxylic acid (or carboxylate salt) and the amine (or amine salt).

$$CH_3-\overset{\overset{\displaystyle O}{\|}}{C}-NH_2 + HCl + H_2O \rightarrow CH_3-\overset{\overset{\displaystyle O}{\|}}{C}-OH + NH_4^+\ Cl^-$$

$$CH_3-\overset{\overset{\displaystyle O}{\|}}{C}-NH_2 + NaOH \rightarrow CH_3-\overset{\overset{\displaystyle O}{\|}}{C}-O^-Na^+ + NH_3$$

◆ Learning Exercise 13.5A

Write the structural formula(s) of the amides formed in each of the following reactions.

1. $CH_3-CH_2-\overset{\overset{\displaystyle O}{\|}}{C}-OH + NH_3 \xrightarrow{\text{Heat}}$

2. 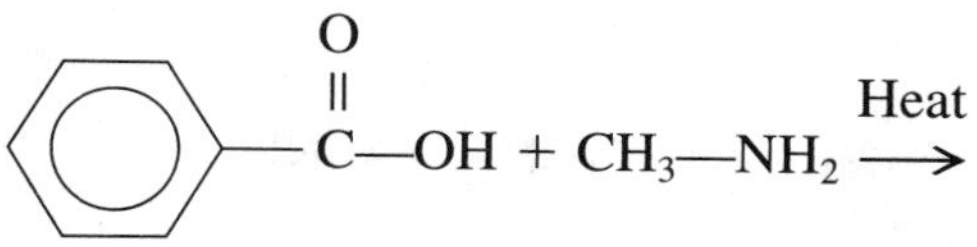 [benzene ring]$-\overset{\overset{\displaystyle O}{\|}}{C}-OH + CH_3-NH_2 \xrightarrow{\text{Heat}}$

3. $CH_3-\overset{\overset{\displaystyle O}{\|}}{C}-OH + \overset{\overset{\displaystyle CH_3}{|}}{NH}-CH_3 \xrightarrow{\text{Heat}}$

Answers

1. $CH_3-CH_2-\overset{\overset{\displaystyle O}{\|}}{C}-NH_2$

2. 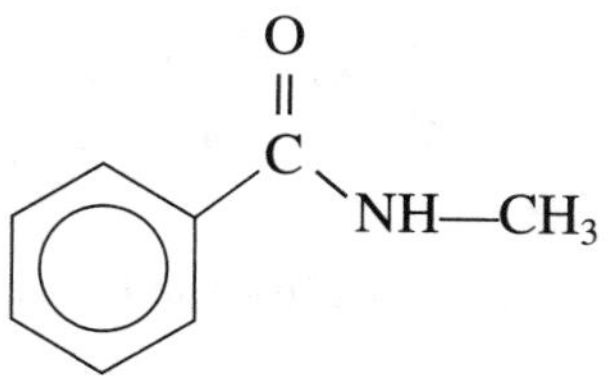 [benzene ring]$-\overset{\overset{\displaystyle O}{\|}}{C}-NH-CH_3$

3. $CH_3-\overset{\overset{\displaystyle O}{\|}}{C}-\overset{\overset{\displaystyle CH_3}{|}}{N}-CH_3$

◆ Learning Exercise 13.5B

Name the following amides.

1. $CH_3-CH_2-\overset{\overset{\displaystyle O}{\|}}{C}-NH_2$ _____

2. _____

$$\overset{\displaystyle O}{\overset{\displaystyle \|}{}}$$

3. $CH_3—CH_2—CH_2—CH_2—C—NH_2$ _____

$$\overset{\displaystyle O}{\overset{\displaystyle \|}{}}$$

4. $CH_3—C—NH_2$ _____

Answers   **1.** propanamide (propionamide)   **2.** benzamide
             3. pentanamide   **4.** ethanamide (acetamide)

◆ **Learning Exercise 13.5C**

Write the structural formulas for each of the following amides.

1. propanamide   **2.** butanamide

3. 3-chloropentanamide   **4.** benzamide

$$\overset{\displaystyle O}{\overset{\displaystyle \|}{}}$$

Answers   **1.** $CH_3—CH_2—C—NH_2$ **2.** $CH_3—CH_2—CH_2—C—NH_2$

$$\overset{Cl}{\overset{|}{}} \quad \overset{O}{\overset{\|}{}}$$

3. $CH_3—CH_2—CH—CH_2—C—NH_2$ **4.** 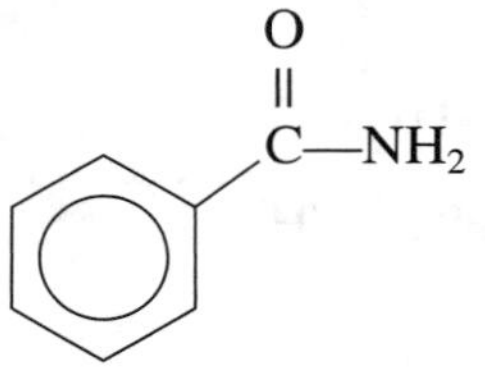 (benzamide structure)

◆ **Learning Exercise 13.5D**

Using structural formulas, write the products for the hydrolysis of each of the following with HCl and NaOH.

$$\overset{\displaystyle O}{\overset{\displaystyle \|}{}}$$

1. $CH_3—CH_2—C—NH_2$

$$\overset{\displaystyle O}{\overset{\displaystyle \|}{}}$$

2. $CH_3—C—NH_2$

Answers

1. (HCl) $CH_3—CH_2—\overset{O}{\overset{\|}{C}}—OH + NH_4{}^+Cl^-$ (NaOH) $CH_3—CH_2—\overset{O}{\overset{\|}{C}}—O^-Na^+ + NH_3$

2. (HCl) $CH_3—\overset{O}{\overset{\|}{C}}—OH + \overset{+}{NH_4}Cl^-$ (NaOH) $CH_3—\overset{O}{\overset{\|}{C}}—O^-Na^+ + NH_3$

Checklist for Chapter 13

You are ready to take the practice test for Chapter 13. Be sure that you have accomplished the following learning goals for this chapter. If you are not sure, review the section listed at the end of the goal. Then apply your new skills and understanding to the practice test.

After studying Chapter 13, I can successfully:

_____ Write the IUPAC and common names, and draw condensed structural formulas, of carboxylic acids (13.1).

_____ Describe the solubility and ionization of carboxylic acids in water (13.2).

_____ Describe the behavior of carboxylic acids as weak acids and write the structural formulas for the products of neutralization (13.2).

_____ Write equations for the preparation, hydrolysis, and saponification of esters (13.3).

_____ Write the IUPAC or common names and draw the condensed structural formulas of esters (13.3).

_____ Classify amines as primary, secondary, or tertiary (13.4).

_____ Write the IUPAC and common names of amines and draw their condensed structural formulas (13.4).

_____ Compare the boiling points and solubility of amines to alkanes and alcohols of similar mass (13.4).

_____ Write equations for the ionization and neutralization of amines (13.4).

_____ Write the IUPAC and common names of amides and draw their condensed structural formulas (13.5).

_____ Write equations for the hydrolysis of amides (13.5).

Practice Test for Chapter 13

Match structures 1 through 5 with functional groups A to E.

A. alcohol   **B.** aldehyde   **C.** carboxylic acid   **D.** ester   **E.** ketone

$$\text{CH}_3$$
$$|$$
1. _____ $\text{CH}_3\text{—CH—CH}_2\text{—OH}$

$$O$$
$$||$$
2. _____ $\text{CH}_3\text{—CH}_2\text{—C—H}$

$$O$$
$$||$$
3. _____ $\text{CH}_3\text{—C—CH}_2\text{—CH}_3$

$$O$$
$$||$$
4. _____ $\text{CH}_3\text{—CH}_2\text{—C—OH}$

$$O$$
$$||$$
5. _____ $\text{CH}_3\text{—C—O—CH}_3$

Match the names of the compounds 6 through 10 with their structures A to E.

$$O$$
$$||$$
A. $\text{CH}_3\text{—C—O—CH}_2\text{—CH}_3$

$$O$$
$$||$$
B. $\text{CH}_3\text{—CH}_2\text{—CH}_2\text{—C—O}^-\text{Na}^+$

$$O$$
$$||$$
C. $\text{CH}_3\text{—C—O}^-\text{Na}^+$

$$\text{CH}_3 \quad O$$
$$| \quad\quad ||$$
D. $\text{CH}_3\text{—CH}_2\text{—CH—C—OH}$

$$O$$
$$||$$
E. $\text{CH}_3\text{—CH}_2\text{—C—O—CH}_3$

6. _____ 2-methylbutanoic acid

7. _____ methyl propanoate

8. _____ sodium butanoate

9. _____ ethyl acetate

10. _____ sodium acetate

11. What is the product when a carboxylic acid reacts with sodium hydroxide?
    A. carboxylic acid salt   **B.** alcohol   **C.** ester   **D.** aldehyde   **E.** no reaction

12. Carboxylic acids are water soluble due to their
    A. nonpolar nature.   **B.** ionic bonds.   **C.** ability to lower pH.
    D. ability to hydrogen bond.   **E.** high melting points.

Questions 13 through 16 refer to the following reactions:

A. $CH_3-\overset{\overset{\textstyle O}{\|}}{C}-OH + CH_3-OH \xrightarrow{H^+} CH_3-\overset{\overset{\textstyle O}{\|}}{C}-O-CH_3 + H_2O$

B. $CH_3-\overset{\overset{\textstyle O}{\|}}{C}-OH + NaOH \rightarrow CH_3-\overset{\overset{\textstyle O}{\|}}{C}-O^-Na^+ + H_2O$

C. $CH_3-\overset{\overset{\textstyle O}{\|}}{C}-O-CH_3 + H_2O \xrightarrow{H^+} CH_3-\overset{\overset{\textstyle O}{\|}}{C}-OH + CH_3-OH$

D. $CH_3-\overset{\overset{\textstyle O}{\|}}{C}-O-CH_3 + NaOH \rightarrow CH_3-\overset{\overset{\textstyle O}{\|}}{C}-O^-Na^+ + CH_3-OH$

13. _____ is an ester hydrolysis.

14. _____ is a neutralization.

15. _____ is a saponification.

16. _____ is an esterification.

17. What is the name of the organic product of reaction A?
    A. methyl acetate   **B.** acetic acid   **C.** methyl alcohol
    D. acetaldehyde   **E.** ethyl methanoate

18. The compound with the highest boiling point is
    A. formic acid.   **B.** acetic acid.   **C.** propanol.
    D. propanoic acid.   **E.** ethyl acetate.

19. The ester produced from the reactions of 1-butanol and propanoic acid is
    A. butyl propanoate.   **B.** butyl propanone.   **C.** propyl butyrate.
    D. propyl butanone.   **E.** heptanoate.

20. The reaction of methyl acetate with NaOH produces
    A. ethanol and formic acid.   **B.** ethanol and sodium formate.
    C. ethanol and sodium ethanoate.   **D.** methanol and acetic acid.
    E. methanol and sodium acetate.

21. Identify the carboxylic acid and alcohol needed to produce

$CH_3-CH_2-CH_2-\overset{\overset{\textstyle O}{\|}}{C}-O-CH_2-CH_3$

    A. propanoic acid and ethanol   **B.** acetic acid and 1-pentanol   **C.** acetic acid and 1-butanol
    D. butanoic acid and ethanol   **E.** hexanoic acid and methanol

22. The name of $CH_3-CH_2-\overset{\overset{\textstyle O}{\|}}{C}-O-CH_2-CH_3$ is
    A. ethyl acetate.   **B.** ethyl ethanoate.   **C.** ethyl propanoate.
    D. propyl ethanoate.   **E.** ethyl butyrate.

23. Soaps are
  A. long-chain fatty acids.   **B.** fatty acid salts.   **C.** esters of acetic acid.
  D. alcohols with 10 carbon atoms.   **E.** aromatic compounds.

24. In a hydrolysis reaction,
  A. an acid reacts with an alcohol.   **B.** an ester reacts with NaOH.
  C. an ester reacts with H_2O.   **D.** an acid neutralizes a base.
  E. water is added to an alkene.

25. Esters
  A. have pleasant odors.
  B. can undergo hydrolysis.
  C. are formed from alcohols and carboxylic acids.
  D. have a lower boiling point than the corresponding acid.
  E. All of the above

26. The products of $\overset{\overset{\displaystyle O}{\|}}{H-C-OH}$ + H_2O are

  A. $\overset{\overset{\displaystyle O}{\|}}{H-C-O^-}$ + H_3O^+ **B.** $\overset{\overset{\displaystyle O}{\|}}{H-C-OH_2^+}$ + OH^- **C.** $\overset{\overset{\displaystyle O}{\|}}{H-C-O-CH_3}$

  D. $\overset{\overset{\displaystyle O}{\|}}{CH_3-C-OH}$ **E.** $\overset{\overset{\displaystyle O}{\|}}{H-C-O^-Na^+}$ + H_2O

Classify the amines in questions 27 through 30 as

  A. primary amine   **B.** secondary amine   **C.** tertiary amine

27. _____ $\overset{\overset{\displaystyle CH_3}{|}}{CH_3-CH-NH_2}$ **28.** _____ $\overset{\overset{\displaystyle CH_3}{|}}{CH_3-CH_2-N-CH_3}$

29. _____ $\overset{\overset{\displaystyle NH_2}{|}}{CH_3-CH_2-CH-CH_2-CH_3}$ **30.** _____ $\overset{\overset{\displaystyle H}{|}}{CH_3-N-CH_2-CH_3}$

Match the amines and amides in questions 31 through 33 with the following names.

  A. ethyl dimethyl amine   **B.** butanamide   **C.** *N*-methylacetamide
  D. benzamide   **E.** *N*-ethylbutyramide

31. $\overset{\overset{\displaystyle CH_3}{|}}{CH_3-CH_2-N-CH_3}$ **32.** $\overset{\overset{\displaystyle O}{\|}}{CH_3-CH_2-CH_2-C-NH_2}$ **33.**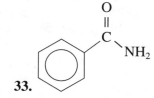

Use the following products to complete the reactions in questions 34 through 37.

  A. $\overset{\overset{\displaystyle O}{\|}}{CH_3-CH_2-C-OH}$ + NH_3 **B.** $CH_3-CH_2-NH_3^+Cl^-$

  C. $CH_3-CH_2-CH_2-NH_3^+OH^-$ **D.** $\overset{\overset{\displaystyle O}{\|}}{CH_3-C-NH_2}$

34. _____ ionization of 1-propanamine in water

35. _____ hydrolysis of propanamide

36. _____ reaction of ethanamine and hydrochloric acid

37. _____ amidation of acetic acid

38. _____ Amines used in drugs are converted to their amine salt because the salt is
 A. a solid at room temperature.   **B.** soluble in water.   **C.** odorless.
 D. soluble in body fluids.   **E.** All of these

Match the following alkaloids with their sources

 A. caffeine   **B.** nicotine   **C.** morphine   **D.** quinine

39. a painkiller from the Oriental poppy plant

40. obtained from the bark of the cinchona tree and used in the treatment of malaria

41. a stimulant obtained from the leaves of tobacco plants

42. a stimulant obtained from coffee beans and tea

Answers to the Practice Test

1. A	**2.** B	**3.** E	**4.** C	**5.** D
6. D	**7.** E	**8.** B	**9.** A	**10.** C
11. A	**12.** D	**13.** C	**14.** B	**15.** D
16. A	**17.** A	**18.** D	**19.** A	**20.** E
21. D	**22.** C	**23.** B	**24.** C	**25.** E
26. A	**27.** A	**28.** C	**29.** A	**30.** B
31. A	**32.** B	**33.** D	**34.** C	**35.** A
36. B	**37.** D	**38.** E	**39.** C	**40.** E
41. B	**42.** A			

Answers and Solutions to Selected Text Problems

13.1   Methanoic acid (formic acid) is the carboxylic acid that is responsible for the pain associated with ant stings.

13.3   Each compound contains three carbon atoms. They differ because propanal, an aldehyde, contains a carbonyl group bonded to a hydrogen. In propanoic acid, the carbonyl group connects to a hydroxyl group.

13.5   **a.** Ethanoic acid (acetic acid) is the carboxylic acid with two carbons.
   b. Propanoic acid (propionic acid) is a three-carbon carboxylic acid.
   c. 4-hydroxybenzoic acid has a carboxylic acid group on benzene and a hydroxyl group on carbon 4.
   d. 4-bromopentanoic acid is a five-carbon carboxylic acid with a –Br atom on carbon 4.

13.7   **a.**
$$\text{CH}_3\text{—CH}_2\text{—}\overset{\overset{\text{O}}{\|}}{\text{C}}\text{—OH}$$
     Propionic acid has three carbons.

   b. 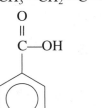   Benzoic acid is the carboxylic acid of benzene.

$$
\text{c.} \quad \underset{\overset{\displaystyle O}{\displaystyle \|}}{\text{Cl—CH}_2\text{—C—OH}}
$$

c. Cl—CH$_2$—C—OH   2-chloroethanoic acid is a carboxylic acid that has a two-carbon chain with a chlorine atom on carbon 2.

d. HO—CH$_2$—CH$_2$—C—OH   3-hydroxypropanoic acid is a carboxylic acid that has a three-carbon chain with a hydroxyl on carbon 3.

e. CH$_3$—CH$_2$—CH$_2$—C—OH   Butyric acid is a carboxylic acid that has a four-carbon chain.

f. CH$_3$—CH$_2$—CH$_2$—CH$_2$—CH$_2$—CH$_2$—C—OH   Heptanoic acid is a carboxylic acid that has a seven-carbon chain.

13.9 a. Butanoic acid has a higher molar mass and would have a higher boiling point.
b. Propanoic acid can form more hydrogen bonds (dimers) and would have a higher boiling point.
c. Butanoic acid can form more hydrogen bonds (dimers) and would have a higher boiling point.

13.11 a. Propanoic acid is the most soluble of the group because it has the fewest number of carbon atoms in its alkyl chain. Solubility of carboxylic acids decreases as the number of carbon atoms in the alkyl chain increases.
b. Propanoic acid is more soluble than 1-butanol because a carboxyl group forms more hydrogen bonds with water than a hydroxyl group. An alkane is not soluble in water.

13.13 a. H—C—OH + H$_2$O $\rightleftarrows$ H—C—O$^-$ + H$_3$O$^+$

b. CH$_3$—CH$_2$—C—OH + H$_2$O $\rightleftarrows$ CH$_3$—CH$_2$—C—O$^-$ + H$_3$O$^+$

c. CH$_3$—C—OH + H$_2$O $\rightleftarrows$ CH$_3$—C—O$^-$ + H$_3$O$^+$

13.15 a. H—C—OH + NaOH $\rightarrow$ H—C—O$^-$Na$^+$ + H$_2$O

b. CH$_3$—CH$_2$—C—OH + NaOH $\rightarrow$ CH$_3$—CH$_2$—C—O$^-$Na$^+$ + H$_2$O

c.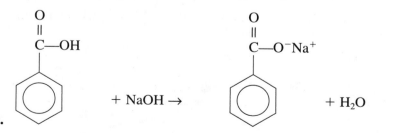

13.17 a. This is an *aldehyde* since it has a carbonyl bonded to carbon and hydrogen.
b. This is an *ester* since it has a carbonyl bonded to oxygen that is also bonded to a carbon.
c. This is a *ketone* since it has a carbonyl bonded to two carbon atoms.
d. This is a *carboxylic acid* since it has a carboxylic group; a carbonyl bonded to a hydroxyl.

13.19 **a.**

$$CH_3-\overset{\overset{\displaystyle O}{\|}}{C}-O-CH_3$$

The carbonyl portion of the ester has two carbons bonded to a one-carbon methyl group.

b.

$$CH_3-CH_2-CH_2-\overset{\overset{\displaystyle O}{\|}}{C}-O-CH_3$$

The carbonyl portion of the ester is a four-carbon chain bonded to a one-carbon methyl group.

c.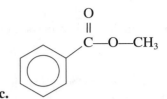

13.21 A carboxylic acid and an alcohol react to give an ester with the elimination of water.

a.

$$CH_3-CH_2-\overset{\overset{\displaystyle O}{\|}}{C}-O-CH_2-CH_2-CH_3$$

b.

$$CH_3-CH_2-CH_2-CH_2-\overset{\overset{\displaystyle O}{\|}}{C}-O-\overset{\overset{\displaystyle CH_3}{|}}{CH}-CH_3$$

13.23 **a.** The name of this ester is methyl methanoate (methyl formate). The carbonyl portion of the ester contains one carbon; the name is derived from methanoic (formic) acid. The alkyl portion has one carbon; it is methyl.
   b. The name of this ester is methyl ethanoate (methyl acetate). The carbonyl portion of the ester contains two carbons; the name is derived from ethanoic (acetic) acid. The alkyl portion has one carbon; it is methyl.
   c. The name of this ester is methyl butanoate (methyl butyrate). The carbonyl portion of the ester contains four carbons; the name is derived from butanoic (butyric) acid. The alkyl portion has one carbon; it is methyl.
   d. The name of this ester is ethyl butanoate (ethyl butyrate). The carbonyl portion of the ester has a four-carbon chain. The alkyl portion with two carbons is an ethyl.

13.25 **a.**

$$CH_3-\overset{\overset{\displaystyle O}{\|}}{C}-O-CH_3$$

Acetic acid is the two-carbon carboxylic acid. Methanol gives a one-carbon alkyl group.

b.

$$H-\overset{\overset{\displaystyle O}{\|}}{C}-O-CH_2-CH_2-CH_2-CH_3$$

Formic acid is the carboxylic acid bonded to the four-carbon 1-butanol.

c.

$$CH_3-CH_2-CH_2-CH_2-\overset{\overset{\displaystyle O}{\|}}{C}-O-CH_2-CH_3$$

Pentanoic acid is the carboxylic acid bonded to ethanol.

d.

$$CH_3-CH_2-\overset{\overset{\displaystyle O}{\|}}{C}-O-CH_2-CH_2-CH_3$$

Propanoic acid is the carboxylic acid bonded to propanol.

13.27 **a.** The flavor and odor of bananas is pentyl ethanoate (pentyl acetate).
b. The flavor and odor of oranges is octyl ethanoate (octyl acetate).
c. The flavor and odor of apricots is pentyl butanoate (pentyl butyrate).

13.29 Acid hydrolysis of an ester adds water in the presence of acid and gives a carboxylic acid and an alcohol.

13.31 Acid hydrolysis of an ester gives the carboxylic acid and the alcohol, which were combined to form the ester; basic hydrolysis of an ester gives the salt of carboxylic acid and the alcohol, which combine to form the ester.

a. $CH_3-CH_2-\overset{\displaystyle O}{\overset{\displaystyle \|}{C}}-O^-Na^+$ and CH_3-OH

b. $CH_3-\overset{\displaystyle O}{\overset{\displaystyle \|}{C}}-OH$ and $CH_3-CH_2-CH_2-OH$

c. $CH_3-CH_2-CH_2-\overset{\displaystyle O}{\overset{\displaystyle \|}{C}}-OH$ and CH_3-CH_2-OH

d. $\overset{\displaystyle O}{\overset{\displaystyle \|}{C}}-O^-Na^+$ and CH_3CH_2OH

13.33 **a.** This is a primary (1°) amine; there is only one alkyl group attached to the nitrogen atom.
b. This is a secondary (2°) amine; there are two alkyl groups attached to the nitrogen atom.
c. This is a tertiary (3°) amine; there are three alkyl groups attached to the nitrogen atom.
d. This is a tertiary (3°) amine; there are three alkyl groups attached to the nitrogen atom.

13.35 The common name of an amine consists of naming the alkyl groups bonding to the nitrogen atom in alphabetical order. In the IUPAC name, the *e* in the alkane chain is replaced with *amine*.

a. An ethyl group attached to $-NH_2$ is ethylamine.
b. Two alkyl groups attach to nitrogen as methyl and propyl for methylpropylamine.
c. Diethylmethylamine
d. Isopropylamine

13.37 **a.** $CH_3-CH_2-NH_2$

b.

c. $CH_3-CH_2-CH_2-CH_2-\overset{\displaystyle H}{\overset{\displaystyle |}{N}}-CH_2-CH_2-CH_3$

13.39 Amines have higher boiling points than hydrocarbons, but lower than alcohols of similar mass.

a. CH_3-CH_2-OH **b.** $CH_3-CH_2-CH_2-NH_2$ **c.** $CH_3-CH_2-CH_2-NH_2$

13.41 Amines with one to five carbon atoms are soluble. The solubility in water of amines with longer carbon chains decreases.

a. yes   **b.** yes   **c.** no   **d.** yes

13.43 Amines, which are weak bases, bond with a proton from water to give a hydroxide ion and an ammonium ion.

a. $CH_3-NH_2 + H_2O \rightleftarrows CH_3-NH_3^+ + OH^-$

b.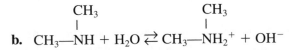

c.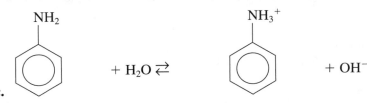

13.45 a. Ethanamide (acetamide) tells us that this amide has two carbon atoms.
b. Butanamide (butyramide) is a chain of four-carbon atoms bonded to an amino group.
c. Methanamide (formamide) is a one-carbon amide.

13.47 a. This is an amide of propionic acid, which has three carbon atoms.

$$CH_3-CH_2-\overset{\displaystyle O}{\overset{\displaystyle \|}{C}}-NH_2$$

b. 2-methyl indicates that a methyl is bonded to carbon 2 in an amide chain of five carbon atoms.

$$CH_3-CH_2-CH_2-\overset{\displaystyle CH_3}{\overset{\displaystyle |}{C}}H-\overset{\displaystyle O}{\overset{\displaystyle \|}{C}}-NH_2$$

c. $CH-\overset{\displaystyle O}{\overset{\displaystyle \|}{C}}-NH_2$

13.49 Acid hydrolysis of amides gives the carboxylic acid and the amine salt.

a. $CH_3-COOH + NH_4^+Cl^-$

b. $CH_3-CH_2-COOH + NH_4^+Cl^-$

c. $CH_3-CH_2-CH_2-COOH + CH_3-NH_3^+Cl^-$

d. a benzene ring $-\overset{\displaystyle O}{\overset{\displaystyle \|}{C}}-OH + NH_4^+Cl^-$

e. $CH_3-CH_2-CH_2-CH_2-COOH + CH_3-CH_2-NH_3^+Cl^-$

13.51 a. $CH_3-CH_2-CH_2-O-\overset{\displaystyle O}{\overset{\displaystyle \|}{C}}-CH_3$

b. $CH_3-CH_2-CH_2-OH + HO-\overset{\displaystyle O}{\overset{\displaystyle \|}{C}}-CH_3 \xrightarrow{H^+} CH_3-CH_2-CH_2-O-\overset{\displaystyle O}{\overset{\displaystyle \|}{C}}-CH_3 + H_2O$

c. $CH_3-CH_2-CH_2-O-\overset{\displaystyle O}{\overset{\displaystyle \|}{C}}-CH_3 + H_2O \xrightarrow{H^+} CH_3-CH_2-CH_2-OH + HO-\overset{\displaystyle O}{\overset{\displaystyle \|}{C}}-CH_3$

d. CH_3—CH_2—CH_2—O—$\overset{\displaystyle O}{\overset{\|}{C}}$—$CH_3$ + NaOH → CH_3—CH_2—CH_2—OH + $Na^{+-}O$—$\overset{\displaystyle O}{\overset{\|}{C}}$—$CH_3$

e. $C_5H_{10}O_2$ molar mass = 102 g/mole

$$1.58 \text{ g} \times \frac{1 \text{ mole}}{102 \text{ g}} \times \frac{1000 \text{ mL NaOH}}{0.208 \text{ mole}} = 74.5 \text{ mL NaOH}$$

13.53 Neo-Synephrine contains aromatic, amine, and alcohol functional groups.

13.55   **a.** 3-methylbutanoic acid   **b.** ethyl benzoate
        c. ethyl propanoate; ethyl propionate   **d.** 2-chlorobenzoic acid
        e. pentanoic acid

13.57   **a.** CH_3—O—$\overset{\displaystyle O}{\overset{\|}{C}}$—$CH_3$

        b. CH_3—CH_2—O—$\overset{\displaystyle O}{\overset{\|}{C}}$—$CH_2$—$CH_2$—$CH_3$

        c. CH_3—CH_2—$\overset{\displaystyle CH_3}{\overset{\|}{CH}}$—$CH_2$—$\overset{\displaystyle O}{\overset{\|}{C}}$—OH

        d. $\overset{\displaystyle O}{\overset{\|}{C}}$—$O$—$CH_2$—$CH_3$ (attached to benzene ring)

13.59   **a.** CH_3—$\overset{\displaystyle O}{\overset{\|}{C}}$—OH

        Ethanoic acid has a higher boiling point than 1-propanol because two molecules of ethanoic acid can hydrogen bond to form a dimer, which gives twice the mass and requires a higher temperature to reach the boiling point.

        b. CH_3—CH_2—$\overset{\displaystyle O}{\overset{\|}{C}}$—OH

        Propanoic acid forms hydrogen bonds and dimers, which require a higher temperature for the boiling point. Butane does not form hydrogen bonds or dimers.

        c. CH_3—CH_2—$\overset{\displaystyle O}{\overset{\|}{C}}$—OH

        Propanoic acid has a higher mass than ethanoic acid and requires a higher temperature to reach the boiling point.

13.61   **a.** CH_3—CH_2—$\overset{\displaystyle O}{\overset{\|}{C}}$—$O^-K^+$ + H_2O

b. $CH_3-CH_2-\overset{\overset{\displaystyle O}{\|}}{C}-O-CH_3 + H_2O$

c.

$\overset{\overset{\displaystyle O}{\|}}{C}-O-CH_2-CH_3 + H_2O$

13.63 a. $CH_3-CH_2-\overset{\overset{\displaystyle O}{\|}}{C}-OH$ and $HO-\overset{\overset{\displaystyle CH_3}{|}}{CH}-CH_3$

b. $CH_3-\overset{\overset{\displaystyle CH_3}{|}}{CH}-\overset{\overset{\displaystyle O}{\|}}{C}-OH$ and $HO-CH_2-CH_2-CH_3$

13.65 a. $CH_3-\overset{\overset{\displaystyle CH_3}{|}}{NH}$

b. $\overset{\displaystyle NH_2}{\bigcirc}$

c. This is an ammonium salt with two methyl groups bonded to the nitrogen atom.

$CH_3-\overset{\overset{\displaystyle CH_3}{|}}{NH_2}{}^+Cl^-$

d. Three ethyl groups are bonded to a nitrogen atom.

$CH_3-CH_2-\overset{\overset{\displaystyle CH_2-CH_3}{|}}{N}-CH_2-CH_3$

e. $\overset{\overset{\displaystyle H}{|}}{N}-CH_3$ (bonded to benzene ring)

13.67 The smaller amines are more soluble in water.

a. Ethylamine has a smaller carbon chain.
b. Trimethylamine has fewer carbon atoms.
c. Butylamine forms hydrogen bonds with water; pentane does not.
d. Butyramide forms hydrogen bonds with water; hexane does not.

13.69 **a.** An amine in water accepts a proton from water, which produces an ammonium ion and OH^-.

$$CH_3—CH_2—NH_3^+OH^-$$

b. The amine accepts a proton to give an ammonium salt: $CH_3—CH_2—NH_3^+ \ Cl^-$

c. $CH_3—CH_2—\overset{+}{N}H_2—CH_3 + OH^-$

d. $CH_3—CH_2—\overset{+}{N}H_2—CH_3 \ Cl^-$

e. An ammonium salt and a strong base produce the amine, a salt, and water.
$$CH_3—CH_2—CH_2—NH_2 + NaCl + H_2O$$

f. $CH_3—CH_2—\underset{\underset{CH_3}{|}}{N}H + NaCl + H_2O$

13.71 carboxylic acid salt, aromatic, amine

13.73 **a.** dimethyl ether, ethyl amine, ethyl alcohol, acetic acid
b. Dimethyl ether and ethyl alcohol are isomers.
c. Acetic acid has the highest boiling point since it can form many hydrogen bonds. Ethanol can also form strong hydrogen bonds. Ethyl amine will form weak hydrogen bonds. Dimethyl ether will not form hydrogen bonds and will have the lowest boiling point.

13.75 **a.**

$$H_2N—\underset{}{\bigcirc}—\overset{\overset{O}{\|}}{C}—O—CH_2—CH_2—\overset{\overset{CH_2—CH_3}{|}}{\underset{\underset{CH_2—CH_3}{|}}{N^+}}—H \quad Cl^-$$

b. The amine salt (Novocain) is more soluble in aqueous body fluids than procaine.

Study Goals

- Identify the common carbohydrates in the diet.
- Distinguish between monosaccharides, disaccharides, and polysaccharides.
- Identify the chiral carbons in a carbohydrate.
- Label the Fischer projection for a monosaccharide as the D or L enantiomer.
- Write Haworth structures for monosaccharides.
- Describe the structural units and bonds in disaccharides and polysaccharides.

Think About It

1. What are some foods you eat that contain carbohydrates?

2. What elements are found in carbohydrates?

3. What carbohydrates are present in table sugar, milk, and wood?

4. What is meant by a "high-fiber" diet?

Key Terms

Match the following key terms with the descriptions shown below.

a. carbohydrate **b.** glucose **c.** disaccharide **d.** Haworth structure **e.** cellulose

1. _____ A simple or complex sugar composed of a carbon chain with an aldehyde or ketone group and several hydroxyl groups

2. _____ A cyclic structure that represents the closed chain form of a monosaccharide

3. _____ An unbranched polysaccharide that cannot be digested by humans

4. _____ An aldohexose that is the most prevalent monosaccharide in the diet

5. _____ A carbohydrate that contains two monosaccharides linked by a glycosidic bond

Answers   **1.** a   **2.** d   **3.** e   **4.** b   **5.** c

14.1 Carbohydrates

- Carbohydrates are classified as monosaccharides (simple sugars), disaccharides (two monosaccharide units), and polysaccharides (many monosaccharide units).
- In a chiral molecule, there is one or more carbon atoms attached to four different atoms or groups.
- Monosaccharides are polyhydroxy aldehydes (aldoses) or ketones (ketoses).
- Monosaccharides are classified by the number of carbon atoms as *trioses, tetroses, pentoses,* or *hexoses.*

◆ **Learning Exercise 14.1A**

Complete and balance the equations for the photosynthesis of

1. glucose: _____ + _____ → $C_6H_{12}O_6$ + _____

2. ribose: _____ + _____ → $C_5H_{10}O_5$ + _____

Answers **1.** $6\ CO_2 + 6\ H_2O \rightarrow C_6H_{12}O_6 + 6\ O_2$
2. $5\ CO_2 + 5\ H_2O \rightarrow C_5H_{10}O_5 + 5\ O_2$

◆ **Learning Exercise 14.1B**

Indicate the number of monosaccharide units (1, 2, or many) in each of the following carbohydrates.

1. Sucrose, a disaccharide _____ **4.** Amylose, a polysaccharide _____

2. Cellulose, a polysaccharide _____ **5.** Maltose, a disaccharide _____

3. Glucose, a monosaccharide _____

Answers **1.** two **2.** many **3.** one **4.** many **5.** two

◆ **Learning Exercise 14.1C**

Identify the following monosaccharides as aldo- or ketotrioses, tetroses, pentoses, or hexoses.

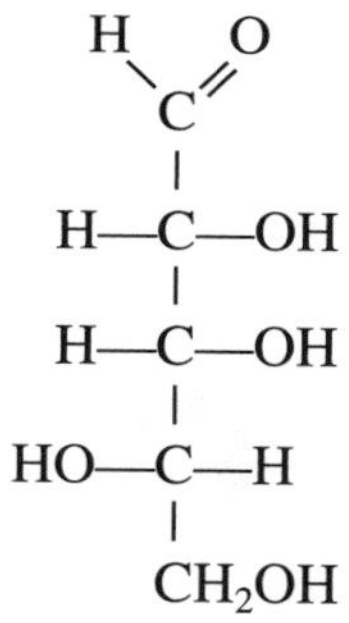

1. _____ **4.** _____

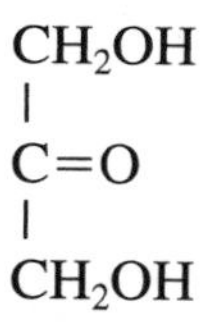

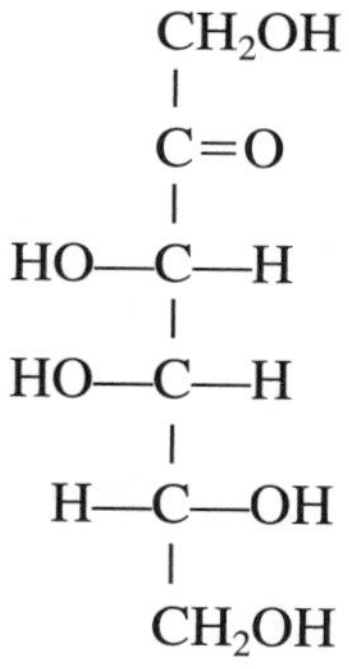

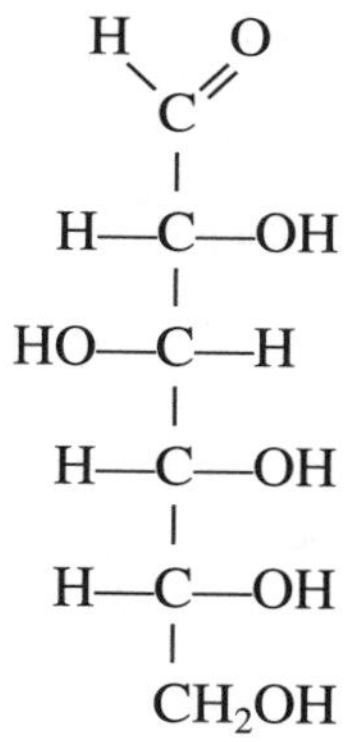

2. _____ **5.** _____

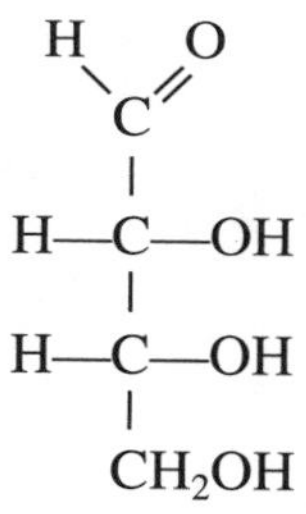

3. _____

Answers **1.** ketotriose **2.** aldopentose **3.** ketohexose
4. aldohexose **5.** aldotetrose

14.2 Fischer Projections of Monosaccharides

- In a Fischer projection (straight chain), the prefixes D and L are used to distinguish between the mirror images. In D-glyceraldehyde, the —OH is on the right of the chiral carbon; it is on the left in L-glyceraldehyde.
- In the Fischer projection of a monosaccharide, the chiral —OH farthest from the carbonyl group (C=O) is on the left side in the L isomer, and on the right side in the D isomer.
- Important monosaccharides are the aldohexoses glucose and galactose and the ketohexose fructose.

◆ Learning Exercise 14.2A

Identify each of the following sugars as the D or L isomer.

1. 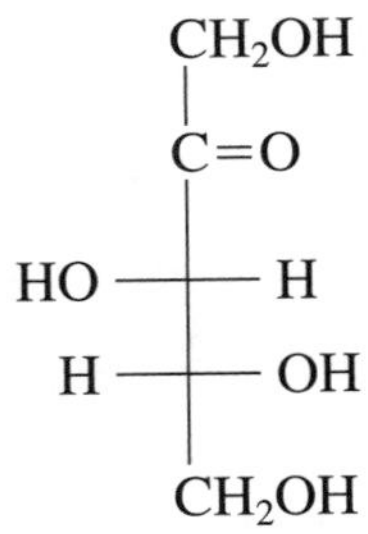
```
      CH₂OH
       |
       C=O
       |
HO ——— H
       |
 H ——— OH
       |
      CH₂OH
```

2. 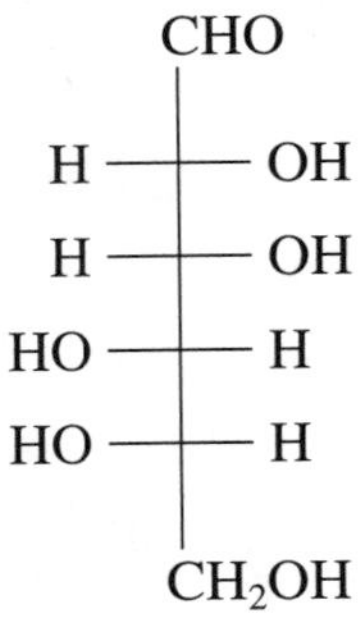
```
       CHO
        |
 H ——— OH
        |
 H ——— OH
        |
HO ——— H
        |
HO ——— H
        |
      CH₂OH
```

3. 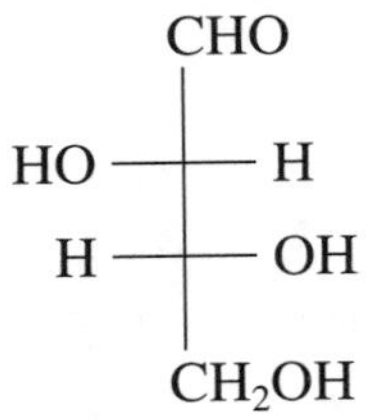
```
        CHO
         |
HO ——— H
         |
 H ——— OH
         |
      CH₂OH
```

4. 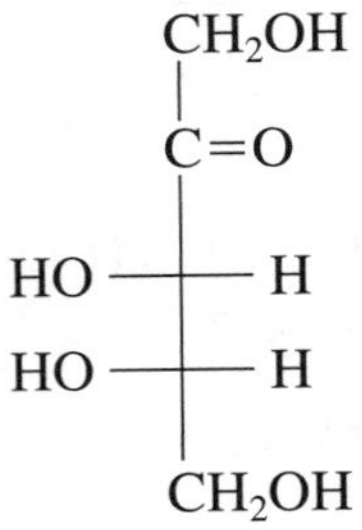
```
      CH₂OH
       |
       C=O
       |
HO ——— H
       |
HO ——— H
       |
      CH₂OH
```

_____ -Xylulose   _____ -Mannose   _____ -Threose   _____ -Ribulose

Answers   **1.** D-Xylulose   **2.** L-Mannose   **3.** D-Threose   **4.** L-Ribulose

◆ Learning Exercise 14.2B

Write the mirror image of each of the sugars in learning exercise 14.2A and give the D or L name.

1.   **2.**   **3.**   **4.**

Answers

1. 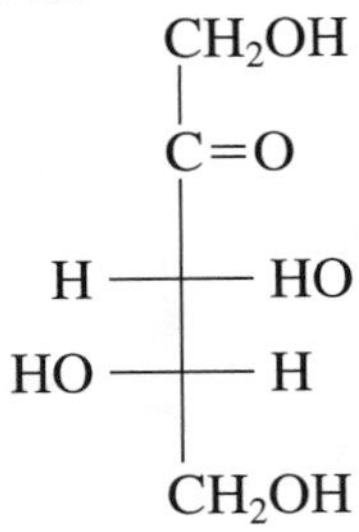
```
      CH₂OH
       |
       C=O
       |
 H ——— HO
       |
HO ——— H
       |
      CH₂OH
```
L-Xylulose

2. 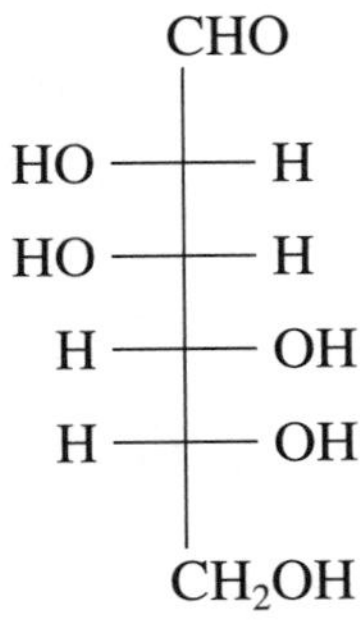
```
       CHO
        |
HO ——— H
        |
HO ——— H
        |
 H ——— OH
        |
 H ——— OH
        |
      CH₂OH
```
D-Mannose

3. 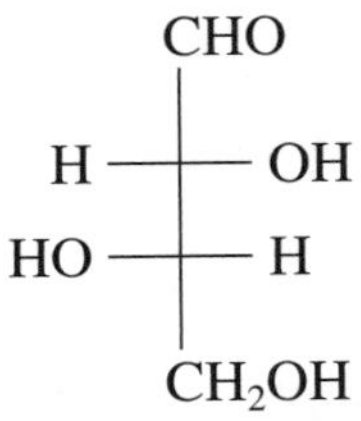
```
        CHO
         |
 H ——— OH
         |
HO ——— H
         |
      CH₂OH
```
L-Threose

4. 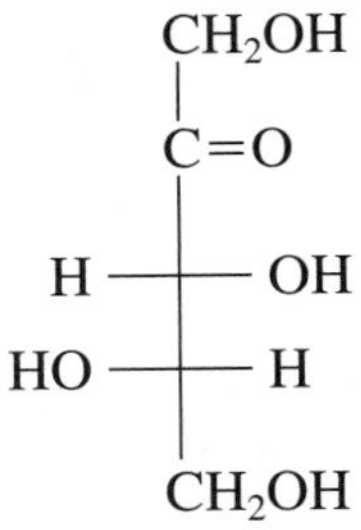
```
      CH₂OH
       |
       C=O
       |
 H ——— OH
       |
HO ——— H
       |
      CH₂OH
```
D-Ribulose

◆ Learning Exercise 14.2C

Identify the monosaccharide (glucose, fructose, or galactose) that fits the following description.

1. a building block in cellulose _____

2. also known as fruit sugar _____

3. accumulates in the disease known as *galactosemia* _____

4. the most common monosaccharide _____

5. the sweetest monosaccharide _____

Answers **1.** glucose **2.** fructose **3.** galactose **4.** glucose **5.** fructose

◆ Learning Exercise 14.2D

Draw the open-chain structure for the following monosaccharides:

D-Glucose   D-Galactose   D-Fructose

Answers

D-Glucose   D-Galactose   D-Fructose

14.3 Haworth Structures of Monosaccharides

• The predominant form of monosaccharides is the cyclic form of five or six atoms called the Haworth structure. The cyclic structure forms by a reaction between an OH on carbon 5 in hexoses with the carbonyl group of the same molecule.

• The formation of a new hydroxyl group on carbon 1 (or 2 in fructose) gives α and β forms of the cyclic monosaccharide. Because the molecule opens and closes continuously in solution, both α and β forms are present.

◆ Learning Exercise 14.3

Write the Haworth structures (α-form) for the following.

1. D-Glucose   **2.** D-Galactose   **3.** D-Fructose

Answers

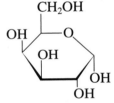

α-D-Glucose   α-D-Galactose   α-D-Fructose

14.4 Chemical Properties of Monosaccharides

• Monosaccharides contain functional groups that undergo oxidation or reduction.

• Monosaccharides are called *reducing sugars* because the aldehyde group (also available in ketoses) is oxidized by a metal ion such as Cu^{2+} in Benedict's solution, which is reduced.

• Monosaccharides are also reduced to give sugar alcohols.

◆ Learning Exercise 14.4

What changes occur when a reducing sugar reacts with Benedict's reagent?

Answers   The carbonyl group of the reducing sugar is oxidized to a carboxylic acid group; the Cu^{2+} ion in Benedict's reagent is reduced to Cu^+, which forms a brick-red solid of Cu_2O.

14.5 Disaccharides

- Disaccharides are two monosaccharide units joined together by a glycosidic bond:

    monosaccharide (1) + monosaccharide $(2) \rightarrow$ disaccharide + H_2O

- In the most common disaccharides, maltose, lactose, and sucrose, there is at least one glucose unit.
- In the disaccharide maltose, two glucose units are linked by an α-1, 4 bond. The α-1, 4 indicates that the —OH of the alpha form at carbon 1 was bonded to the —OH on carbon 4 of the other glucose molecule.
- When a disaccharide is hydrolyzed by water, the products are a glucose unit and one other monosaccharide.

$$Maltose + H_2O \rightarrow Glucose + Glucose$$
$$Lactose + H_2O \rightarrow Glucose + Galactose$$
$$Sucrose + H_2O \rightarrow Glucose + Fructose$$

◆ Learning Exercise 14.5

For the following disaccharides, state (a) the monosaccharide units, (b) the type of glycosidic bond, and (c) the name of the disaccharide.

	a. Monosaccharide(s)	b. Type of glycosidic bond	c. Name of disaccharide
1.			
2.			

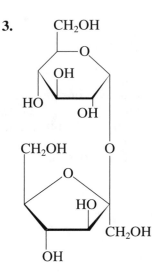

3.

4.

	a. Monosaccharide units	b. Type of glycosidic bond	c. Name of disaccharide
3.			
4.			

Answers

1. **a.** two glucose units   **b.** α-1, 4-glycosidic bond   **c.** β-maltose
2. **a.** galactose + glucose   **b.** α-1, 4-glycosidic bond   **c.** α-lactose
3. **a.** fructose + glucose   **b.** α-1, β-2 glycosidic bond   **c.** sucrose
4. **a.** two glucose units   **b.** α-1, 4-glycosidic bond   **c.** α-maltose

14.6 Polysaccharides

- Polysaccharides are polymers of monosaccharide units.
- Starches consist of amylose, an unbranched chain of glucose; amylopectin is a branched polymer of glucose. Glycogen, the storage form of glucose in animals, is similar to amylopectin with more branching.
- Cellulose is also a polymer of glucose, but in cellulose the glycosidic bonds are β bonds rather than α bonds as in the starches. Humans can digest starches to obtain energy, but not cellulose. However, cellulose is important as a source of fiber in our diets.

◆ Learning Exercise 14.6

List the monosaccharides and describe the glycosidic bonds in each of the following carbohydrates:

	Monosaccharides	Type(s) of glycosidic bonds
1. amylose	_____	_____
2. amylopectin	_____	_____
3. glycogen	_____	_____
4. cellulose	_____	_____

Answers
   1. glucose; α-1, 4-glycosidic bonds   **2.** glucose; α-1, 4- and α-1, 6-glycosidic bonds
   3. glucose; α-1, 4- and α-1, 6-glycosidic bonds **4.** glucose; β-1, 4-glycosidic bonds

Checklist for Chapter 14

You are ready to take the practice test for Chapter 14. Be sure that you have accomplished the following learning goals for this chapter. If you are not sure, review the section listed at the end of the goal. Then apply your new skills and understanding to the practice test.

After studying Chapter 14, I can successfully:

_____ Classify carbohydrates as monosaccharides, disaccharides, and polysaccharides (14.1).

_____ Classify a monosaccharide as aldose or ketose and indicate the number of carbon atoms (14.1).

_____ Draw and identify D- and L-Fischer projections for carbohydrate molecules (14.2).

_____ Draw the open-chain structures for D-glucose, D-galactose, and D-fructose (14.2).

_____ Draw or identify the cyclic structures of monosaccharides (14.3).

_____ Describe some chemical properties of carbohydrates (14.4).

_____ Describe the monosaccharide units and linkages in disaccharides (14.5).

_____ Describe the structural features of amylose, amylopectin, glycogen, and cellulose (14.6).

Practice Test for Chapter 14

1. The requirements for photosynthesis are

   A. sun.   **B.** sun and water.   **C.** water and carbon dioxide.
   D. sun, water, and carbon dioxide.   **E.** carbon dioxide and sun.

2. What are the products of photosynthesis?

   A. carbohydrates   **B.** carbohydrates and oxygen
   C. carbon dioxide and oxygen   **D.** carbohydrates and carbon dioxide
   E. water and oxygen

3. The name "carbohydrate" came from the fact that

   A. carbohydrates are hydrates of water.
   B. carbohydrates contain hydrogen and oxygen in a 2:1 ratio.
   C. carbohydrates contain a great quantity of water.
   D. all plants produce carbohydrates.
   E. carbon and hydrogen atoms are abundant in carbohydrates.

4. What functional groups are in the open chains of monosaccharides?

    A. hydroxyl groups
    B. aldehyde groups
    C. ketone groups
    D. hydroxyl and aldehyde or ketone groups
    E. hydroxyl and ether groups

5. What is the classification of the following sugar?

CH_2OH
|
$C=O$
|
CH_2OH

    A. aldotriose   **B.** ketotriose   **C.** aldotetrose   **D.** ketotetrose   **E.** ketopentose

Questions 6 through 10 refer to

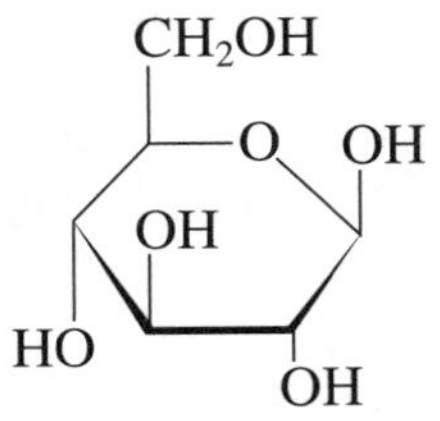

6. It is the cyclic structure of an

    A. aldotriose.   **B.** ketopentose.   **C.** aldopentose.
    D. aldohexose.   **E.** aldoheptose.

7. This is a Haworth structure of

    A. fructose.   **B.** glucose.   **C.** ribose.
    D. glyceraldehyde. **E.** galactose.

8. It is at least one of the products of the complete hydrolysis of

    A. maltose.   **B.** sucrose.   **C.** lactose.
    D. glycogen.   **E.** All of these

9. A Benedict's test with this sugar would

    A. be positive.   **B.** be negative.   **C.** produce a blue precipitate.
    D. give no color change.   **E.** produce a silver mirror.

10. It is the monosaccharide unit used to build polymers of

    A. amylose.   **B.** amylopectin.   **C.** cellulose.
    D. glycogen   **E.** All of these

Identify each of the carbohydrates described in 11 through 15 as:

    A. maltose   **B.** sucrose   **C.** cellulose
    D. amylopectin   **E.** glycogen

11. _____ A disaccharide that is not a reducing sugar

12. _____ A disaccharide that occurs as a breakdown product of amylose

13. _____ A carbohydrate that is produced as a storage form of energy in plants

14. _____ The storage form of energy in humans

15. _____ A carbohydrate that is used for structural purposes by plants

For questions 16 through 20 select answers from:

    A. amylose   **B.** cellulose   **C.** glycogen
    D. lactose   **E.** sucrose

16. _____ A polysaccharide composed of many glucose units linked by α-1, 4-glycosidic bonds.

17. _____ A sugar containing both glucose and galactose.

18. _____ A sugar composed of glucose units joined by both α-1, 4- and α-1, 6-glycosidic bonds.

19. _____ A disaccharide that is not a reducing sugar.

20. _____ A carbohydrate composed of glucose units joined by β-1, 4-glycosidic bonds.

For questions 21 through 25, select answers from the following:

    A. glucose   **B.** lactose   **C.** sucrose   **D.** maltose

21. _____ A sugar composed of glucose and fructose

22. _____ Also called table sugar

23. _____ Found in milk and milk products

24. _____ Gives sorbitol upon reduction

25. _____ Gives galactose upon hydrolysis

Answers to the Practice Test

1. D	**2.** B	**3.** B	**4.** D	**5.** B
6. D	**7.** B	**8.** E	**9.** A	**10.** E
11. B	**12.** A	**13.** D	**14.** E	**15.** C
16. A	**17.** D	**18.** C	**19.** E	**20.** B
21. C	**22.** C	**23.** B	**24.** A	**25.** B

Answers and Solutions to Selected Text Problems

14.1 Photosynthesis requires CO_2, H_2O, and the energy from the sun. Respiration requires O_2 from the air and glucose from our foods.

14.3 Monosaccharides can be a chain of three to eight carbon atoms, one in a carbonyl group as an aldehyde or ketone, and the rest attached to hydroxyl groups. A monosaccharide cannot be split or hydrolyzed into smaller carbohydrates. A disaccharide consists of two monosaccharide units joined together that can be split.

14.5 Hydroxyl groups and a carbonyl are found in all monosaccharides.

14.7 The name ketopentose tells us that the compound contains a ketone functional group and has five carbon atoms. In addition, all monosaccharides contain hydroxyl groups.

14.9   **a.** This monosaccharide is a ketose; it has a carbonyl between two carbon atoms.
     b. This monosaccharide is an aldose; it has a CHO, an aldehyde group.
     c. This monosaccharide is a ketose; it has a carbonyl group between two carbon atoms.
     d. This monosaccharide is an aldose; it has a CHO, an aldehyde group.
     e. This monosaccharide is an aldose; it has a CHO, an aldehyde group.

14.11 In the α isomer, the —OH on the chiral carbon atom at the bottom of the chain is on the right side, whereas in the β isomer the —OH appears on the left side.

14.13 **a.** This structure is a D isomer since the hydroxyl on the chiral carbon farthest from the carbonyl is on the right.
b. This structure is a D isomer since the hydroxyl on the chiral carbon farthest from the carbonyl is on the right.
c. This structure is an L isomer since the hydroxyl on the chiral carbon farthest from the carbonyl is on the left.
d. This structure is a D isomer since the hydroxyl on the chiral carbon farthest from the carbonyl is on the right.

14.15 **a.** 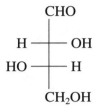 **b.**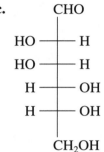

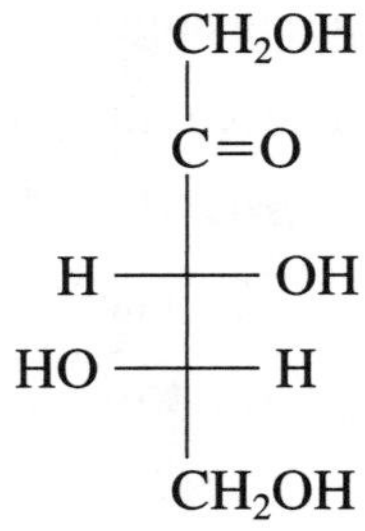

c. 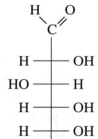 **d.**

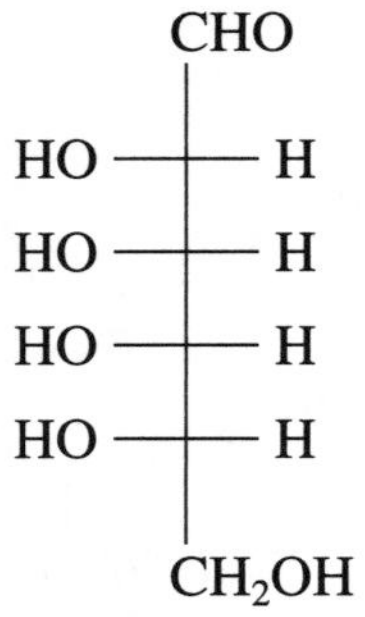

14.17 L-glucose is the mirror image of D-glucose.

D-glucose 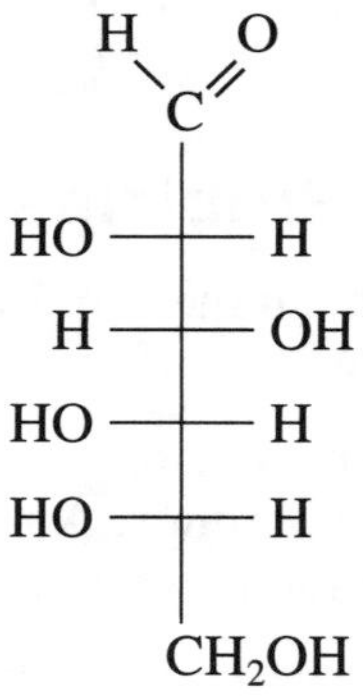   L-glucose

14.19 In D-galactose the hydroxyl on carbon four extends to the left; in glucose this hydroxyl group goes to the right.

14.21 **a.** Glucose is also called dextrose.
b. Galactose is not metabolized in the condition called galactosemia.
c. Another name for fructose is fruit sugar.

14.23 In the cyclic structure of glucose, there are five carbon atoms and an oxygen atom in the ring.

14.25 In the α-form, the hydroxyl (—OH) on carbon 1 is down; the β-form has the hydroxyl (—OH) on carbon 1 up.

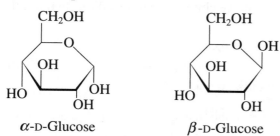

α-D-Glucose β-D-Glucose

14.27 **a.** This is the α-form because the —OH on carbon 1 is down.
b. This is the α-form because the —OH on carbon 1 is down.

14.29

```
        CH₂OH
          |
   H——C——OH
          |
  HO——C——H
          |
   H——C——OH
          |
        CH₂OH
```

D-Xylitol

14.31 Oxidation product:

```
        O
        ‖
        C——OH
        |
  HO——C——H
        |
   H——C——OH
        |
   H——C——OH
        |
      CH₂OH
```

Reduction product:

```
       CH₂OH
        |
 HO——C——H
        |
  H——C——OH
        |
  H——C——OH
        |
      CH₂OH
```

D-arabitol

14.33 **a.** When this disaccharide is hydrolyzed, galactose and glucose are produced. The glycosidic bond is a β-1, 4 bond since the ether bond is up from the 1 carbon of the galactose, which is on the left in the drawing, to the 4 carbon of the glucose on the right. β-Lactose is the name of this disaccharide since the free hydroxyl is up.

b. When this disaccharide is hydrolyzed, two molecules of glucose are produced. The glycosidic bond is an α-1, 4 bond since the ether bond is down from the 1 carbon of the glucose on the left to the 4 carbon of the glucose on the right. α-Maltose is the name of this disaccharide since the free hydroxyl is down.

14.35 **a.** Can be oxidized.   **b.** Can be oxidized.

14.37 **a.** Another name for table sugar is sucrose.
   b. Lactose is the disaccharide found in milk and milk products.
   c. Maltose is also called malt sugar.
   d. When lactose is hydrolyzed, the products are the monosaccharides galactose and glucose.

14.39 **a.** Amylose is an unbranched polymer of glucose units joined by α-1, 4 bonds; amylopectin is a branched polymer of glucose joined by α-1, 4 and α-1, 6 bonds.
   b. Amylopectin, produced by plants, is a branched polymer of glucose joined by α-1, 4 and α-1, 6 bonds. Glycogen, which is made by animals, is a highly branched polymer of glucose joined by α-1, 4 and α-1, 6 bonds.

14.41 **a.** Cellulose is not digestible by humans since we do not have the enzymes necessary to break the β-1, 4-glycosidic bonds in cellulose.
   b. Amylose and amylopectin are the storage forms of carbohydrates in plants.
   c. Amylose is the polysaccharide, which contains only α-1, 4-glycosidic bonds.
   d. Glycogen contains many α-1, 4 and α-1, 6 bonds and is the most highly branched polysaccharide.

14.43 **a.** Isomaltose is a disaccharide.
   b. Isomaltose consists of two glucose molecules.
   c. The glycosidic link in isomaltose is an α-1, 6 bond.
   d. The structure shown is α-isomaltose.
   e. Isomaltose is a reducing sugar.

14.45 **a.** Melezitose is a trisaccharide.
   b. Melezitose contains two aldohexose molecules of glucose.

14.47 D-Galactose has a carbonyl group on carbon 1 and a hydroxyl group on carbon 2. In D-fructose, the carbonyl group is on carbon 2 and the hydroxyl group is on carbon 1. The hydroxyl group on carbon 4 is on the left in D-galactose and on the right in D-fructose.

14.49 D-Galactose is the mirror image of L-galactose. In D-galactose, the —OH on carbons 2 and 5 are on the right side, but on the left for carbons 3 and 4. In L-galactose, the —OH are reversed; carbons 2 and 5 have —OH on the left, and carbons 3 and 4 have —OH on the right.

14.51 **a.**

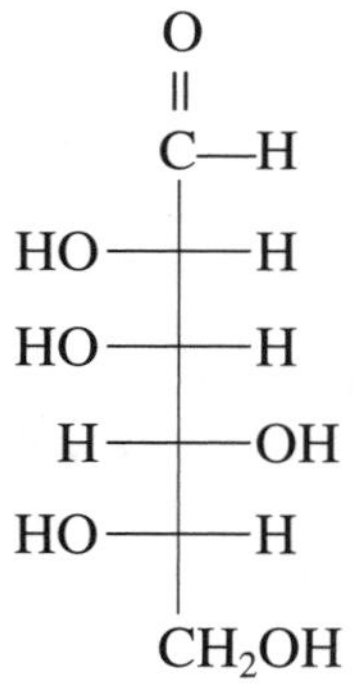

L-Gulose

b.

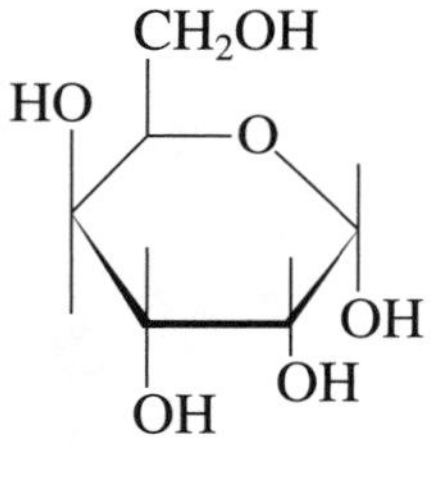

α-D-Gulose

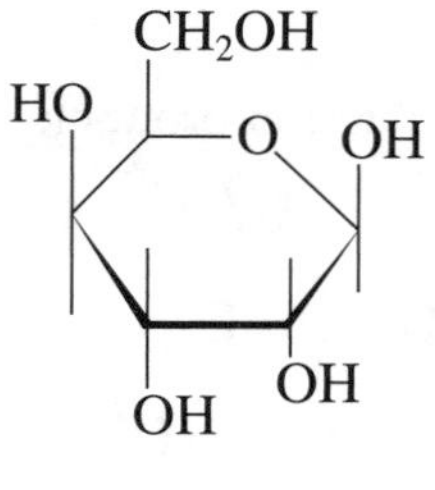

β-D-Gulose

14.53 Since sorbitol can be oxidized to D-glucose, it must contain the same number of carbons with the same groups attached as glucose. The difference is that sorbitol has only hydroxyls while glucose has an aldehyde group. In sorbitol, the aldehyde group is changed to a hydroxyl.

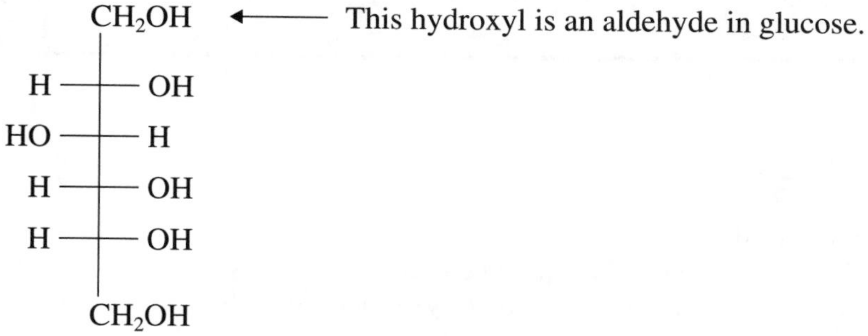

CH₂OH ← This hydroxyl is an aldehyde in glucose.

14.55 When α-galactose forms an open chain structure it can close to form either α- or β-galactose.

14.57

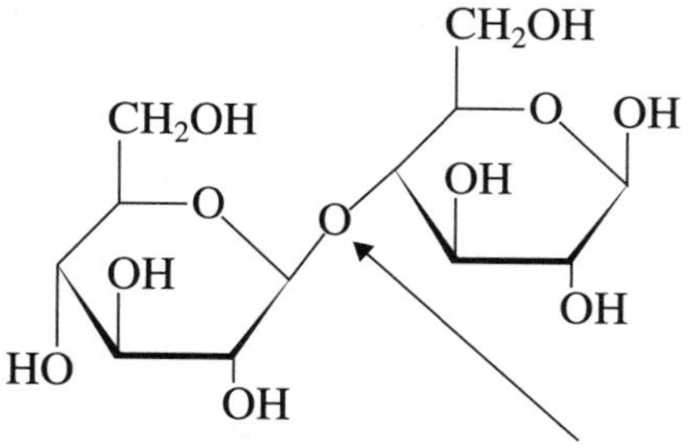

β-1,4-glycosidic bond. The bond from the glucose on the left is up (β).

14.59 a.

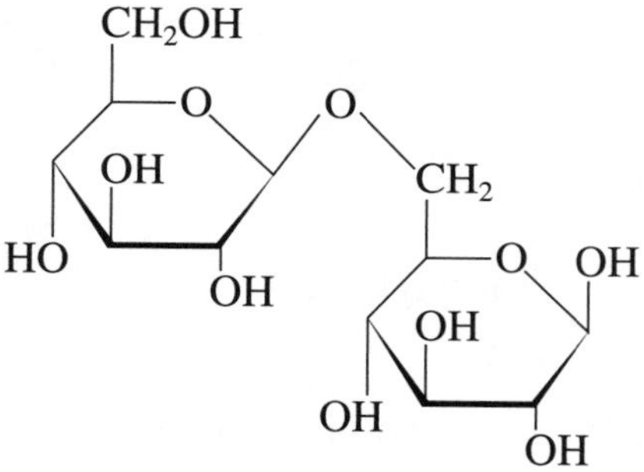

b. Yes. The ring on the right side can open up to form an aldehyde.

Study Goals

- Describe the properties and types of lipids.
- Write the structures of triacylglycerols obtained from glycerol and fatty acids.
- Draw the structure of the product from hydrogenation, hydrolysis, and saponification of triacylglycerols.
- Describe glycerophospholipids.
- Describe steroids and their role in bile salts, vitamins, and hormones.
- Describe the lipid bilayer in a cell.

Think About It

1. What are fats used for in the body?

2. What foods are high in fat?

3. What oils are used to produce margarines?

4. What kind of lipid is cholesterol?

Key Terms

Match the key terms with the correct statement shown below.

a. lipid	**b.** fatty acid	**c.** triacylglycerol
d. saponification	**e.** glycerophospholipid	**f.** steroid

1. _____ A lipid consisting of glycerol bonded to two fatty acids and a phosphate group attached to an amino group

2. _____ A type of compound that is not soluble in water but is soluble in nonpolar solvents

3. _____ The hydrolysis of a triacylglycerol with a strong base producing salts called soaps and glycerol

4. _____ A lipid consisting of glycerol bonded to three fatty acids

5. _____ A lipid composed of a multicyclic ring system

6. _____ Long-chain carboxylic acid found in triacylglycerols

Answers   **1.** e   **2.** a   **3.** d   **4.** c   **5.** f   **6.** b

15.1 Lipids

- Lipids are nonpolar compounds that are not soluble in water.
- Classes of lipids include waxes, triacylglycerols, glycerophospholipids, and steroids.

◆ Learning Exercise 15.1

Match one of the classes of lipids with the composition of lipids below:

    a. wax   **b.** triacylglycerol   **c.** glycerophospholipid   **d.** steroid

1. _____ A fused structure of four cycloalkanes

2. _____ A long chain alcohol and a fatty acid

3. _____ Glycerol and three fatty acids

4. _____ Glycerol, two fatty acids, phosphate, and choline

Answers   **1.** d   **2.** a   **3.** b   **4.** c

15.2 Fatty Acids

- Fatty acids are unbranched carboxylic acids that typically contain an even number (12–18) of carbon atoms.
- Fatty acids may be saturated, monounsaturated with one double bond, or polyunsaturated with two or more double bonds. The double bonds in unsaturated fatty acids are almost always cis.

◆ Learning Exercise 15.2

Draw the structural formulas of linoleic acid, stearic acid, and oleic acid.

    a. linoleic acid

    b. stearic acid

    c. oleic acid

Which of these three fatty acids

1. _____ is the most saturated?   **4.** _____ has the highest melting point?

2. _____ is the most unsaturated?   **5.** _____ is found in vegetables?

3. _____ has the lowest melting point?   **6.** _____ is from animal sources?

Answers

linoleic acid $CH_3-(CH_2)_4-CH=CH-CH_2-CH=CH-(CH_2)_7-\overset{\overset{\displaystyle O}{\|}}{C}-OH$

stearic acid $CH_3-(CH_2)_{16}-\overset{\overset{\displaystyle O}{\|}}{C}-OH$

oleic acid $CH_3-(CH_2)_7-CH=CH-(CH_2)_7-\overset{\overset{\displaystyle O}{\|}}{C}-OH$

1. b   **2.** a   **3.** a   **4.** b   **5.** a and c   **6.** b

15.3 Waxes, Fats, and Oils

- A wax is an ester of a long-chain fatty acid and a long-chain alcohol.
- The triacylglycerols in fats and oils are esters of glycerol with three long-chain fatty acids.
- Fats from animal sources contain more saturated fatty acids and have higher melting points than fats found in most vegetable oils.

◆ Learning Exercise 15.3A

Write the formula of the wax formed by the reaction of palmitic acid, CH_3—$(CH_2)_{14}$—COOH, and cetyl alcohol, CH_3—$(CH_2)_{14}$—CH_2—OH.

Answer

$$CH_3\text{—}(CH_2)_{14}\text{—}\overset{\overset{\displaystyle O}{\|}}{C}\text{—}O\text{—}CH_2\text{—}(CH_2)_{14}\text{—}CH_3$$

◆ Learning Exercise 15.3B

Consider the following fatty acid called oleic acid.

1. Why is the compound an acid?

2. Is it a saturated or unsaturated compound? Why?

3. Is the double bond cis or trans?

4. Is it likely to be a solid or a liquid at room temperature?

5. Why is it not soluble in water?

Answers
1. contains a carboxylic acid group
2. unsaturated; double bond
3. cis
4. liquid
5. it has a long hydrocarbon chain

◆ Learning Exercise 15.3C

Write the structure and name of the triacylglycerol formed from the following.

1. Glycerol and three palmitic acids, $CH_3-(CH_2)_{14}-COOH$

2. Glycerol and three myristic acids, $CH_3-(CH_2)_{12}-COOH$

Answers

1.

$$CH_2-O-\overset{\overset{\displaystyle O}{\|}}{C}-(CH_2)_{14}-CH_3$$
$$HC-O-\overset{\overset{\displaystyle O}{\|}}{C}-(CH_2)_{14}-CH_3$$
$$CH_2-O-\overset{\overset{\displaystyle O}{\|}}{C}-(CH_2)_{14}-CH_3$$

Glyceryl tripalmitate
(Tripalmitin)

2.

$$CH_2-O-\overset{\overset{\displaystyle O}{\|}}{C}-(CH_2)_{12}-CH_3$$
$$HC-O-\overset{\overset{\displaystyle O}{\|}}{C}-(CH_2)_{12}-CH_3$$
$$CH_2-O-\overset{\overset{\displaystyle O}{\|}}{C}-(CH_2)_{12}-CH_3$$

Glyceryl trimyristate
(Trimyristin)

◆ Learning Exercise 15.3D

Write the structural formulas of the following triacylglycerols:

1. Glyceryl tristearate (tristearin)

2. Glyceryl trioleate (triolein)

Answers

1.

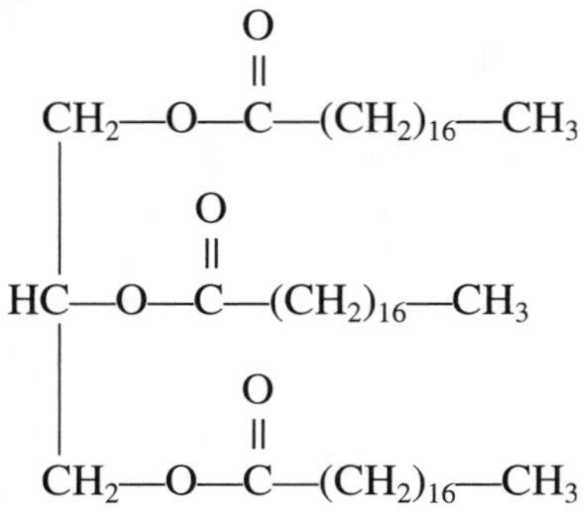

$$CH_2\text{—}O\text{—}\overset{\displaystyle O}{\overset{\|}{C}}\text{—}(CH_2)_{16}\text{—}CH_3$$

$$HC\text{—}O\text{—}\overset{\displaystyle O}{\overset{\|}{C}}\text{—}(CH_2)_{16}\text{—}CH_3$$

$$CH_2\text{—}O\text{—}\overset{\displaystyle O}{\overset{\|}{C}}\text{—}(CH_2)_{16}\text{—}CH_3$$

Glyceryl tristearate
(Tristearin)

2.

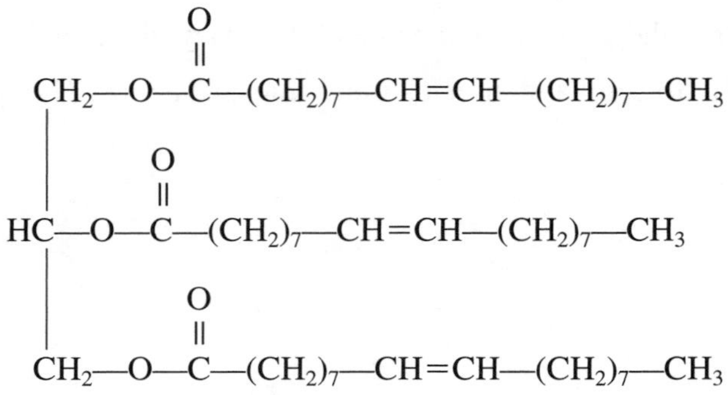

$$CH_2\text{—}O\text{—}\overset{\displaystyle O}{\overset{\|}{C}}\text{—}(CH_2)_7\text{—}CH{=}CH\text{—}(CH_2)_7\text{—}CH_3$$

$$HC\text{—}O\text{—}\overset{\displaystyle O}{\overset{\|}{C}}\text{—}(CH_2)_7\text{—}CH{=}CH\text{—}(CH_2)_7\text{—}CH_3$$

$$CH_2\text{—}O\text{—}\overset{\displaystyle O}{\overset{\|}{C}}\text{—}(CH_2)_7\text{—}CH{=}CH\text{—}(CH_2)_7\text{—}CH_3$$

Glyceryl trioleate
(Triolein)

15.4 Chemical Properties of Triacylglycerols

- The hydrogenation of unsaturated fatty acids converts double bonds to single bonds.
- The hydrolysis of the ester bonds in fats or oils produces glycerol and fatty acids.
- In saponification, a fat heated with a strong base produces glycerol and the salts of the fatty acids or soaps. The dual polarity of soap permits its solubility in both water and oil.

◆ Learning Exercise 15.4

Write the equations for the following reactions of glyceryl trioleate (triolein).

1. hydrogenation with a nickel catalyst

2. acid hydrolysis with HCl

3. saponification with NaOH

Answers

1.

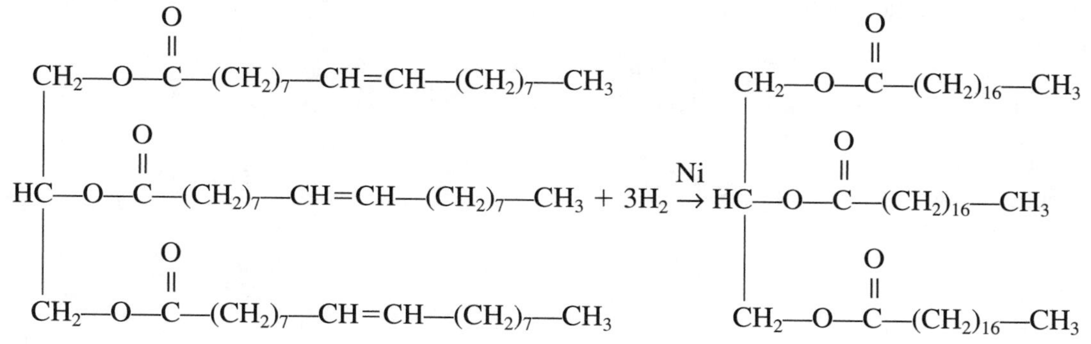

$$\begin{array}{c}
CH_2-O-\overset{\overset{O}{\|}}{C}-(CH_2)_7-CH=CH-(CH_2)_7-CH_3 \\
HC-O-\overset{\overset{O}{\|}}{C}-(CH_2)_7-CH=CH-(CH_2)_7-CH_3 \\
CH_2-O-\overset{\overset{O}{\|}}{C}-(CH_2)_7-CH=CH-(CH_2)_7-CH_3
\end{array} + 3H_2 \xrightarrow{Ni}
\begin{array}{c}
CH_2-O-\overset{\overset{O}{\|}}{C}-(CH_2)_{16}-CH_3 \\
HC-O-\overset{\overset{O}{\|}}{C}-(CH_2)_{16}-CH_3 \\
CH_2-O-\overset{\overset{O}{\|}}{C}-(CH_2)_{16}-CH_3
\end{array}$$

2.

$$\begin{array}{c}
CH_2-O-\overset{\overset{O}{\|}}{C}-(CH_2)_7-CH=CH-(CH_2)_7-CH_3 \\
HC-O-\overset{\overset{O}{\|}}{C}-(CH_2)_7-CH=CH-(CH_2)_7-CH_3 \\
CH_2-O-\overset{\overset{O}{\|}}{C}-(CH_2)_7-CH=CH-(CH_2)_7-CH_3
\end{array} + 3H_2O \xrightarrow{H^+}
\begin{array}{c}
CH_2-OH \\
HC-OH \\
CH_2-OH
\end{array}$$

$$+\ 3\ HO-\overset{\overset{O}{\|}}{C}-(CH_2)_7-CH=CH-(CH_2)_7-CH_3$$

3.

$$\begin{array}{c}
CH_2-O-\overset{\overset{O}{\|}}{C}-(CH_2)_7-CH=CH-(CH_2)_7-CH_3 \\
HC-O-\overset{\overset{O}{\|}}{C}-(CH_2)_7-CH=CH-(CH_2)_7-CH_3 \\
CH_2-O-\overset{\overset{O}{\|}}{C}-(CH_2)_7-CH=CH-(CH_2)_7-CH_3
\end{array} + 3NaOH \rightarrow
\begin{array}{c}
CH_2-OH \\
HC-OH \\
CH_2-OH
\end{array}$$

$$+\ 3\ Na^{+-}O-\overset{\overset{O}{\|}}{C}-(CH_2)_7-CH=CH-(CH_2)_7-CH_3$$

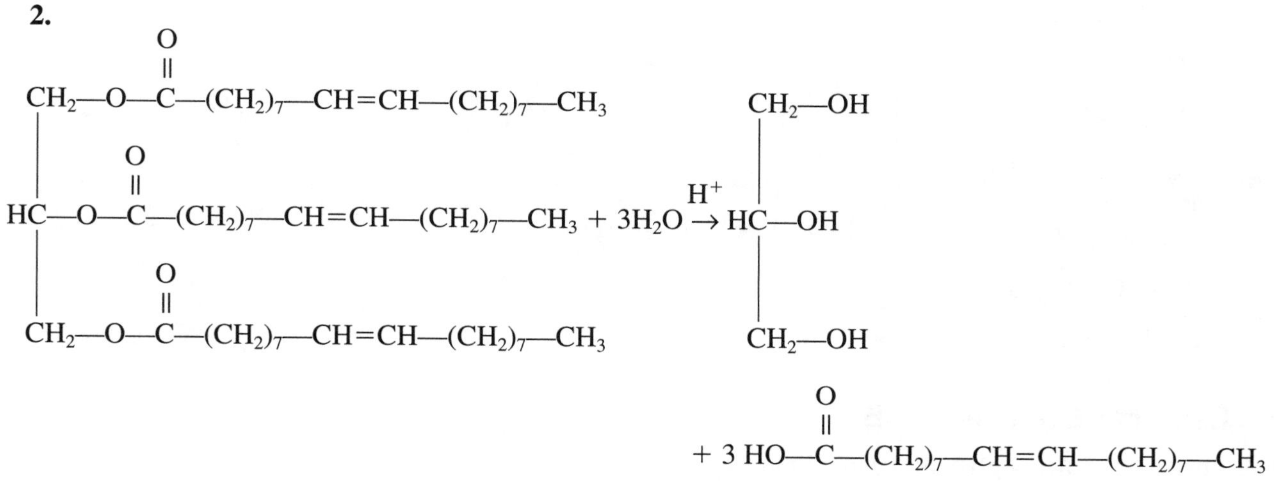

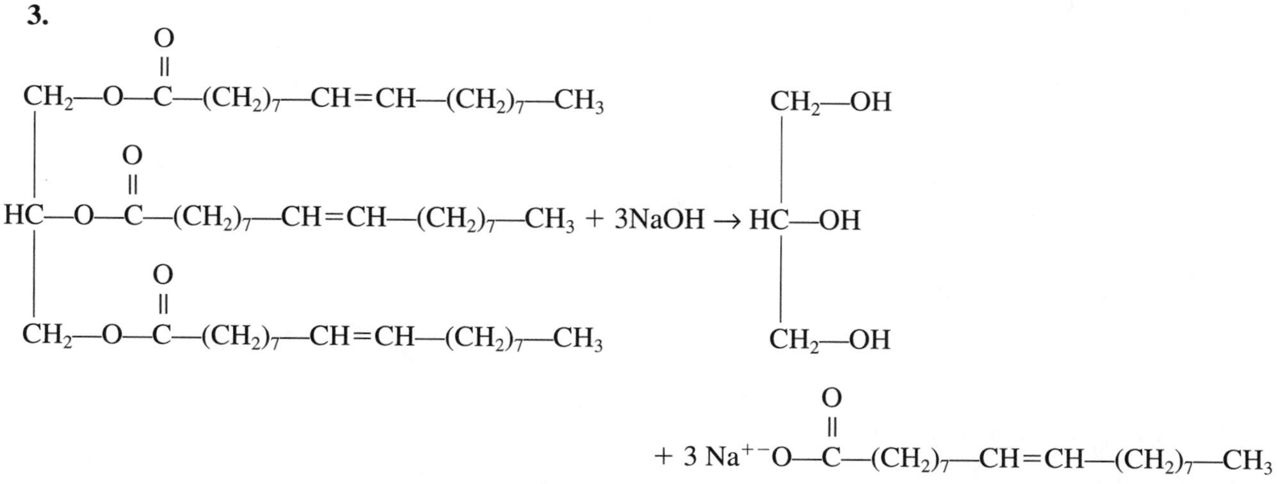

15.5 Glycerophospholipids

- Glycerophospholipids are esters of glycerol with two fatty acids and a phosphate group attached to an amino alcohol.
- The fatty acids are a nonpolar region, whereas the phosphate group and the amino alcohol make up a polar region.

◆ Learning Exercise 15.5A

Draw the structure of a glycerophospholipid that is formed from two molecules of palmitic acid and serine, an amino alcohol.

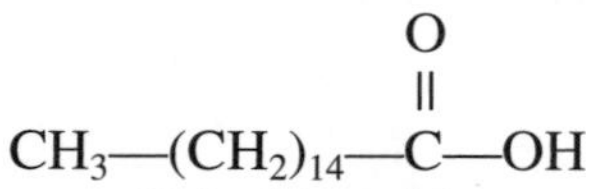

$$
\begin{array}{cc}
\overset{\displaystyle O}{\overset{\displaystyle \|}{}} & \overset{+}{NH_3}\ \overset{\displaystyle O}{\overset{\displaystyle \|}{}} \\
CH_3—(CH_2)_{14}—C—OH & HO—CH_2—CH—C—O^- \\
\text{Palmitic acid} & \text{Serine}
\end{array}
$$

Answer

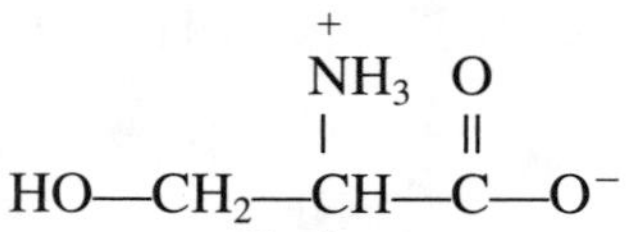

$$
\begin{array}{l}
\qquad\quad \overset{\displaystyle O}{\overset{\displaystyle \|}{}} \\
CH_2—O—C—(CH_2)_{14}—CH_3 \\
|\qquad\quad \overset{\displaystyle O}{\overset{\displaystyle \|}{}} \\
HC—O—C—(CH_2)_{14}—CH_3 \\
|\qquad\quad \overset{\displaystyle O}{\overset{\displaystyle \|}{}}\qquad\quad \overset{+}{NH_3}\ \overset{\displaystyle O}{\overset{\displaystyle \|}{}} \\
CH_2—O—P—O—CH_2—CH—C—O^- \\
\qquad\quad\ |\\
\qquad\quad O^-
\end{array}
$$

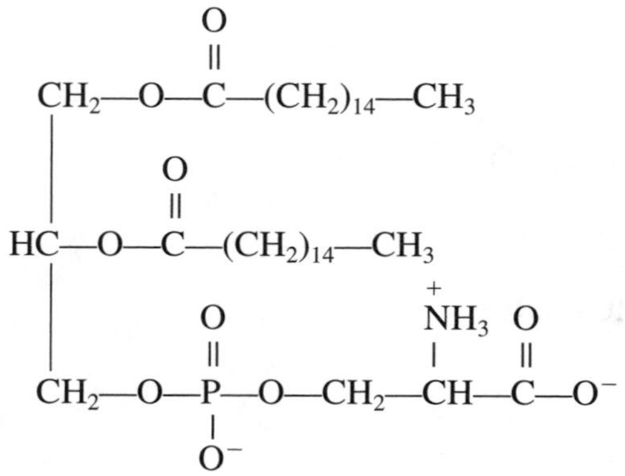

◆ Learning Exercise 15.5B

Consider the following glycerophospholipid.

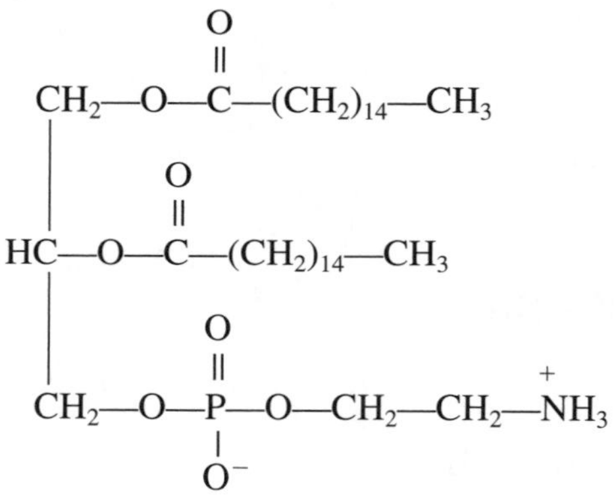

$$
\begin{array}{l}
\qquad\quad \overset{\displaystyle O}{\overset{\displaystyle \|}{}} \\
CH_2—O—C—(CH_2)_{14}—CH_3 \\
|\qquad\quad \overset{\displaystyle O}{\overset{\displaystyle \|}{}} \\
HC—O—C—(CH_2)_{14}—CH_3 \\
|\qquad\quad \overset{\displaystyle O}{\overset{\displaystyle \|}{}} \\
CH_2—O—P—O—CH_2—CH_2—\overset{+}{N}H_3 \\
\qquad\quad\ |\\
\qquad\quad O^-
\end{array}
$$

On the above structure, indicate the

A. two fatty acids.	**B.** the part from the glycerol molecule.
C. the phosphate section.	**D.** the amino alcohol group.
E. the nonpolar region.	**F.** the polar region.

1. What is the name of the amino alcohol group? _____

2. Why is a glycerophospholipid more soluble in water than most lipids? _____

Answers

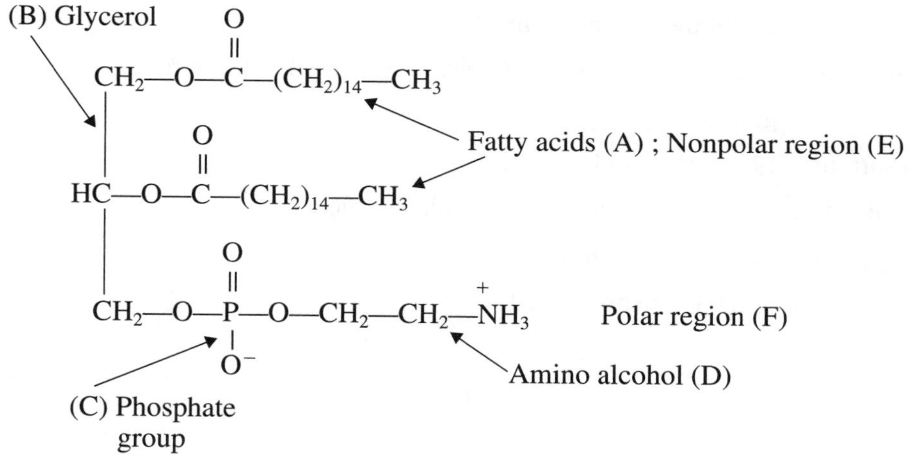

(B) Glycerol

$$CH_2-O-\overset{\overset{O}{\|}}{C}-(CH_2)_{14}-CH_3$$

$$HC-O-\overset{\overset{O}{\|}}{C}-(CH_2)_{14}-CH_3$$

Fatty acids (A) ; Nonpolar region (E)

$$CH_2-O-\overset{\overset{O}{\|}}{\underset{\underset{O^-}{|}}{P}}-O-CH_2-CH_2-\overset{+}{N}H_3$$

Polar region (F)

Amino alcohol (D)

(C) Phosphate group

1. ethanolamine
2. The polar portion is attracted to water, which makes this type of lipid more soluble in water than other lipids.

15.6 Steroids: Cholesterol and Steroid Hormones

- Steroids are lipids containing the steroid nucleus, which is a fused structure of four rings.
- Steroids include cholesterol, hormones, and vitamin D.
- The steroid hormones are closely related in structure to cholesterol and depend on cholesterol for their synthesis. The sex hormones such as estrogen in females and testosterone in males are responsible for sexual characteristics and reproduction. The adrenal corticosteroid aldosterone regulates reabsorption of Na^+ and water retention in the kidneys. The adrenal corticosteroid cortisone regulates blood glucose levels.

◆ Learning Exercise 15.6A

1. Write the structure of the steroid nucleus. 2. Write the structure of cholesterol.

Answers

1.

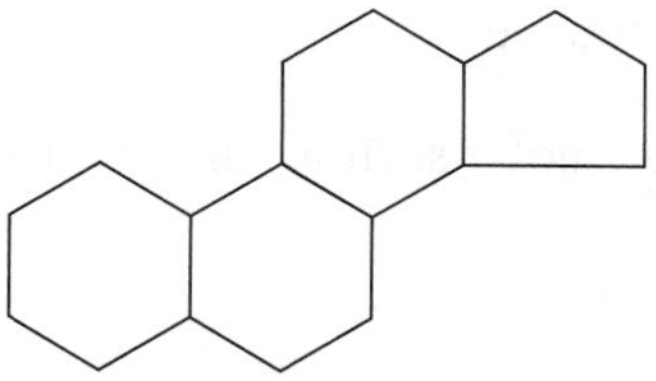

2.

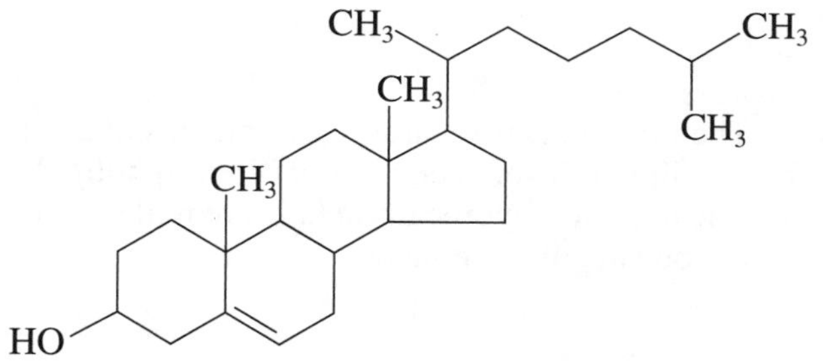

◆ **Learning Exercise 15.6B**

Identify one of these compounds with the following statements.

    a. estrogen   **b.** testosterone   **c.** cortisone   **d.** aldosterone

1. _____ Increases the blood level of glucose

2. _____ Increases the reabsorption of Na^+ in the kidneys

3. _____ Stimulates development of secondary sex characteristics in females

4. _____ Stimulates reabsorption of water by the kidneys

5. _____ Stimulates the secondary sex characteristics in males

Answers   **1.** c   **2.** d   **3.** a   **4.** d   **5.** b

15.7 Cell Membranes

- Cell membranes surround all of our cells and separate the cellular contents from the external liquid environment.
- A cell membrane is a lipid bilayer composed of two rows of phospholipids such that the nonpolar hydrocarbon tails are in the center and the polar sections are aligned along the outside.
- The inner portion of the lipid bilayer consists of nonpolar chains of the fatty acids, with the polar heads at the outer and inner surfaces.
- Molecules of cholesterol, proteins, and carbohydrates are embedded in the lipid bilayer.

◆ **Learning Exercise 15.7**

  a. What is the function of the lipid bilayer in cell membranes?

  b. What type of lipid makes up the lipid bilayer?

  c. What is the general arrangement of the lipids in a lipid bilayer?

Answers
  a. The lipid bilayer separates the contents of a cell from the surrounding aqueous environment.
  b. The lipid bilayer is composed of phospholipids.
  c. The nonpolar hydrocarbon tails are in the center of the bilayer, while the polar sections are aligned along the outside of the bilayer.

Checklist for Chapter 15

You are ready to take the practice test for Chapter 15. Be sure that you have accomplished the following learning goals for this chapter. If you are not sure, review the section listed at the end of the goal. Then apply your new skills and understanding to the practice test.
After studying Chapter 15, I can successfully:

_____ Describe the classes of lipids (15.1).

_____ Identify a fatty acid as saturated or unsaturated (15.2).

_____ Write the structural formula of a wax or triacylglycerol produced by the reaction of a fatty acid and an alcohol or glycerol (15.3).

_____ Draw the structure of the product from the reaction of a triacylglycerol with hydrogen, an acid, or a base (15.4).

_____ Describe the components of glycerophospholipids (15.5).

_____ Describe the structure of a steroid and cholesterol (15.6).

_____ Describe the function of the lipid bilayer in cell membranes (15.7).

Practice Test for Chapter 15

1. An ester of a fatty acid is called a
   A. carbohydrate   **B.** lipid   **C.** protein   **D.** oxyacid   **E.** soap

2. A fatty acid that is unsaturated is usually from
   A. animal sources and liquid at room temperature.
   B. animal sources and solid at room temperature.
   C. vegetable sources and liquid at room temperature.
   D. vegetable sources and solid at room temperature.
   E. both vegetable and animal sources and solid at room temperature.

3. $CH_3-(CH_2)_{16}-\overset{\displaystyle O}{\overset{\displaystyle \|}{C}}-OH$ is a
   A. unsaturated fatty acid   **B.** saturated fatty acid   **C.** wax
   D. triacylglycerol   **E.** steroid

For questions 4 through 7, consider the following compound.

$$CH_2-O-\overset{\displaystyle O}{\overset{\displaystyle \|}{C}}-(CH_2)_{16}-CH_3$$
$$HC-O-\overset{\displaystyle O}{\overset{\displaystyle \|}{C}}-(CH_2)_{16}-CH_3$$
$$CH_2-O-\overset{\displaystyle O}{\overset{\displaystyle \|}{C}}-(CH_2)_{16}-CH_3$$

4. This compound belongs in the family called
    A. wax.   **B.** triacylglycerol.   **C.** glycerophospholipid.
    D. fatty acid.   **E.** steroid.

5. The molecule shown above was formed by
    A. esterification.   **B.** hydrolysis (acid).   **C.** saponification.
    D. emulsification.   **E.** oxidation.

6. If this molecule were reacted with a strong base such as NaOH, the products would be
    A. glycerol and fatty acids.   **B.** glycerol and water.
    C. glycerol and soap.   **D.** an ester and salts of fatty acids.
    E. an ester and fatty acids.

7. The compound would be expected to be
    A. saturated, and a solid at room temperature.
    B. saturated, and a liquid at room temperature.
    C. unsaturated, and a solid at room temperature.
    D. unsaturated, and a liquid at room temperature.
    E. supersaturated, and a liquid at room temperature.

8. Which are found in glycerophospholipids?
    A. fatty acids   **B.** glycerol   **C.** a nitrogen compound
    D. phosphate   **E.** All of these

For questions 9 and 10, consider the following reaction:

$$\text{Triacylglycerol} + 3\text{NaOH} \rightarrow 3 \text{ sodium salts of fatty acids and glycerol}$$

9. The reaction of a triacylglycerol with a strong base such as NaOH is called
    A. esterification.   **B.** lipogenesis.   **C.** hydrolysis.
    D. saponification.   **E.** oxidation.

10. What is another name for the sodium salts of the fatty acids?
    A. margarines   **B.** fat substitutes   **C.** soaps
    D. perfumes   **E.** vitamins

For questions 11 through 16, consider the following glycerophospholipid.

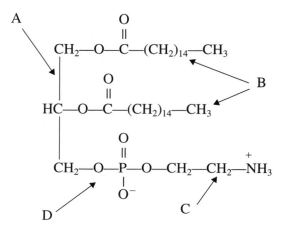

Match the labels in the glycerophospholipid with the following:

11. _____ The glycerol portion   **12.** _____ The phosphate portion

13. _____ The amino alcohol   **14.** _____ The polar region

15. _____ The nonpolar region

16. Type of glycerophospholipid
    A. choline   **B.** cephalin   **C.** fat   **D.** steroid   **E.** cerebroside

Classify the lipids in 17 through 20 as

    A. wax   **B.** triacylglycerol   **C.** glycerophospholipid
    D. steroid   **E.** fatty acid

17. _____ cholesterol

18. _____ $CH_3—(CH_2)_{14}—\overset{\displaystyle O}{\overset{\|}{C}}—OH$

19. _____ $CH_3—(CH_2)_{14}—\overset{\displaystyle O}{\overset{\|}{C}}—O—(CH_2)_{29}—CH_3$

20. _____ an ester of glycerol with three palmitic acids

Select answers for 21 through 25 from the following:

    A. testosterone   **B.** estrogen   **C.** prednisone
    D. cortisone   **E.** aldosterone

21. _____ Stimulates the female sexual characteristics

22. _____ Increases the retention of water by the kidneys

23. _____ Stimulates the male sexual characteristics

24. _____ Increases the blood glucose level

25. _____ Used medically to reduce inflammation and treat asthma

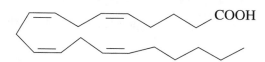

26. This compound is a
    A. cholesterol.   **B.** triacylglycerol.   **C.** fatty acid.
    D. glycerophospholipid.   **E.** steroid.

27. The lipid bilayer of a cell is composed of
    A. cholesterol.   **B.** glycerophospholipids.   **C.** proteins.
    D. carbohydrate side chains.   **E.** All of these

28. The type of lipoprotein that transports cholesterol to the liver for elimination is called
    A. chylomicron.   **B.** high density lipoprotein. **C.** low density lipoprotein.
    D. very low density lipoprotein.   **E.** All of these

259

Answers to the Practice Test

1. B	**2.** C	**3.** B	**4.** B	**5.** A
6. C	**7.** A	**8.** E	**9.** D	**10.** C
11. A	**12.** D	**13.** C	**14.** C, D	**15.** B
16. B	**17.** D	**18.** E	**19.** A	**20.** B
21. B	**22.** E	**23.** A	**24.** D	**25.** C
26. C	**27.** E	**28.** B		

Answers and Solutions to Selected Text Problems

15.1 Lipids provide energy, protection, and insulation for the organs in the body.

15.3 Since lipids are not soluble in water, they are nonpolar molecules.

15.5 All fatty acids contain a long chain of carbon atoms with a carboxylic acid group. Saturated fatty acids contain only carbon-to-carbon single bonds; unsaturated fatty acids contain one or more double bonds.

15.7 **a.** palmitic acid

     b. oleic acid

15.9 **a.** Lauric acid has only carbon–carbon single bonds; it is saturated.
    b. Linolenic acid has three carbon–carbon double bonds; it is unsaturated.
    c. Palmitoleic acid has one carbon–carbon double bond; it is unsaturated.
    d. Stearic acid has only carbon–carbon single bonds; it is saturated.

15.11 In a cis double bond, the alkyl groups are on the same side of the double bond, whereas in trans fatty acids, the alkyl groups are on opposite sides.

15.13 In an omega-3 fatty acid, the first double bond occurs at carbon 3 counting from the methyl. In an omega-6 fatty acid, the first double bond occurs at carbon 6.

15.15 Arachidonic acid contains four double bonds and no side groups. In PGE_1, a part of the chain forms cyclopentane and there are hydroxyl and ketone functional groups.

15.17 Prostaglandins raise blood pressure, stimulate contraction and relaxation of smooth muscle, and may cause inflammation and pain.

15.19 Palmitic acid is the 16-carbon saturated fatty acid. $CH_3-(CH_2)_{14}-\overset{\displaystyle O}{\overset{\displaystyle \|}{C}}-O-(CH_2)_{29}-CH_3$

15.21 Fats are composed of fatty acids and glycerol. In this case, the fatty acid is stearic acid, an 18-carbon saturated fatty acid.

$$
\begin{array}{l}
\quad\quad\quad \overset{\displaystyle O}{\overset{\displaystyle \|}{}} \\
CH_2-O-C-(CH_2)_{16}-CH_3 \\
\mid \quad\quad \overset{\displaystyle O}{\overset{\displaystyle \|}{}} \\
HC-O-C-(CH_2)_{16}-CH_3 \\
\mid \quad\quad \overset{\displaystyle O}{\overset{\displaystyle \|}{}} \\
CH_2-O-C-(CH_2)_{16}-CH_3
\end{array}
$$

15.23 Glyceryl tripalmitate (tripalmitin) has three palmitic acids (16-carbon saturated fatty acid) forming ester bonds with glycerol.

$$\begin{array}{l}
\text{CH}_2\text{—O—}\overset{\displaystyle \text{O}}{\overset{\|}{\text{C}}}\text{—(CH}_2)_{14}\text{—CH}_3 \\[1em]
\text{HC—O—}\overset{\displaystyle \text{O}}{\overset{\|}{\text{C}}}\text{—(CH}_2)_{14}\text{—CH}_3 \\[1em]
\text{CH}_2\text{—O—}\overset{\displaystyle \text{O}}{\overset{\|}{\text{C}}}\text{—(CH}_2)_{14}\text{—CH}_3
\end{array}$$

15.25 Safflower oil contains fatty acids with two or three double bonds; olive oil contains a large amount of oleic acid, which has a single (monounsaturated) double bond.

15.27 Although coconut oil comes from a vegetable source, it has large amounts of saturated fatty acids and small amounts of unsaturated fatty acids. Since coconut oil contains the same kinds of fatty acids as animal fat, coconut oil has a melting point similar to the melting point of animal fats.

15.29

$$\begin{array}{l}
\text{CH}_2\text{—O—}\overset{\text{O}}{\overset{\|}{\text{C}}}\text{—(CH}_2)_7\text{—CH=CH—(CH}_2)_7\text{—CH}_3 \\[1em]
\text{HC—O—}\overset{\text{O}}{\overset{\|}{\text{C}}}\text{—(CH}_2)_7\text{—CH=CH—(CH}_2)_7\text{—CH}_3 + 3\text{H}_2 \overset{\text{Ni}}{\rightarrow} \\[1em]
\text{CH}_2\text{—O—}\overset{\text{O}}{\overset{\|}{\text{C}}}\text{—(CH}_2)_7\text{—CH=CH—(CH}_2)_7\text{—CH}_3
\end{array}$$

$$\begin{array}{l}
\text{CH}_2\text{—O—}\overset{\text{O}}{\overset{\|}{\text{C}}}\text{—(CH}_2)_{16}\text{—CH}_3 \\[1em]
\text{HC—O—}\overset{\text{O}}{\overset{\|}{\text{C}}}\text{—(CH}_2)_{16}\text{—CH}_3 \\[1em]
\text{CH}_2\text{—O—}\overset{\text{O}}{\overset{\|}{\text{C}}}\text{—(CH}_2)_{16}\text{—CH}_3
\end{array}$$

15.31 Acid hydrolysis of a fat gives glycerol and the fatty acids. Basic hydrolysis (saponification) of fat gives glycerol and the salts of the fatty acids.

a.

$$\begin{array}{l}
\text{CH}_2\text{—O—}\overset{\text{O}}{\overset{\|}{\text{C}}}\text{—(CH}_2)_{12}\text{—CH}_3 \\[1em]
\text{CH—O—}\overset{\text{O}}{\overset{\|}{\text{C}}}\text{—(CH}_2)_{12}\text{—CH}_3 + 3\text{H}_2\text{O} \overset{\text{H}^+}{\rightarrow} \\[1em]
\text{CH}_2\text{—O—}\overset{\text{O}}{\overset{\|}{\text{C}}}\text{—(CH}_2)_{12}\text{—CH}_3
\end{array}$$

$$\begin{array}{l}
\text{CH}_2\text{OH} \\[1em]
\text{CHOH} + 3\text{HO}\overset{\text{O}}{\overset{\|}{\text{C}}}\text{—(CH}_2)_{12}\text{—CH}_3 \\[1em]
\text{CH}_2\text{OH}
\end{array}$$

b.

$$\begin{array}{l}
\text{CH}_2\text{—O—}\overset{\text{O}}{\overset{\|}{\text{C}}}\text{—(CH}_2)_{12}\text{—CH}_3 \\[1em]
\text{CH—O—}\overset{\text{O}}{\overset{\|}{\text{C}}}\text{—(CH}_2)_{12}\text{—CH}_3 + 3\text{NaOH} \rightarrow \\[1em]
\text{CH}_2\text{—O—}\overset{\text{O}}{\overset{\|}{\text{C}}}\text{—(CH}_2)_{12}\text{—CH}_3
\end{array}$$

$$\begin{array}{l}
\text{CH}_2\text{OH} \\[1em]
\text{CHOH} + 3\ \text{Na}^+\,{}^-\text{O—}\overset{\text{O}}{\overset{\|}{\text{C}}}\text{—(CH}_2)_{12}\text{—CH}_3 \\[1em]
\text{CH}_2\text{OH}
\end{array}$$

15.33 A triacylglycerol is a combination of three fatty acids bonded to glycerol by ester bonds. Olestra is sucrose bonded to six to eight fatty acids by ester bonds. Olestra cannot be digested because digestive enzymes cannot break down the molecule.

15.35

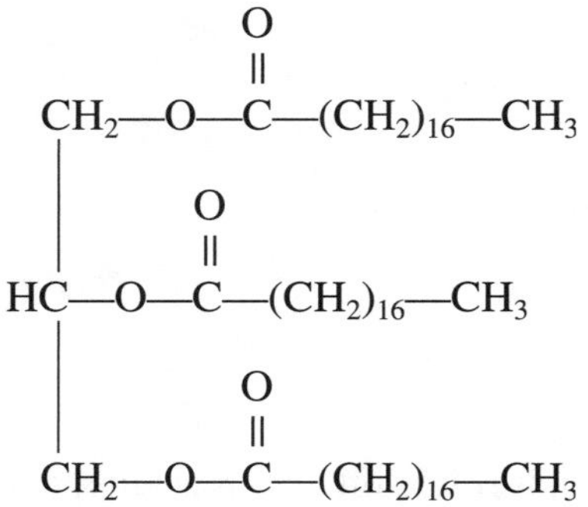

$$CH_2-O-\overset{\overset{\displaystyle O}{\|}}{C}-(CH_2)_{16}-CH_3$$

$$HC-O-\overset{\overset{\displaystyle O}{\|}}{C}-(CH_2)_{16}-CH_3$$

$$CH_2-O-\overset{\overset{\displaystyle O}{\|}}{C}-(CH_2)_{16}-CH_3$$

15.37 A triacylglycerol consists of glycerol and three fatty acids. A glycerophospholipid consists of glycerol, two fatty acids, a phosphate group, and an amino alcohol.

15.39

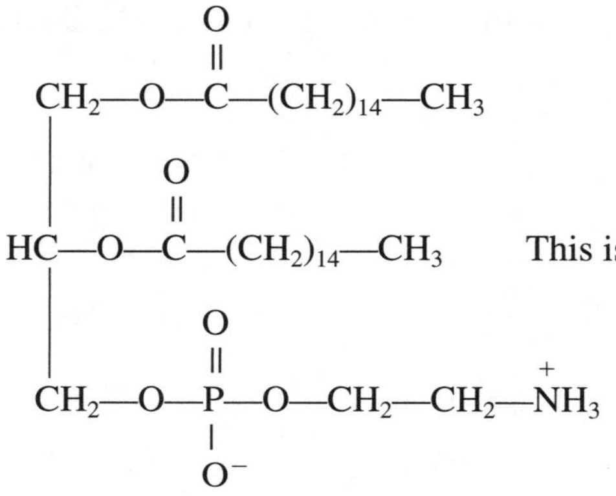

$$CH_2-O-\overset{\overset{\displaystyle O}{\|}}{C}-(CH_2)_{14}-CH_3$$

$$HC-O-\overset{\overset{\displaystyle O}{\|}}{C}-(CH_2)_{14}-CH_3 \qquad \text{This is a cephalin.}$$

$$CH_2-O-\underset{\underset{\displaystyle O^-}{|}}{\overset{\overset{\displaystyle O}{\|}}{P}}-O-CH_2-CH_2-\overset{+}{N}H_3$$

15.41 This glycerophospholipid is a cephalin. It contains glycerol, oleic acid, stearic acid, phosphate, and ethanolamine.

15.43

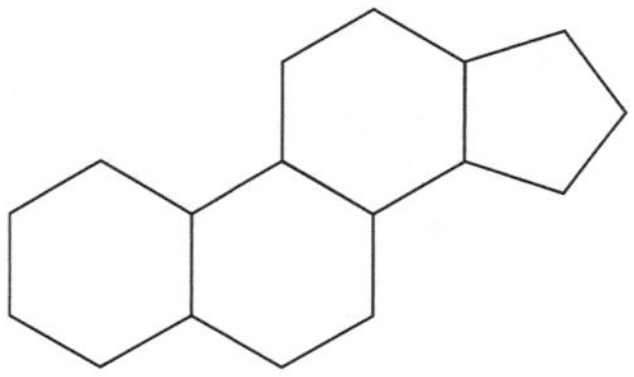

15.45 Lipoproteins are large, spherically shaped molecules that transport lipids in the bloodstream. They consist of an outer layer of glycerophospholipids and proteins surrounding an inner core of hundreds of nonpolar lipids and cholesteryl esters.

15.47 Chylomicrons have a lower density than VLDLs. They pick up triacylglycerols from the intestine, whereas VLDLs transport triacylglycerols synthesized in the liver.

15.49 "Bad" cholesterol is the cholesterol carried by LDLs to the tissues where it can form deposits called plaque, which can narrow the arteries.

15.51 Both estradiol and testosterone contain the steroid nucleus and a hydroxyl group. Testosterone has a ketone group, a double bond, and an extra methyl group. Estradiol has a benzene ring and a second hydroxyl group.

15.53 Testosterone is a male sex hormone.

15.55 Cell membranes contain glycerophospholipids that contain fatty acids with cis double bonds and smaller amounts of cholesterol.

15.57 The function of the lipid bilayer in the plasma membrane is to keep the cell contents separated from the outside environment and to allow the cell to regulate the movement of substances into and out of the cell.

15.59 The peripheral proteins in the membrane emerge on the inner or outer surface only, whereas the integral proteins extend through the membrane to both surfaces.

15.61 The carbohydrates on the surface of cells act as receptors for cell recognition and chemical messengers such as neurotransmitters.

15.63

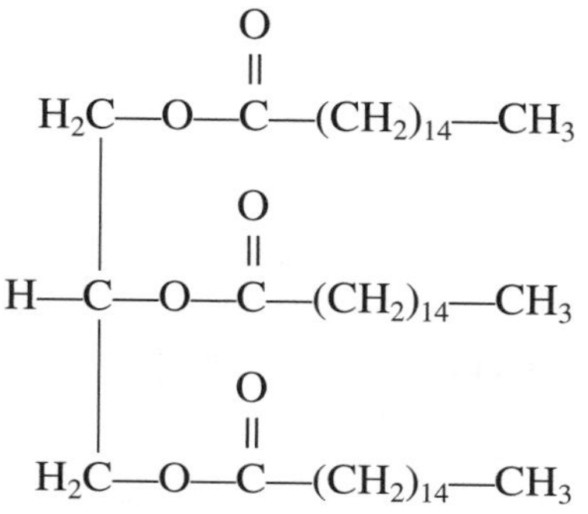

$$H_2C-O-\overset{\overset{\displaystyle O}{\|}}{C}-(CH_2)_{14}-CH_3$$

$$H-\overset{|}{C}-O-\overset{\overset{\displaystyle O}{\|}}{C}-(CH_2)_{14}-CH_3$$

$$H_2C-O-\overset{\overset{\displaystyle O}{\|}}{C}-(CH_2)_{14}-CH_3$$

15.65 a.

$$H_2C-O-\overset{\overset{\displaystyle O}{\|}}{C}-(CH_2)_7-CH=CH-CH_2-CH=CH-(CH_2)_4-CH_3$$

$$H-\overset{|}{C}-O-\overset{\overset{\displaystyle O}{\|}}{C}-(CH_2)_7-CH=CH-CH_2-CH=CH-(CH_2)_4-CH_3$$

$$H_2C-O-\overset{\overset{\displaystyle O}{\|}}{C}-(CH_2)_7-CH=CH-CH_2-CH=CH-(CH_2)_4-CH_3$$

$$H_2C-O-\overset{\overset{\displaystyle O}{\|}}{C}-(CH_2)_7-CH=CH-CH_2-CH=CH-(CH_2)_4-CH_3$$

$$H-\overset{|}{C}-O-\overset{\overset{\displaystyle O}{\|}}{C}-(CH_2)_7-CH=CH-(CH_2)_7-CH_3$$

$$H_2C-O-\overset{\overset{\displaystyle O}{\|}}{C}-(CH_2)_7-CH=CH-CH_2-CH=CH-(CH_2)_4-CH_3$$

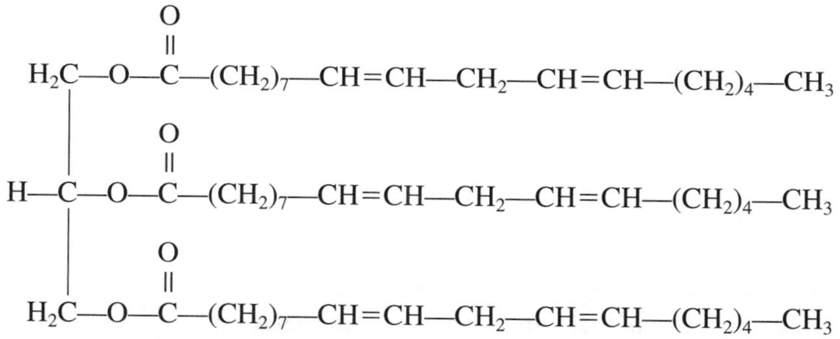

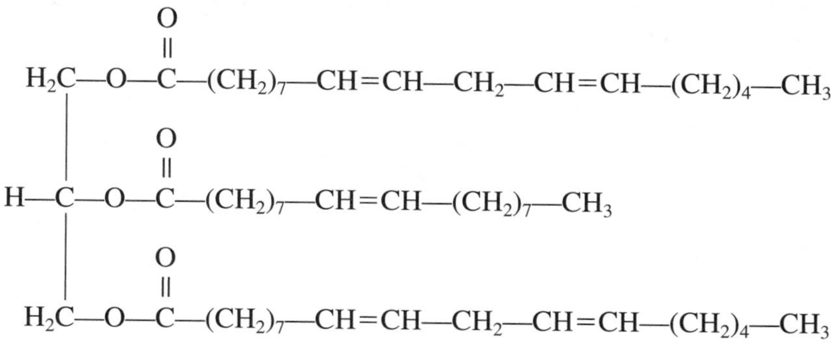

b.

$$H_2C\text{—}O\text{—}\overset{\overset{\displaystyle O}{\|}}{C}\text{—}(CH_2)_7\text{—}CH=CH\text{—}CH_2\text{—}CH=CH\text{—}(CH_2)_4\text{—}CH_3$$

$$H\text{—}\overset{|}{C}\text{—}O\text{—}\overset{\overset{\displaystyle O}{\|}}{C}\text{—}(CH_2)_7\text{—}CH=CH\text{—}(CH_2)_7\text{—}CH_3 + 5H_2 \xrightarrow{Ni}$$

$$H_2C\text{—}O\text{—}\overset{\overset{\displaystyle O}{\|}}{C}\text{—}(CH_2)_7\text{—}CH=CH\text{—}CH_2\text{—}CH=CH\text{—}(CH_2)_4\text{—}CH_3$$

$$H_2C\text{—}O\text{—}\overset{\overset{\displaystyle O}{\|}}{C}\text{—}(CH_2)_{16}\text{—}CH_3$$

$$H\text{—}\overset{|}{C}\text{—}O\text{—}\overset{\overset{\displaystyle O}{\|}}{C}\text{—}(CH_2)_{16}\text{—}CH_3 \xrightarrow{H^+}$$

$$H_2C\text{—}O\text{—}\overset{\overset{\displaystyle O}{\|}}{C}\text{—}(CH_2)_{16}\text{—}CH_3$$

15.67   Beeswax and carnauba are waxes. Vegetable oil and glyceryl tricaprate are triacylglycerols.

$$CH_2\text{—}O\text{—}\overset{\overset{\displaystyle O}{\|}}{C}\text{—}(CH_2)_8\text{—}CH_3$$

$$CH\text{—}O\text{—}\overset{\overset{\displaystyle O}{\|}}{C}\text{—}(CH_2)_8\text{—}CH_3 \qquad \text{Glyceryl tricaprate}$$

$$CH_2\text{—}O\text{—}\overset{\overset{\displaystyle O}{\|}}{C}\text{—}(CH_2)_8\text{—}CH_3$$

15.69   **a.** A typical monounsaturated fatty acid has a cis double bond.
   b. A cis unsaturated fatty acid contains hydrogen atoms on the same side of each double bond. A trans unsaturated fatty acid has the hydrogen atoms on the opposite sides of the each double bond that forms during hydrogenation.
   c.

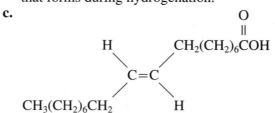

15.71

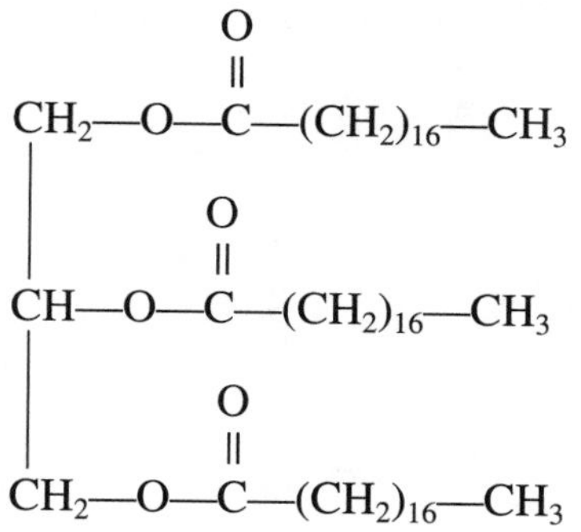

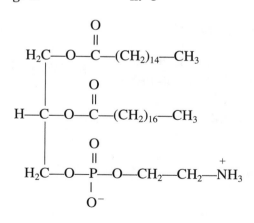

O
‖
CH₂—O—C—(CH₂)₁₆—CH₃

O
‖
CH—O—C—(CH₂)₁₆—CH₃   Glyceryl tristearate

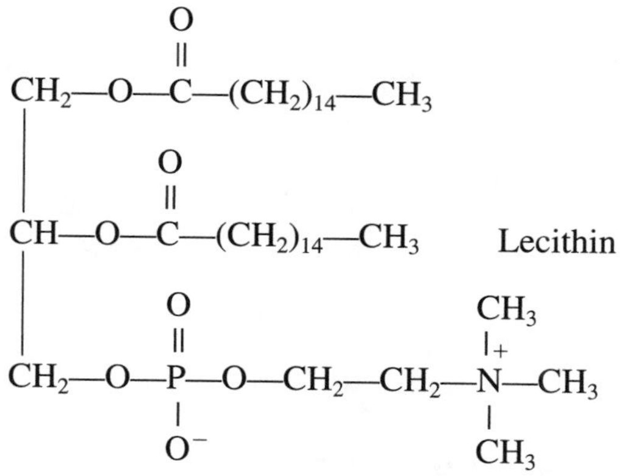

O
‖
CH₂—O—C—(CH₂)₁₆—CH₃

O
‖
CH₂—O—C—(CH₂)₁₄—CH₃

O
‖
CH—O—C—(CH₂)₁₄—CH₃   Lecithin

$$CH_2-O-\overset{\overset{\textstyle O}{\|}}{P}-O-CH_2-CH_2-\overset{\overset{\textstyle CH_3}{|+}}{\underset{\underset{\textstyle CH_3}{|}}{N}}-CH_3$$
O⁻

15.73 **a.** wax   **b.** steroid   **c.** glycerophospholipid   **d.** triacylglycerol
      e. soap   **f.** triacylglycerol   **g.** triacylglycerol   **h.** triacylglycerol
      i. steroid   **j.** steroid   **k.** fatty acid

15.75 **a.** 5   **b.** 1, 2, 3, 4   **c.** 2
      d. 1, 2   **e.** 1, 2, 3, 4

15.77 **a.** 4   **b.** 3   **c.** 1
      d. 4   **e.** 4   **f.** 3
      g. 2   **h.** 1

15.79

O
‖
H₂C—O—C—(CH₂)₁₄—CH₃

O
‖
H—C—O—C—(CH₂)₁₆—CH₃

O
‖
H₂C—O—P—O—CH₂—CH₂—NH₃⁺
O⁻

15.81 **a.** Adding NaOH would hydrolyze the tristearin and break it up to wash down the drain.

b.

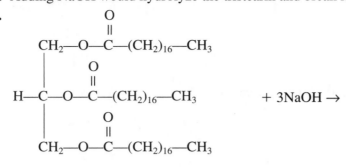

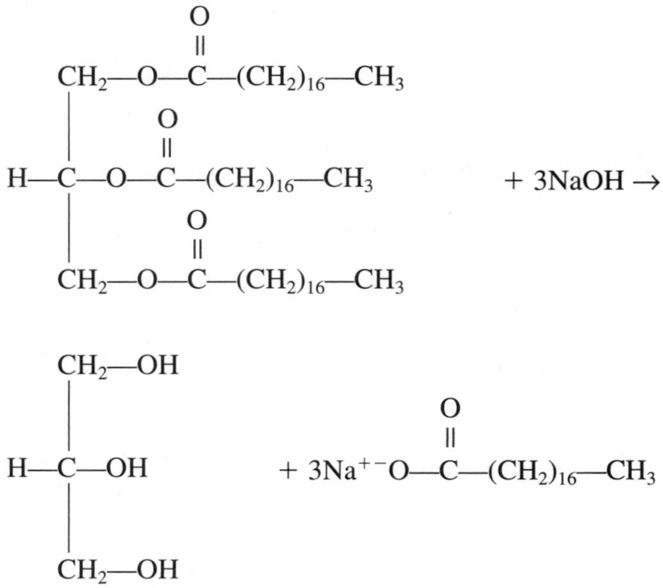

$$CH_2\!-\!OH$$

$$H\!-\!C\!-\!OH \qquad + \; 3Na^{+\,-}O\!-\!\underset{\displaystyle \overset{O}{\|}}{C}\!-\!(CH_2)_{16}\!-\!CH_3$$

$$CH_2\!-\!OH$$

Glycerol   Salts of stearic acid

Answers to Combining Ideas from Chapters 13 to 15

CI.25 **a.**

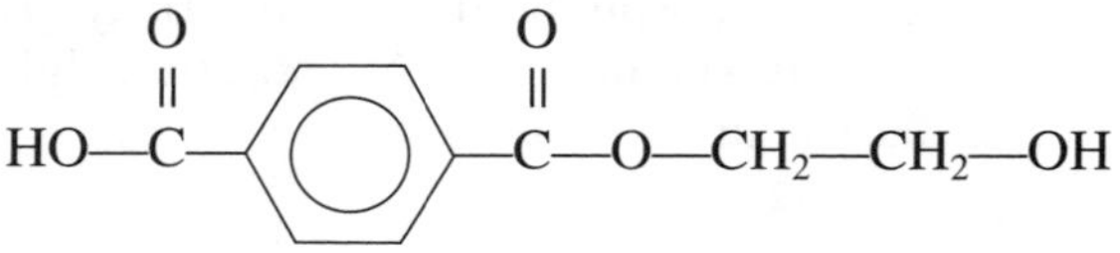

$$HO\!-\!\underset{\displaystyle \overset{O}{\|}}{C}\!-\!\bigcirc\!-\!\underset{\displaystyle \overset{O}{\|}}{C}\!-\!O\!-\!CH_2\!-\!CH_2\!-\!OH$$

b.

$$HO\!-\!CH_2\!-\!CH_2\!-\!O\!-\!\underset{\displaystyle \overset{O}{\|}}{C}\!-\!\bigcirc\!-\!\underset{\displaystyle \overset{O}{\|}}{C}\!-\!O\!-\!CH_2\!-\!CH_2\!-\!OH$$

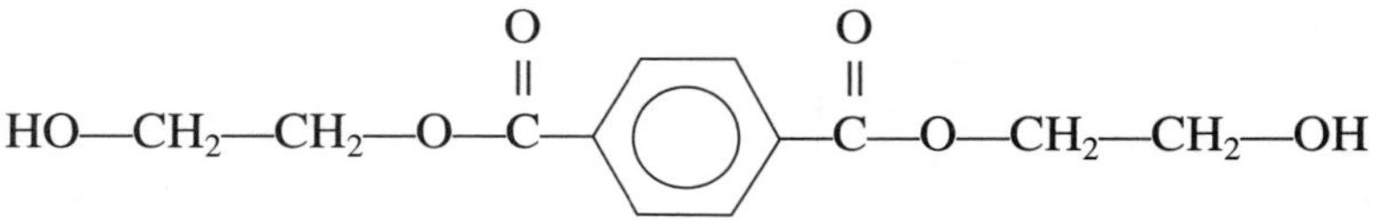

c. $1.7 \times 10^9 \; \text{lb PETE} \times \dfrac{1 \text{ kg PETE}}{2.20 \text{ lb PETE}} = 7.7 \times 10^8 \text{ kg of PETE}$

d. $1.7 \times 10^9 \; \text{lb PETE} \times \dfrac{454 \text{ g PETE}}{1 \text{ lb PETE}} \times \dfrac{1 \text{ mL PETE}}{1.38 \text{ g PETE}} \times \dfrac{1 \text{ L PETE}}{1000 \text{ mL PETE}} = 5.6 \times 10^8 \text{ L of PETE}$

e. $5.6 \times 10^8 \; \text{L PETE} \times \dfrac{1 \text{ landfill}}{2.7 \times 10^7 \text{ L PETE}} = 21 \text{ landfills}$

CI.27 **a.**

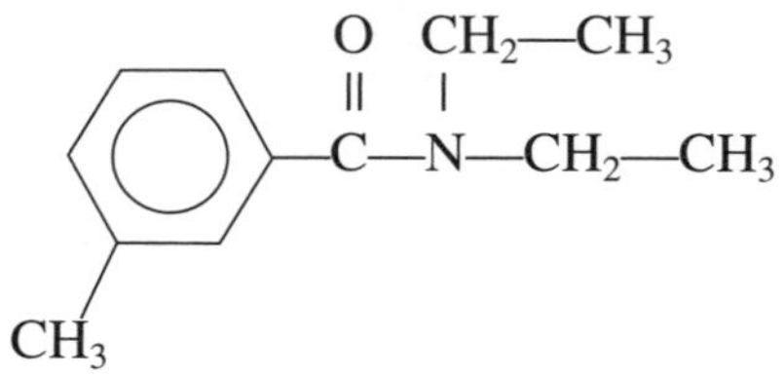

$$\bigcirc\!-\!\underset{\displaystyle \overset{O}{\|}}{C}\!-\!\underset{\displaystyle \underset{CH_3}{\,}}{N}\!-\!CH_2\!-\!CH_3$$
with $CH_2\!-\!CH_3$ on N and CH_3 on ring.

b. $C_{12}H_{17}NO$

c. $12 \times C(12.0) + 17 \times H(1.0) + 1 \times N(14.0) + 1 \times O(16.0) = 191$ g/mole

d. $6.0 \text{ fl oz} \times \dfrac{1 \text{ qt}}{32 \text{ fl oz}} \times \dfrac{1000 \text{ mL}}{1.06 \text{ qt}} \times \dfrac{1.0 \text{ g solution}}{1 \text{ mL}} \times \dfrac{25.0 \text{ g DEET}}{100 \text{ g solution}} = 44$ g of DEET

e. $6.0 \text{ fl oz} \times \dfrac{1 \text{ qt}}{32 \text{ fl oz}} \times \dfrac{1000 \text{ mL}}{1.06 \text{ qt}} \times \dfrac{1.0 \text{ g solution}}{1 \text{ mL}} \times \dfrac{25.0 \text{ g DEET}}{100 \text{ g solution}}$

$\times \dfrac{1 \text{ mole DEET}}{191 \text{ g DEET}} \times \dfrac{6.02 \times 10^{23} \text{ molecules DEET}}{1 \text{ mole DEET}} = 1.4 \times 10^{23}$ molecules of DEET

CI.29 a. A, B, and C are all glucose.

b. An α-1, 6-glycosidic bond links A and B.

c. An α-1, 4-glycosidic bond links B and C.

d. β-panose

e. Panose is a reducing sugar because it has a free hydroxyl on C1 of structure C, which allows glucose (C) to form the aldehyde.

16
Amino Acids, Proteins, and Enzymes

Study Goals

- Classify proteins by their functions in the cells.
- Draw the structures of amino acids.
- Draw the ionized structures of amino acids.
- Write the structural formulas of di- and tripeptides.
- Identify the structural levels of proteins as primary, secondary, tertiary, and quaternary.
- Describe the effects of denaturation on the structure of proteins.
- Classify enzymes according to the type of reaction they catalyze.
- Describe the lock-and-key and induced-fit models of enzyme action.
- Discuss the effect of changes in temperature, pH, and concentration of substrate on enzyme action.
- Describe the competitive and noncompetitive inhibition of enzymes.
- Identify the types of cofactors that are necessary for enzyme action.

Think About It

1. What are some uses of protein in the body?

2. What are the units that make up a protein?

3. How do you obtain protein in your diet?

4. What are some functions of enzymes in the cells of the body?

5. Why are enzymes sensitive to high temperatures and low or high pH levels?

Key Terms

1. Match the following key terms with the correct statement shown below.

   a. amino acid   **b.** peptide bond   **c.** denaturation
   d. primary structure   **e.** zwitterion

   1. _____ The order of amino acids in a protein

   2. _____ The ionized structure of an amino acid has a net charge of zero

   3. _____ The bond that connects amino acids in peptides and proteins

   4. _____ The loss of secondary and tertiary protein structure caused by agents such as heat and acid

   5. _____ The building block of proteins

Answers   **1.** d   **2.** e   **3.** b   **4.** c   **5.** a

2. Match the following key terms with the correct statement shown below.

a. lock-and-key theory   **b.** vitamin   **c.** inhibitor   **d.** enzyme   **e.** active site

1. _____ The portion of an enzyme structure where a substrate undergoes reaction

2. _____ A protein that catalyzes a biological reaction in the cells

3. _____ A model of enzyme action in which the substrate exactly fits the shape of an enzyme like a key fits into a lock

4. _____ A substance that makes an enzyme inactive by interfering with its ability to react with a substrate

5. _____ An organic compound essential for normal health and growth that must be obtained from the diet

Answers   **1.** e   **2.** d   **3.** a   **4.** c   **5.** b

16.1 Functions of Proteins

- Some proteins are enzymes or hormones, while others are important in structure, transport, protection, storage, and contraction of muscles.

◆ Learning Exercise 16.1

Match one of the following functions of a protein with the examples below.

a. structural   **b.** contractile   **c.** storage   **d.** transport
e. hormonal   **f.** enzyme   **g.** protection

1. _____ hemoglobin carries oxygen in blood   **5.** _____ collagen makes up connective tissue

2. _____ amylase hydrolyzes starch   **6.** _____ immunoglobulin

3. _____ egg albumin, a protein in egg white   **7.** _____ keratin, a major protein of hair

4. _____ hormone that controls growth   **8.** _____ lipoprotein carries lipids in blood

Answers   **1.** d   **2.** f   **3.** c   **4.** e
             5. a   **6.** g   **7.** a   **8.** d

16.2 Amino Acids

- A group of 20 amino acids provides the molecular building blocks of proteins.
- In an amino acid, a central (alpha) carbon is attached to an amino group, a carboxyl group, and a side chain, which is a characteristic group for each amino acid.
- The particular side group makes each amino acid polar, nonpolar, acidic, or basic. Nonpolar amino acids contain hydrocarbon side chains, while polar amino acids contain electronegative atoms such as oxygen ($-OH$) or sulfur ($-SH$). Acidic side chains contain a carboxylic acid group and basic side chains contain an amino group ($-NH_2$).

◆ Learning Exercise 16.2

Using the side chain shown in parentheses, complete the structural formula of each of the following amino acids. Indicate whether the amino acid would be polar, nonpolar, acidic, or basic.

Glycine (—H)

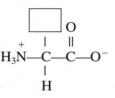

Alanine (—CH₃)

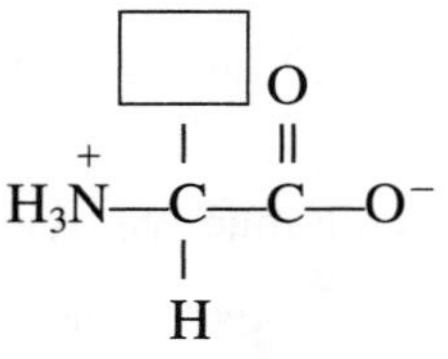

$$\begin{array}{c} \square \quad O \\ + \quad | \quad || \\ H_3N—C—C—O^- \\ | \\ H \end{array}$$

_____

Serine (—CH₂—OH)

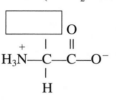

Aspartic acid (—CH₂—C—O⁻ with O double bond)

$$\begin{array}{c} \square \quad O \\ + \quad | \quad || \\ H_3N—C—C—O^- \\ | \\ H \end{array}$$

_____

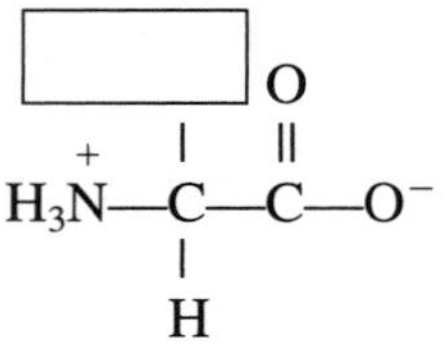

Answers

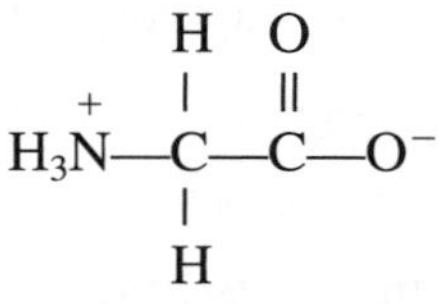

$$\begin{array}{c} H \quad O \\ + \quad | \quad || \\ H_3N—C—C—O^- \\ | \\ H \end{array}$$

Nonpolar

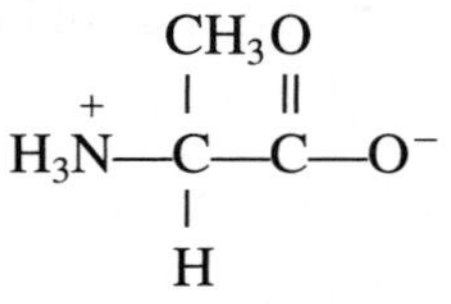

$$\begin{array}{c} CH_3 \quad O \\ + \quad | \quad || \\ H_3N—C—C—O^- \\ | \\ H \end{array}$$

Nonpolar

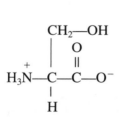

Polar

$$\begin{array}{c} \quad\quad\quad O \\ \quad\quad\quad || \\ CH_2—C—O^- \\ \quad O \\ + \quad | \quad || \\ H_3N—C—C—O^- \\ | \\ H \end{array}$$

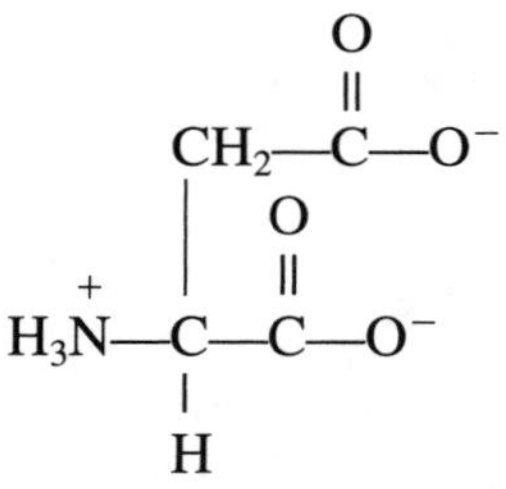

Acidic

16.3 Amino Acids as Acids and Bases

• Amino acids exist as dipolar ions called zwitterions, which are neutral at the isoelectric point (pI).
• A zwitterion has a positive charge at pH levels below its pI and a negative charge at pH levels higher than its pI.

Study Note

Example: Glycine has an isoelectric point at a pH of 6.0. Write the zwitterion of glycine at its iso-electric point (pI), and at pH levels above and below its isoelectric point.

Solution: In more acidic solutions, glycine has a net positive charge, and in more basic solutions, a net negative charge.

$$H_3N^+—CH_2—COOH \xleftarrow{H^+} H_3N^+—CH_2—COO^- \xrightarrow{OH^-} H_2N—CH_2—COO^-$$

below pI   *zwitterion of glycine*   *above pI*

◆ **Learning Exercise 16.3**

Write the structure of the amino acids under the given conditions.

Zwitterion	H^+	OH^-
Alanine		
Serine		

Answers

Zwitterion	H^+	OH^-
Alanine $\overset{\displaystyle CH_3}{\underset{\displaystyle H_3\overset{+}{N}—CH—COO^-}{\mid}}$	$\overset{\displaystyle CH_3}{\underset{\displaystyle H_3\overset{+}{N}—CH—COOH}{\mid}}$	$\overset{\displaystyle CH_3}{\underset{\displaystyle H_2N—CH—COO^-}{\mid}}$
Serine $\overset{\displaystyle CH_2OH}{\underset{\displaystyle H_3\overset{+}{N}—CH—COO^-}{\mid}}$	$\overset{\displaystyle CH_2OH}{\underset{\displaystyle H_3\overset{+}{N}—CH—COOH}{\mid}}$	$\overset{\displaystyle CH_2OH}{\underset{\displaystyle H_2N—CH—COO^-}{\mid}}$

16.4 Formation of Peptides

- A peptide bond is an amide bond between the carboxyl group of one amino acid and the amino group of the second.

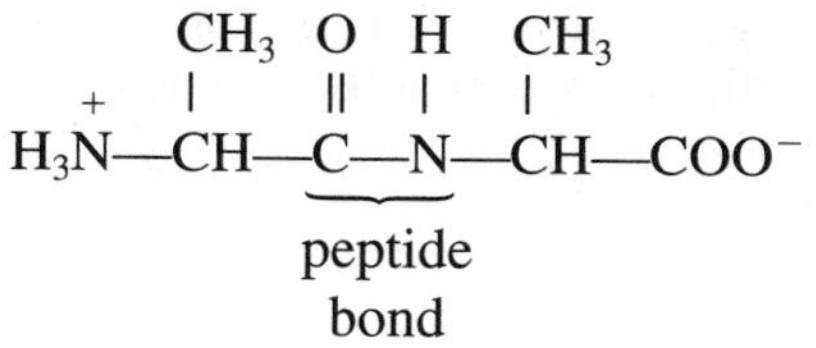

$$H_3\overset{+}{N}—\overset{\overset{\displaystyle CH_3}{\mid}}{C}H—\underset{\underbrace{\quad\quad}_{\substack{peptide\\bond}}}{\overset{\overset{\displaystyle O}{\|}}{C}—\overset{\overset{\displaystyle H}{\mid}}{N}}—\overset{\overset{\displaystyle CH_3}{\mid}}{C}H—COO^-$$

- Short chains of amino acids are called peptides. Long chains of amino acids are called proteins.

◆ Learning Exercise 16.4

Draw the structural formulas of the following di- and tripeptides.

1. serylglycine

2. cystylvaline

3. Gly-Ser-Cys

Answers

1.   HOCH$_2$ O H
$$\overset{+}{H_3N}-\underset{}{CH}-\underset{}{C}-\underset{}{N}-CH_2-COO^-$$

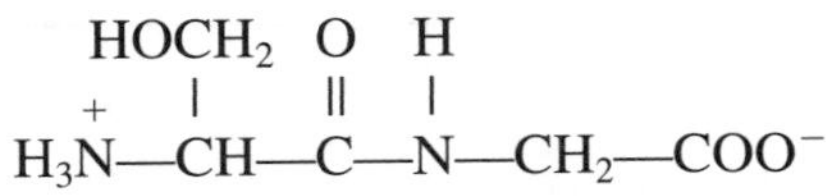

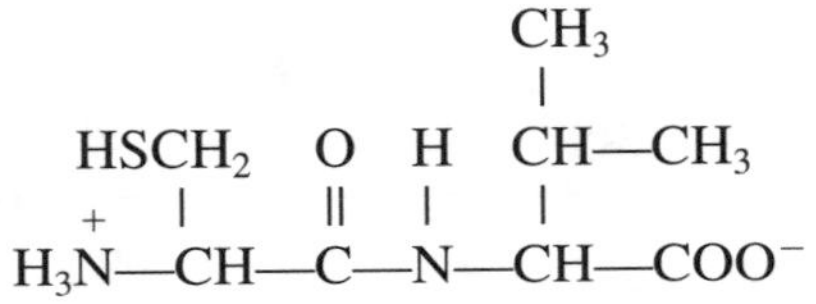

2.   CH$_3$
                            |
     HSCH$_2$ O H CH—CH$_3$
$$\overset{+}{H_3N}-CH-C-N-CH-COO^-$$

3.   O HOCH$_2$ O   CH$_2$SH
$$\overset{+}{H_3N}-CH_2-\underset{}{C}-\underset{|}{N}-CH-\underset{}{C}-\underset{|}{N}-CH-COO^-$$
                       | |
                       H   H

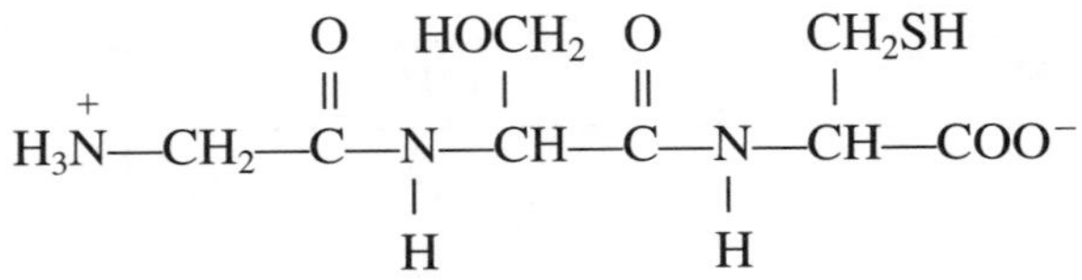

16.5 Levels of Protein Structure

- The primary structure of a protein is the sequence of amino acids.
- In the secondary structure, hydrogen bonds between different sections of the peptide produce a characteristic shape such as an α helix, β-pleated sheet, or a triple helix.
- In globular proteins, the polypeptide chain including its α-helical and β-pleated sheet regions folds upon itself to form a tertiary structure.
- In a tertiary structure, amino acids with hydrophobic side chains are found on the inside and amino acids with hydrophilic side chains are found on the outside surface. The tertiary structure is stabilized by interactions between the side chains.
- In a quaternary structure, two or more subunits must combine for biological activity. They are held together by the same interactions found in tertiary structures.
- Denaturation of a protein occurs when heat or other denaturing agents destroys the secondary and tertiary structure (but not the primary structure) of the protein until biological activity is lost.

◆ Learning Exercise 16.5A

Identify the following descriptions of protein structure as primary or secondary structure.

1. _____ Hydrogen bonding forms an alpha (α) helix.

2. _____ Hydrogen bonding occurs between C=O and N—H within a peptide chain.

3. _____ The order of amino acids, which are linked by peptide bonds.

4. _____ Hydrogen bonds between protein chains form a pleated-sheet structure.

Answers   **1.** secondary   **2.** secondary   **3.** primary   **4.** secondary

◆ Learning Exercise 16.5B

Identify the following descriptions of protein structure as tertiary or quaternary.

1. _____ A disulfide bond joining distant parts of a peptide.

2. _____ The combination of four protein subunits.

3. _____ Hydrophilic side groups seeking contact with water.

4. _____ A salt bridge forms between two oppositely charged side chains.

5. _____ Hydrophobic side groups forming a nonpolar center.

Answers   **1.** tertiary   **2.** quaternary   **3.** tertiary   **4.** tertiary   **5.** tertiary

◆ Learning Exercise 16.5C

Indicate the denaturing agent in the following examples.

   a. heat or UV light   **b.** pH change   **c.** organic solvent
   d. heavy metal ions   **e.** agitation

1. _____ Placing surgical instruments in a 120 °C autoclave

2. _____ Whipping cream to make a desert topping

3. _____ Applying tannic acid to a burn

4. _____ Placing $AgNO_3$ drops in the eyes of newborns

5. _____ Using alcohol to disinfect a wound

6. _____ Using lactobacillus bacteria culture to produce acid that converts milk to yogurt.

Answers   **1.** a   **2.** e   **3.** b   **4.** d   **5.** c   **6.** b

16.6 Enzymes

- Enzymes are globular proteins that act as biological catalysts.
- Enzymes accelerate the rate of biological reactions by lowering the activation energy of a reaction.
- The names of most enzymes are indicated by their *ase* endings.
- Enzymes are classified by the type of reaction they catalyze: oxidoreductase, hydrolase, isomerase, transferase, lyase, or ligase.

◆ **Learning Exercise 16.6A**

Indicate whether each of the following characteristics of an enzyme is *true* or *false*.

An enzyme

1. _____ is a biological catalyst.

2. _____ functions at a low pH.

3. _____ does not change the equilibrium position of a reaction.

4. _____ is obtained from the diet.

5. _____ greatly increases the rate of a cellular reaction.

6. _____ is needed for every reaction that takes place in the cell.

7. _____ catalyzes at a faster rate at higher temperatures.

8. _____ functions best at mild conditions of pH 7.4 and 37 °C.

9. _____ lowers the activation energy of a biological reaction.

10. _____ increases the rate of the forward reaction, but not the reverse.

| *Answers* | **1.** T | **2.** F | **3.** T | **4.** F | **5.** T |
| | **6.** T | **7.** F | **8.** T | **9.** T | **10.** F |

◆ **Learning Exercise 16.6B**

Match the common name of each of the following enzymes with the description of the reaction.

| **1.** dehydrogenase | **2.** oxidase | **3.** peptidase |
| **4.** decarboxylase | **5.** esterase | **6.** transaminase |

a. _____ hydrolyzes the ester bonds in triacylglycerols to yield fatty acids and glycerol

b. _____ removes hydrogen from a substrate

c. _____ removes CO_2 from a substrate

d. _____ decomposes hydrogen peroxide to water and oxygen

e. _____ hydrolyzes peptide bonds during the digestion of proteins

f. _____ transfers an amino (NH_2) group from an amino acid to an α-keto acid

| *Answers* | **a.** 5 | **b.** 1 | **c.** 4 | **d.** 2 | **e.** 3 | **f.** 6 |

◆ **Learning Exercise 16.6C**

Match the IUPAC classification for enzymes with each of the following types of reactions.

| **1.** oxidoreductase | **2.** transferase | **3.** hydrolase |
| **4.** lyase | **5.** isomerase | **6.** ligase |

a. _____ combines small molecules using energy from ATP

b. _____ transfers phosphate groups

c. _____ hydrolyzes a disaccharide into two glucose units

d. _____ converts a substrate to an isomer of the substrate

e. _____ adds hydrogen to a substrate

f. _____ removes H_2O from a substrate

g. _____ adds oxygen to a substrate

h. _____ converts a cis structure to a trans structure

Answers   **a.** 6   **b.** 2   **c.** 3   **d.** 5
                    e. 1   **f.** 4   **g.** 1   **h.** 5

16.7 Enzyme Action

- Within the structure of the enzyme, there is a small pocket called the active site, which has a specific shape that fits a specific substrate.
- In the lock-and-key model or the induced-fit model, an enzyme and substrate form an enzyme-substrate complex so the reaction of the substrate can be catalyzed at the active site.

◆ Learning Exercise 16.7A

Match the terms (A) active site, (B) substrate, (C) enzyme-substrate complex, (D) lock-and-key, and (E) induced-fit with the following descriptions:

1. _____ the combination of an enzyme with a substrate

2. _____ a model of enzyme action in which the rigid shape of the active site exactly fits the shape of the substrate

3. _____ has a tertiary structure that fits the structure of the active site

4. _____ a model of enzyme action in which the shape of the active site adjusts to fit the shape of a substrate

5. _____ the portion of an enzyme that binds to the substrate and catalyzes the reaction

Answers   **1.** C   **2.** D   **3.** B   **4.** E   **5.** A

◆ Learning Exercise 16.7B

Write an equation to illustrate the following:

1. The formation of an enzyme-substrate complex.

2. The conversion of enzyme-substrate complex to product.

Answers   **1.** $E + S \rightleftarrows ES$ **2.** $ES \rightarrow E + P$

16.8 Factors Affecting Enzyme Activity

- Enzymes are most effective at optimum temperature and pH. The rate of an enzyme reaction decreases considerably at temperatures and pH above or below the optimum.
- An enzyme can be made inactive by changes in pH, temperature, or by chemical compounds called inhibitors.
- An increase in substrate concentration increases the reaction rate of an enzyme-catalyzed reaction until all the enzyme molecules combine with substrate.
- A competitive inhibitor has a structure similar to the substrate and competes for the active site. When the active site is occupied by a competitive inhibitor, the enzyme cannot catalyze the reaction of the substrate.

- A noncompetitive inhibitor attaches elsewhere on the enzyme changing the shape of both the enzyme and the active site. As long as the noncompetitive inhibitor is attached to the enzyme, the altered active site cannot bind with substrate.

◆ **Learning Exercise 16.8A**

Urease, which has an optimum pH of 5, catalyzes the hydrolysis of urea to ammonia and CO_2 in the liver.

$$H_2N\text{—}\overset{\overset{\displaystyle O}{\|}}{C}\text{—}NH_2 + H_2O \xrightarrow{\text{Urease}} 2NH_3 + CO_2$$

Draw a graph to represent the effects of each of the following on the reaction rate of an enzyme. Indicate the optimum pH and optimum temperature.

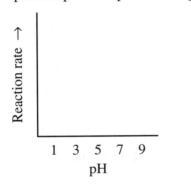
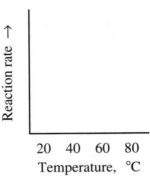

Indicate the effect of each of the following on the reaction rate of the urease-catalyzed reaction.

a. increases   **b.** decreases   **c.** not changed

1. _____ adding more urea when an excess of enzyme is present

2. _____ running the reaction at pH 8

3. _____ lowering the temperature to 0 °C

4. _____ running the reaction at 85 °C

5. _____ increasing the concentration of urea for a specific amount of urease

6. _____ adjusting pH to the optimum

Answers

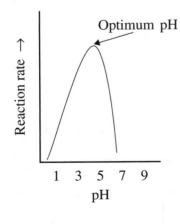
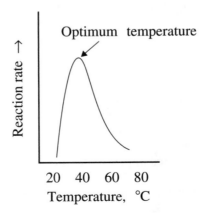

1. a   **2.** b   **3.** b   **4.** b   **5.** a   **6.** a

◆ **Learning Exercise 16.8B**

Identify each of the following as characteristic of competitive inhibition (C) or noncompetitive inhibition (N).

1. _____ An inhibitor binds to the surface of the enzyme away from the active site.

2. _____ An inhibitor resembling the substrate molecule blocks the active site on the enzyme.

3. _____ The action of this inhibitor can be reversed by adding more substrate.

4. _____ Increasing substrate concentration does not change the effect of this inhibition.

5. _____ Sulfanilamide stops bacterial infections because its structure is similar to PABA (*p*-aminobenzoic acid), which is essential for bacterial growth.

Answers   **1.** N   **2.** C   **3.** C   **4.** N   **5.** C

16.9 Enzyme Cofactors

- Simple enzymes are biologically active as a protein only, whereas other enzymes require a cofactor.
- A cofactor may be a metal ion such as Cu^{2+} or Fe^{2+}, or an organic compound called a coenzyme, usually a vitamin.
- Vitamins are organic molecules that are essential for proper health.
- Vitamins must be obtained from the diet because they are not synthesized in the body.
- Vitamins B and C are classified as water-soluble; vitamins A, D, E, and K are fat-soluble vitamins.
- Many water-soluble vitamins function as coenzymes.

◆ **Learning Exercise 16.9**

Indicate whether each statement describes a simple enzyme or a protein that requires a cofactor.

1. _____ An enzyme consisting only of protein

2. _____ An enzyme requiring magnesium ions for activity

3. _____ An enzyme containing a sugar group

4. _____ An enzyme that gives only amino acids upon hydrolysis

5. _____ An enzyme that requires zinc ions for activity

Answers   **1.** simple   **2.** requires a cofactor   **3.** requires a cofactor
             4. simple   **5.** requires a cofactor

Checklist for Chapter 16

You are ready to take the practice test for Chapter 16. Be sure that you have accomplished the following learning goals for this chapter. If you are not sure, review the section listed at the end of the goal. After studying Chapter 16, I can successfully:

_____ Classify proteins by their functions in the cells (16.1).

_____ Draw the structure for an amino acid (16.2).

_____ Draw the zwitterion at the isoelectric point (pI) above and below the pI (16.3).

_____ Describe a peptide bond; draw the structure for a peptide (16.4).

_____ Distinguish between the primary and secondary structures of a protein (16.5).

_____ Distinguish between the tertiary and quaternary structures of a protein (16.5).

_____ Describe the ways that denaturation affects the structure of a protein (16.5).

_____ Classify enzymes according to the type of reaction they catalyze (16.6).

_____ Describe the lock-and-key and induced fit models of enzyme action (16.7).

_____ Discuss the effect of changes in temperature, pH, and concentration of substrate on enzyme action (16.8).

_____ Describe the competitive and noncompetitive inhibition of enzymes (16.8).

_____ Identify the types of cofactors that are necessary for enzyme action (16.9).

Practice Test for Chapter 16

1. Which amino acid is nonpolar?
   A. serine   **B.** aspartic acid   **C.** valine   **D.** cysteine   **E.** glutamine

2. Which amino acid will form disulfide cross-links in a tertiary structure?
   A. serine   **B.** aspartic acid   **C.** valine   **D.** cysteine   **E.** glutamine

3. Which amino acid has a basic side chain?
   A. serine   **B.** aspartic acid   **C.** valine   **D.** cysteine   **E.** glutamine

4. All amino acids
   A. have the same side chains.   **B.** form zwitterions.
   C. have the same isoelectric points.   **D.** show hydrophobic tendencies.
   E. are essential amino acids.

5. Essential amino acids
   A. are the amino acids that must be supplied by the diet.
   B. are not synthesized by the body.
   C. are missing in incomplete proteins.
   D. are present in proteins from animal sources.
   E. All of the above

Use the following to answer questions for the amino acid alanine in questions 6 through 9.

$$\textbf{A. } \overset{+}{H_3}N-\underset{\underset{CH_3}{|}}{CH}-COO^- \qquad \textbf{B. } H_2N-\underset{\underset{CH_3}{|}}{CH}-COO^- \qquad \textbf{C. } \overset{+}{H_3}N-\underset{\underset{CH_3}{|}}{CH}-COOH$$

6. _____ Alanine in its zwitterion form

7. _____ Alanine at a low pH

8. _____ Alanine at a high pH

9. _____ Alanine at its isoelectric point

10. The sequence Tyr-Ala-Gly
   A. is a tripeptide.   **B.** has two peptide bonds.   **C.** has tyrosine with a free —NH_2 end.
   D. has glycine with a free —COOH end.   **E.** All of these

11. The type of bonding expected between lysine and aspartic acid is a(n)
   A. ionic bond.   **B.** hydrogen bond.   **C.** disulfide bond.
   D. hydrophobic attraction.   **E.** hydrophilic attraction.

12. What type of bond is used to form the α helix structure of a protein?
  A. peptide bond   **B.** hydrogen bond   **C.** ionic bond
  D. disulfide bond   **E.** hydrophobic attraction

13. What type of bonding places portions of the protein chain in the center of a tertiary structure?
  A. peptide bonds   **B.** ionic bonds   **C.** disulfide bonds
  D. hydrophobic attraction **E.** hydrophilic attraction

In questions 14 through 18, identify the protein structural levels that each of the following statements describe.

  A. primary   **B.** secondary   **C.** tertiary   **D.** quaternary   **E.** pentenary

14. _____ peptide bonds   **15.** _____ a pleated sheet

16. _____ two or more protein subunits   **17.** _____ an α helix

18. _____ disulfide bonds

In questions 19 through 22, match the function of a protein with each example.

19. _____ enzyme

20. _____ structural

21. _____ transport

22. _____ storage

  A. myoglobin in the muscles
  B. α-keratin in skin
  C. peptidase for protein hydrolysis
  D. casein in milk

23. Denaturation of a protein
  A. occurs at a pH of 7.   **B.** causes a change in protein structure.
  C. hydrolyzes a protein.   **D.** oxidizes the protein.
  E. adds amino acids to a protein.

24. Which of the following will <u>not</u> cause denaturation?
  A. 0 °C   **B.** $AgNO_3$ **C.** 80 °C   **D.** ethanol   **E.** pH 1

25. Enzymes are
  A. biological catalysts.   **B.** polysaccharides.   **C.** insoluble in water.
  D. always contain a cofactor.   **E.** named with an "ose" ending.

Classify the enzymes described in questions 26 through 29 as (A) simple or (B) requiring a cofactor.

26. _____ An enzyme that yields amino acids and a glucose molecule on analysis

27. _____ An enzyme consisting of protein only

28. _____ An enzyme requiring zinc ion for activation

29. _____ An enzyme containing vitamin K

For problems 30 through 34, select answers from the following: ($E = $ enzyme; $S = $ substrate; $P = $ product).

  A. $S \rightarrow P$ **B.** $EP \rightarrow E + P$ **C.** $E + S \rightleftarrows ES$
  D. $ES \rightarrow EP$ **E.** $EP \rightarrow ES$

30. _____ The enzymatic reaction occurring at the active site

31. _____ The release of product from the enzyme

32. _____ The first step in the lock-and-key theory of enzyme action

33. _____ The formation of the enzyme-substrate complex

34. _____ The final step in the lock-and-key theory of enzyme action

In problems 35 through 39, match the names of enzymes with a reaction they each catalyze.

    A. decarboxylase   **B.** isomerase   **C.** dehydrogenase
    D. lipase   **E.** sucrase

35. _____
$$\underset{\text{OH}}{\text{CH}_3-\overset{|}{\text{C}}\text{H}-\text{COOH}} \rightarrow \underset{\text{O}}{\text{CH}_3-\overset{\|}{\text{C}}-\text{COOH}}$$

36. _____ sucrose $+$ H_2O $\rightarrow$ glucose and fructose

37. _____
$$\text{CH}_3-\overset{\overset{\text{O}}{\|}}{\text{C}}-\text{COOH} \rightarrow \text{CH}_3-\text{COOH} + \text{CO}_2$$

38. _____ fructose $\rightarrow$ glucose

39. _____ triglyceride $+$ 3 H_2O $\rightarrow$ fatty acids and glycerol

For problems 40 through 44 select your answers from A, B, or C.

    A. increases the rate of reaction   **B.** decreases the rate of reaction
    C. denatures the enzyme and no reaction occurs

40. _____ Setting the reaction tube in a beaker of water at 100 °C

41. _____ Adding substrate to the reaction vessel

42. _____ Running the reaction at 10 °C

43. _____ Adding ethanol to the reaction system

44. _____ Adjusting the pH to optimum pH

For problems 45 through 47, identify each description of inhibition as

    A. competitive   **B.** noncompetitive

45. _____ An alteration in the conformation of the enzyme

46. _____ A molecule closely resembling the substrate interferes with activity

47. _____ The inhibition can be reversed by increasing substrate concentration.

48. _____ The presence of zinc in an enzyme called alcohol dehydrogenase classifies a protein as

    A. simple.   **B.** requiring a cofactor.   **C.** hormonal.
    D. structural.   **E.** secondary.

Answers to the Practice Test

1. C	**2.** D	**3.** E	**4.** B	**5.** E
6. A	**7.** C	**8.** B	**9.** A	**10.** E
11. A	**12.** B	**13.** D	**14.** A	**15.** B
16. D	**17.** B	**18.** C	**19.** C	**20.** B
21. A	**22.** D	**23.** B	**24.** A	**25.** A
26. B	**27.** A	**28.** B	**29.** B	**30.** D
31. B	**32.** C	**33.** C	**34.** B	**35.** C
36. E	**37.** A	**38.** B	**39.** D	**40.** C
41. A	**42.** B	**43.** C	**44.** A	**45.** B
46. A	**47.** A	**48.** B		

Answers and Solutions to Selected Text Problems

16.1 **a.** Hemoglobin, which carries oxygen in the blood, is a transport protein.
b. Collagen, which is a major component of tendon and cartilage, is a structural protein.
c. Keratin, which is found in hair, is a structural protein.
d. Amylase, which catalyzes the breakdown of starch, is an enzyme.

16.3 All amino acids contain a carboxylic acid group and an amino group on the alpha carbon.

16.5 **a.**

$$H_3\overset{+}{N}-\underset{\underset{CH_3}{|}}{CH}-\overset{O}{\underset{||}{C}}-O^-$$

b.

$$H_3\overset{+}{N}-\underset{\underset{\underset{CH_3}{|}}{\overset{|}{CH}-OH}}{CH}-\overset{O}{\underset{||}{C}}-O^-$$

c.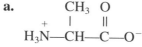

d.

16.7 **a.** Alanine, which has a methyl (hydrocarbon) side group, is nonpolar and hydrophobic.
b. Threonine has a side group that contains the polar —OH. Threonine is polar, hydrophilic, and neutral.
c. Glutamic acid has a side group containing a polar carboxylic acid. Glutamic acid is acidic and hydrophilic.
d. Phenylalanine has a side group with a nonpolar benzene ring. Phenylalanine is nonpolar and hydrophobic.

16.9 The abbreviations of most amino acids are derived from the first three letters in the name.

    a. alanine   **b.** valine   **c.** lysine   **d.** cysteine

16.11 In the L isomer, the —NH_3^+ is on the left side of the horizontal line of the Fischer projection; in the D isomer, the —NH_3^+ group is on the right.

a.

$$H_3\overset{+}{N}-\!\!\!\!\!-\overset{\overset{COO^-}{|}}{\underset{\underset{\underset{H_3C\quad CH_3}{\diagup \diagdown}}{CH}}{|}}\!\!\!\!\!-H$$

b.

$$H-\!\!\!\!\!-\overset{\overset{COO^-}{|}}{\underset{\underset{CH_2SH}{|}}{|}}\!\!\!\!\!-\overset{+}{N}H_3$$

16.13 A zwitterion is formed when the H$^+$ from the acid part of the amino acid is transferred to the amine portion of the amino acid. The resulting dipolar ion has an overall zero charge.

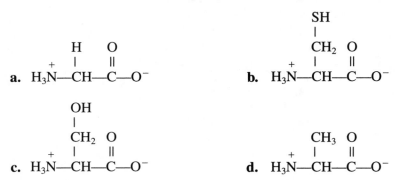

16.15 At low pH (highly acidic), the —COO$^-$ of the zwitterion accepts a proton and the amino acid has a positive charge overall.

OH
|
CH$_2$ O
+ | ||
b. 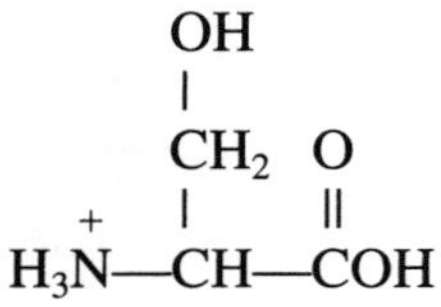 H$_3$N—CH—COH

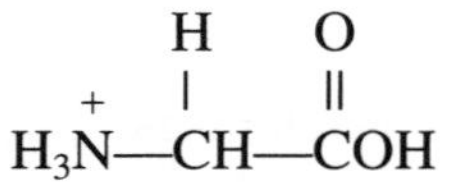

H O
+ | ||
a. H$_3$N—CH—COH

SH
|
CH$_2$ O
+ | ||
c. 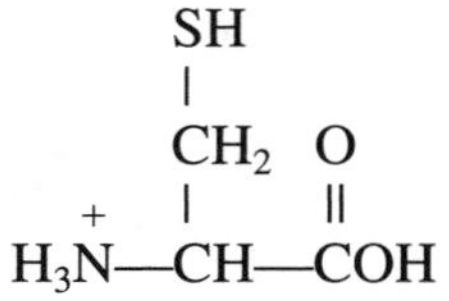 H$_3$N—CH—COH

CH$_3$ O
+ | ||
d. 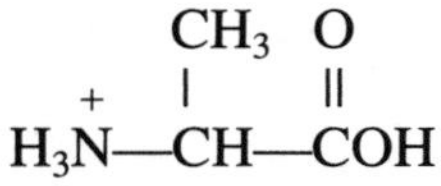 H$_3$N—CH—COH

16.17 **a.** A negative charge means the zwitterion donated a proton (H$^+$) from —NH$_3{}^+$, which occurs at pH levels above the isoelectric point.

b. A positive charge means the zwitterion accepted a proton (H$^+$) from an acidic solution, which occurs at a pH level below the isoelectric point.

c. A zwitterion with a net charge of zero means that the pH is at the isoelectric point.

16.19 In a peptide, the amino acids are joined by peptide bonds (amide bonds). The first amino acid has a free amine group, and the last one has a free carboxyl group.

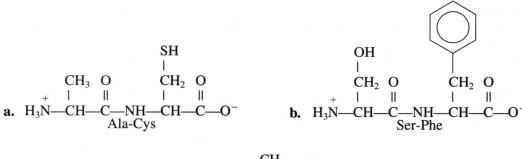

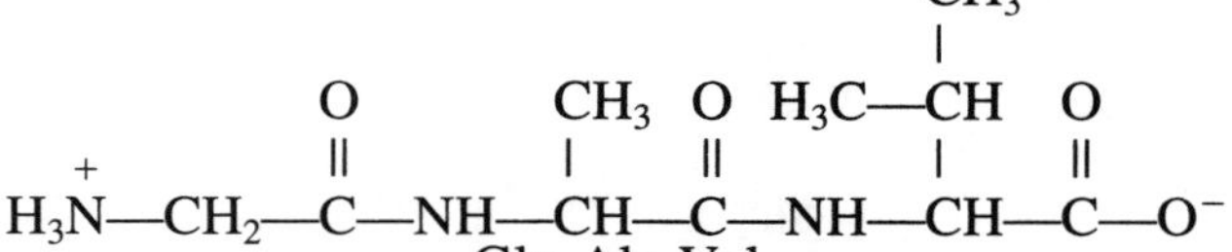

d. H₃N⁺—CH—C—NH—CH—C—NH—CH—C—O⁻
Val-Ile-Trp

16.21 Peptide bonds (amide bonds) connect the amino acids in the primary structure of a protein.

16.23 The possible primary structure of a tripeptide of one valine and two serines are:

Val-Ser-Ser, Ser-Val-Ser, and Ser-Ser-Val.

16.25 The primary structure remains unchanged and intact as hydrogen bonds form secondary structures.

16.27 In an alpha helix, there are hydrogen bonds between the different turns of the helix, which preserves the helical shape of the protein. In a beta-pleated sheet, the hydrogen bonds occur between two protein chains that are side by side or between different parts of a long protein.

16.29 **a.** The two cysteine residues have —SH groups, which react to form a disulfide bond.
b. Glutamic acid is acidic, and lysine is basic; the two groups form an ionic bond, or salt bridge.
c. Serine has a polar —OH group that can form a hydrogen bond with the carboxyl group of aspartic acid.
d. Two leucine residues are hydrocarbon and nonpolar. They would have a hydrophobic interaction.

16.31 **a.** The side chain of cysteine with the —SH group can form disulfide cross-links.
b. Leucine and valine are found on the inside of the protein since they have nonpolar side groups and are hydrophobic.
c. The cysteine and aspartic acid are on the outside of the protein since they are polar.
d. The order of the amino acids (the primary structure) provides the side chains, whose interactions determine the tertiary structure of the protein.

16.33 **a.** Placing an egg in boiling water disrupts hydrogen bonds and hydrophobic interactions, which cause a loss of overall shape.
b. Using an alcohol swab disrupts hydrogen bonds and hydrophobic interactions, which changes the tertiary structure.
c. The heat from an autoclave will disrupt hydrogen bonds and hydrophobic interactions in bacterial protein.
d. Cauterization (heating) of a wound helps to close the wound by breaking hydrogen bonds and disrupting hydrophobic interactions.

16.35 The chemical reactions can occur without enzymes, but the rates are too slow. Catalyzed reactions, which are many times faster, provide the amounts of products needed by the cell at a particular time.

16.37 **a.** galactose   **b.** lipid   **c.** aspartic acid

16.39 **a.** A hydrolase enzyme would catalyze the hydrolysis of sucrose.
b. An oxidoreductase would catalyze the addition of oxygen (oxidation).
c. An isomerase enzyme would catalyze converting glucose to fructose.
d. A transaminase enzyme would catalyze moving an amino group.

16.41 **a.** An enzyme has a tertiary structure that recognizes the substrate.
b. The combination of the enzyme and substrate is the enzyme-substrate complex.
c. The substrate has a structure that complements the structure of the enzyme.

16.43 **a.** The equation for an enzyme-catalyzed reaction is:

$$E + S \rightleftarrows ES \rightarrow E + P$$

E = enzyme, S = substrate, ES = enzyme-substrate complex, P = products

b. The active site is a region or pocket within the tertiary structure of an enzyme that accepts the substrate, aligns the substrate for reaction, and catalyzes the reaction.

16.45 Isoenzymes are slightly different forms of an enzyme that catalyze the same reaction in different organs and tissues of the body.

16.47 A doctor might run tests for the enzymes CK, LDH, and AST to determine if the patient had a heart attack.

16.49 **a.** Decreasing the substrate concentration decreases the rate of reaction.
b. Running the reaction at a pH below optimum pH will decrease the rate of reaction.
c. Temperature above 37 °C (optimum pH) will denature the enzymes and decrease the rate of reaction.

16.51 pepsin, pH 2; urease, pH 5; trypsin, pH 8

16.53 **a.** If the inhibitor has a structure similar to the structure of the substrate, the inhibitor is competitive.
b. If adding more substrate cannot reverse the effect of the inhibitor, the inhibitor is noncompetitive.
c. If the inhibitor competes with the substrate for the active site, it is a competitive inhibitor.
d. If the structure of the inhibitor is not similar to the structure of the substrate, the inhibitor is noncompetitive.
e. If adding more substrate reverses inhibition, the inhibitor is competitive.

16.55 **a.** Methanol has the structural formula CH_3—OH whereas ethanol is CH_3—CH_2—OH.
b. Ethanol has a structure similar to methanol and could compete for the active site.
c. Ethanol is a competitive inhibitor of methanol oxidation.

16.57 **a.** The active form of this enzyme requires a cofactor.
b. The active form of this enzyme requires a cofactor.
c. A simple enzyme is active as a protein.

16.59 **a.** Pantothenic acid (vitamin B_5) is part of coenzyme A.
b. Tetrahydrofolate (THF) is a reduced form of folic acid.
c. Niacin (vitamin B_3) is a component of NAD^+.

16.61 **a.** The oxidation of glycol to an aldehyde and carboxylic acid is catalyzed by an oxidoreductase.
b. At high concentration, ethanol, which acts as a competitive inhibitor of ethylene glycol, would saturate the enzyme to allow ethylene glycol to be removed from the body without producing oxalic acid.

16.63 **a.** aspartic acid, phenylalanine   **b.** aspartylphenylalanine

16.65 **a.** asparagine and serine; hydrogen bond
b. aspartic acid and lysine; salt bridge
c. cysteine and cysteine; disulfide bond
d. Leucine and alanine: hydrophobic interaction

16.67 **a.** Yes; a combination of rice and garbanzo beans contains all three essential amino acids.
b. Yes; a combination of lima beans and cornmeal contains all three essential amino acids.
c. No; a salad of garbanzo beans and lima beans does not contain all three essential amino acids; it is deficient in tryptophan.
d. Yes; a combination of rice and lima beans contains all three essential amino acids in lysine.
e. No; a combination of rice and oatmeal does not contain all the essential amino acids; it is deficient in lysine.
f. Yes; a combination of oatmeal and lima beans contains the three essential amino acids.

16.69 **a.** The secondary structure of a protein depends on hydrogen bonds to form a helix or a beta pleated sheet. The tertiary structure is determined by the interaction of side chains and determines the three-dimensional structure of the protein.

284

b. Nonessential amino acids are synthesized by the body, but essential amino acids must be supplied by the diet.

c. Polar amino acids have hydrophilic side groups while nonpolar amino acids have hydrophobic side groups.

d. Dipeptides contain two amino acids, while tripeptides contain three.

e. An ionic bond is an interaction between a basic and acidic side group; a disulfide bond links the sulfides of two cysteines.

f. Fibrous proteins consist of three to seven α helices coiled like a rope. Globular proteins form a compact spherical shape.

g. The α helix is the secondary shape, like a spiral staircase or corkscrew. The β-pleated sheet is a secondary structure that is formed by many proteins side by side, like a pleated sheet.

h. The tertiary structure of a protein is its three-dimensional structure. In the quaternary structure, two or more peptide subunits are grouped.

16.71 a.

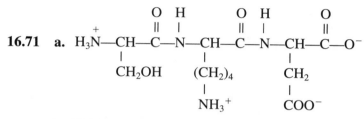

b. This segment contains polar side chains, which would be found on the surface of a globular protein where they can hydrogen bond with water.

16.73 Serine is a polar amino acid, whereas valine is nonpolar. Valine would be in the center of the tertiary structure. However, serine would pull that part of the chain to the outside surface of the protein where valine forms hydrogen bonds with water.

16.75 In chemical laboratories, reactions are often run at high temperatures using catalysts that are strong acids or bases. Enzymes, which function at physiological temperatures and pH, are denatured rapidly if high temperatures or acids or bases are used.

16.77 a. The reactant is lactose and the products are glucose and galactose.

b.

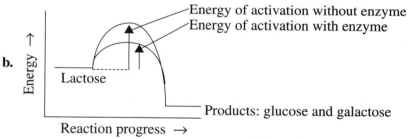

c. By lowering the energy of activation, the enzyme furnishes a lower energy pathway by which the reaction can take place.

16.79 a. The disaccharide lactose is a substrate.

b. The *-ase* in lactase indicates that it is an enzyme.

c. The *-ase* in urease indicates that it is an enzyme.

d. Trypsin is an enzyme, which hydrolyzes polypeptides.

e. Pyruvate is a substrate.

f. The *-ase* in transaminase indicates that it is an enzyme.

16.81 a. Urea is the substrate of urease.

b. Lactose is the substrate of lactase.

c. Aspartate is the substrate of aspartate transaminase.

d. Tyrosine is the substrate of tyrosine synthetase.

16.83 Sucrose fits the shape of the active site in sucrase, but lactose does not.

16.85   A heart attack may be the cause. Normally the enzymes LDH and CK are present only in low levels in the blood.

16.87   **a.** An enzyme is saturated if adding more substrate does not increase the rate.
b. An enzyme is unsaturated when increasing the substrate increases the rate.

16.89   **a.** The Mg^{2+} is a cofactor, which is required by this enzyme.
b. A protein that is catalytically active is a simple enzyme.
c. The folic acid is a coenzyme, which is required by this enzyme.

16.91   **1.** a   **2.** d   **3.** c   **4.** a   **5.** b   **6.** e

Nucleic Acids and Protein Synthesis

Study Goals

- Draw the structures of the nitrogen bases, sugars, and nucleotides in DNA and RNA.
- Describe the structures of DNA and RNA.
- Explain the process of DNA replication.
- Describe the transcription process during the synthesis of mRNA.
- Use the codons in the genetic code to describe protein synthesis.
- Explain how an alteration in the DNA sequence can lead to mutations in proteins.

Think About It

1. Where is DNA in your cells?

2. How does DNA determine your height, or the color of your hair or eyes?

3. What is the genetic code?

4. How does a mutation occur?

Key Terms

Match the statements shown below with the following key terms.

    a. DNA   **b.** RNA   **c.** double helix   **d.** mutation   **e.** transcription

1. _____ The formation of mRNA to carry genetic information from DNA to protein synthesis

2. _____ The genetic material containing nucleotides and nitrogenous bases adenine, cytosine, guanine, and thymine

3. _____ The shape of DNA with a sugar-phosphate backbone and base pairs linked in the center

4. _____ A change in the DNA base sequence that may alter the shape and function of a protein

5. _____ A type of nucleic acid with a single strand of nucleotides of adenine, cytosine, guanine, and uracil

Answers   **1.** e   **2.** a   **3.** c   **4.** d   **5.** b

17.1 Components of Nucleic Acids

- Nucleic acids are composed of four nitrogenous bases, five-carbon sugars, and a phosphate group.
- In DNA, the nitrogen bases are adenine, thymine, guanine, or cytosine. In RNA, uracil replaces thymine.
- In DNA, the sugar is deoxyribose; in RNA the sugar is ribose.
- A nucleoside is composed of a nitrogen base and a sugar.
- A nucleotide is composed of three parts: a nitrogen base, a sugar, and a phosphate group.
- Deoxyribonucleic acid (DNA) and ribonucleic acid (RNA) are polymers of nucleotides.

◆ **Learning Exercise 17.1A**

1. Write the names and abbreviations for the nitrogen bases in each of the following.

DNA _____

RNA _____

2. Write the name of the sugar in each of the following nucleotides.

DNA _____

RNA _____

Answers 1. DNA: adenine (A), thymine (T), guanine (G), cytosine (C)
RNA: adenine (A), uracil (U), guanine (G), cytosine (C)
2. DNA: deoxyribose   RNA: ribose

◆ **Learning Exercise 17.1B**

Name each of the following and classify it as a purine or a pyrimidine.

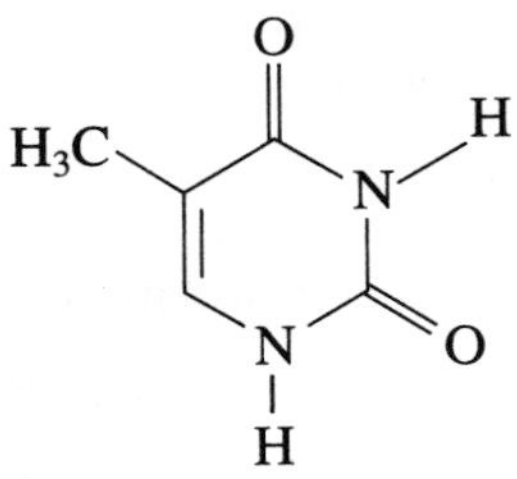

Answers **a.** cytosine; pyrimidine   **b.** adenine; purine
c. guanine; purine   **d.** thymine, pyrimidine

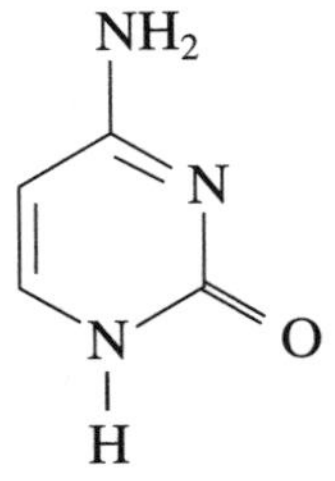

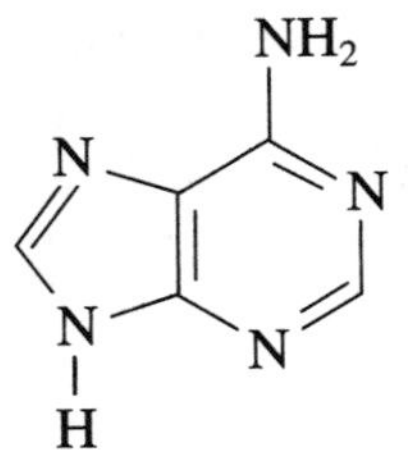

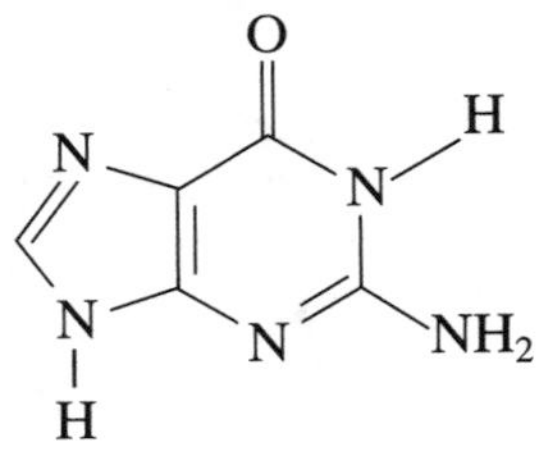

◆ **Learning Exercise 17.1C**

Identify the nucleic acid (DNA or RNA) in which each of the following are found.

1. _____ adenosine 5′-monophosphate   5. _____ guanosine 5′-monophosphate

2. _____ dCMP   6. _____ cytidine 5′-monophosphate

3. _____ deoxythymidine 5′-monophosphate   7. _____ UMP

4. _____ dGMP   8. _____ deoxyadenosine 5′-monophosphate

Answers 1. RNA   2. DNA   3. DNA   4. DNA
5. RNA   6. RNA   7. RNA   8. DNA

◆ Learning Exercise 17.1D

Write the structural formula for deoxyadenosine 5'-monophosphate. Indicate the 5'- and the 3'-carbon atoms on the sugar.

Answer

Deoxyadenosine 5'-monophosphate (dAMP)

17.2 Primary Structure of Nucleic Acids

- Nucleic acids are polymers of nucleotides in which the —OH group on the 3'-carbon of a sugar in one nucleotide bonds to the phosphate group attached to the 5'-carbon of a sugar in the adjacent nucleotide.

◆ Learning Exercise 17.2A

Write the structure of a dipeptide that consists of cytosine 5'-monophosphate (free 5'-phosphate) bonded to guanosine 5'-monophosphate (3'-hydroxyl end). Identify each nucleotide, the phosphodiester bond, the 5'-free phosphate group, and the free 3'-hydroxyl group.

Answer

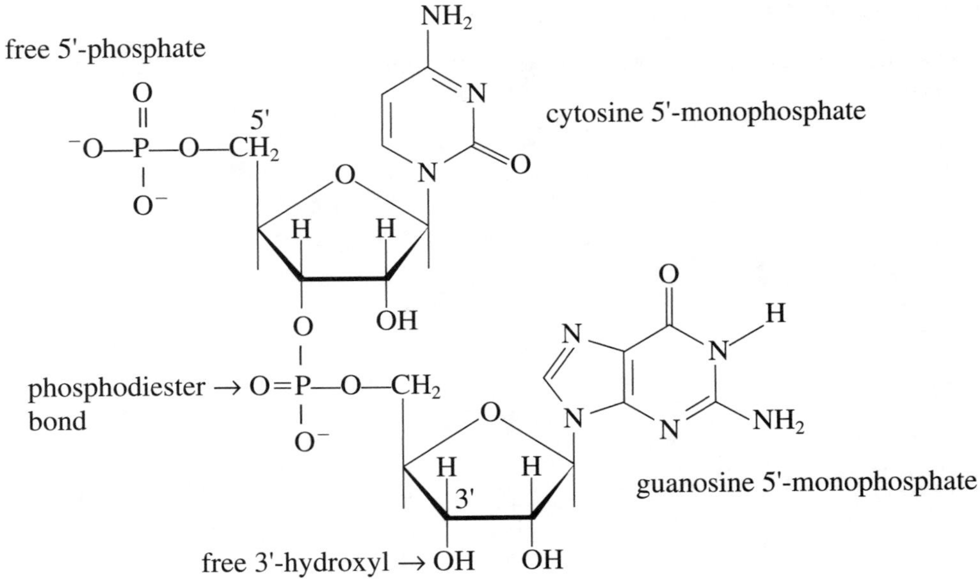

free 5'-phosphate

cytosine 5'-monophosphate

phosphodiester → O=P—O—CH₂
bond

guanosine 5'-monophosphate

free 3'-hydroxyl → OH

◆ Learning Exercise 17.2B

Consider the following sequence of nucleotides in RNA: —A—G—U—C—

1. What are the names of the nucleotides in this sequence?_____

2. Which nucleotide has the free 5'-phosphate group? _____

3. Which nucleotide has the free 3'-hydroxyl group? _____

Answers
1. adenosine 5'-monophosphate, guanosine 5'-monophosphate, uridine 5'-monophosphate, cytidine 5'-monophosphate
2. adenosine 5'-monophosphate (AMP)
3. cytidine 5'-monophosphate (CMP)

17.3 DNA Double Helix

- The two strands in DNA are held together by hydrogen bonds between complementary base pairs, A with T, and G with C.
- During DNA replication, DNA polymerase makes new DNA strands along each of the original DNA strands that serve as templates.
- Complementary base pairing ensures the correct pairing of bases to give identical copies of the original DNA.

◆ Learning Exercise 17.3A

Complete the following statements.

1. The structure of the two strands of nucleotides in DNA is called a _____.

2. The only base pairs that connect the two DNA strands are _____ and _____.

3. The base pairs along one DNA strand are _____ to base pairs on the opposite strand.

Answers   1. double helix   2. A—T; G—C   3. complementary

◆ **Learning Exercise 17.3B**

Complete each DNA section by writing the complementary strand:

1. —A—T—G—C—T—T—G—G—C—T—C—C—

2. —A—A—A—T—T—T—C—C—C—G—G—G—

3. —G—C—G—C—T—C—A—A—A—T—G—C—

Answers
    1. —T—A—C—G—A—A—C—C—G—A—G—G—
    2. —T—T—T—A—A—A—G—G—G—C—C—C—
    3. —C—G—C—G—A—G—T—T—T—A—C—G

◆ **Learning Exercise 17.3C**

Essay: Describe the process by which DNA replicates to produce identical copies.

Answer   In the replication process, the bases on each strand of the separated parent DNA are paired with their complementary bases. Because each complementary base is specific for a base in DNA, the new DNA strands exactly duplicate the original strands of DNA.

17.4 RNA and the Genetic Code

- The three types of RNA differ by function in the cell: ribosomal RNA makes up most of the structure of the ribosomes, messenger RNA carries genetic information from the DNA to the ribosomes, and transfer RNA places the correct amino acids in the protein.
- Transcription is the process by which RNA polymerase produces mRNA from one strand of DNA.
- The bases in the mRNA are complementary to the DNA, except U is paired with A in DNA.
- The production of mRNA occurs when certain proteins are needed in the cell.
- The genetic code consists of a sequence of three bases (triplet) that specifies the order for the amino acids in a protein.
- There are 64 codons for the 20 amino acids, which means there are several codons for most amino acids.
- The AUG codon signals the start of transcription and UAG, UGA, and UAA codons signal stop.

◆ **Learning Exercise 17.4A**

Match each of the following characteristics with a specific type of RNA: mRNA, tRNA, or rRNA.

1. The most abundant type of RNA in a cell _____

2. The RNA that has the shortest chain of nucleotides _____

3. The RNA that carries information from DNA to the ribosomes for protein synthesis _____

4. The RNA that is the major component of ribosomes _____

5. The RNA that carries specific amino acids to the ribosome for protein synthesis _____

6. The RNA that consists of a large and a small subunit _____

Answers   **1.** rRNA   **2.** tRNA   **3.** mRNA
                4. rRNA   **5.** tRNA   **6.** rRNA

◆ **Learning Exercise 17.4B**

Fill in the blanks with a word or phrase that answers each of the following questions:

1. Where in the cell does transcription take place? _____

2. How many strands of the DNA molecules are involved? _____

3. The abbreviations for the four nucleotides in mRNA are _____

4. Write the corresponding section of mRNA produced from each of the following questions.

    a. —C—A—T—T—C—G—G—T—A—

    b. —G—T—A—C—C—T—A—A—C—G—T—C—C—G—

Answers   1. nucleus   2. one   3. A, U, G, C
4. a. —G—U—A—A—G—C—C—A—U—
    b. —C—A—U—G—G—A—U—U—G—C—A—G—G—C—

◆ **Learning Exercise 17.4C**

Indicate the amino acid coded for by the following mRNA codons.

1. UUU _____   6. ACA _____

2. GCG _____   7. AUG _____

3. AGC _____   8. CUC _____

4. CCA _____   9. CAU _____

5. GGA _____   10. GUU _____

Answers   1. Phe   2. Ala   3. Ser   4. Pro   5. Gly
6. Thr   7. Start/Met   8. Leu   9. His   10. Val

17.5 Protein Synthesis

- Proteins are synthesized at the ribosomes in a translation process that includes three steps: initiation, elongation, and termination.
- During translation, the different tRNA molecules bring the appropriate amino acids to the ribosome where the amino acid is bonded by a peptide bond to the growing peptide chain.
- When the polypeptide is released, it takes on its secondary and tertiary structures to become a functional protein in the cell.

◆ **Learning Exercise 17.5A**

Match the components of the translation process a–e with the statements 1–5.

    a. initiation   **b.** activation   **c.** anticodon   **d.** translocation   **e.** termination

1. _____ The three bases in each tRNA that complement a codon on the mRNA

2. _____ The combining of an amino acid with a specific tRNA

3. _____ The placement of methionine on the large ribosome

4. _____ The shift of the ribosome from one codon on mRNA to the next

5. _____ The process that occurs when the ribosome reaches a UAA or UGA codon on mRNA

Answers   **1.** c   **2.** b   **3.** A   **4.** d   **5.** e

◆ Learning Exercise 17.5B

Write the mRNA that would form for the following section of DNA. For each codon in the mRNA, write the amino acid that would be placed in the protein by a tRNA.

1. DNA strand: —CCC—TCA—GGG—CGC—

   mRNA: — ____ — ____ — ____ — ____ —

   amino acid order: ____ — ____ — ____ — ____

2. DNA: —ATA—GCC—TTT—GGC—AAC—

   mRNA: — ____ — ____ — ____ — ____ — ____ —

   amino acid order: ____ — ____ — ____ — ____ — ____

Answers   **1.** mRNA: —GGG—AGU—CCC—GCG—
                      —Gly—Ser—Pro—Ala—
            2. mRNA: —UAU—CGG—AAA—CCG—UUG—
                      —Tyr—Arg—Lys—Pro—Leu—

◆ Learning Exercise 17.5C

A segment of DNA that codes for a protein contains 270 nucleic acids. How many amino acids would be present in the protein for this DNA segment?

Answer   Assuming that the entire segment codes for a protein, there would be 90 (270 ÷ 3) amino acids in the protein produced.

17.6 Genetic Mutations

- A genetic mutation is a change of one or more bases in the DNA sequence that may alter the structure and ability of the resulting protein to function properly.
- In a substitution, one base is altered, which codes for a different amino acid.
- In a frame shift mutation, the insertion or deletion of one base alters all the codons following the base change, which affects the amino acid sequence that follows the mutation.

◆ **Learning Exercise 17.6**

Consider the DNA template of —AAT—CCC—GGG—.

1. Write the mRNA produced.

_____ — _____ — _____

2. Write the amino acid order for the mRNA codons.

_____ — _____ — _____

3. Suppose thymine in the DNA template is replaced with a guanine. Write the mRNA it produces.

_____ — _____ — _____

4. What is the new amino acid order?

_____ — _____ — _____

5. What are some possible causes of genetic mutations?

Answers   **1.** —UUA—GGG—CCC—   **2.** Leu-Gly-Pro
                    3. —UUC—GGG—CCC—   **4.** Phe-Gly-Pro
                    5. X-rays, UV light, chemicals called mutagens, and some viruses are possible causes of mutations.

17.7 Viruses

- Viruses are small particles of 3–200 genes that cannot replicate unless they invade a host cell.
- A viral infection involves using the host cell machinery to replicate the viral DNA.
- A retrovirus contains RNA and a reverse transcriptase enzyme that synthesizes a viral DNA in a host cell.

◆ **Learning Exercise 17.7**

Match the key terms with the statements shown below

    a. host cell   **b.** retrovirus   **c.** vaccine   **d.** protease   **e.** virus

1. _____ The enzyme inhibited by drugs that prevent the synthesis of viral proteins

2. _____ A small, disease-causing particle that contains either DNA or RNA as its genetic material

3. _____ A type of virus that must use reverse transcriptase to make a viral DNA

4. _____ Required by viruses to replicate

5. _____ Inactive form of viruses that boosts the immune response by causing the body to produce antibodies

Answers   **1.** d   **2.** e   **3.** b   **4.** a   **5.** c

Checklist for Chapter 17

You are ready to take the practice test for Chapter 17. Be sure that you have accomplished the following learning goals for this chapter. If you are not sure, review the section listed at the end of the goal. Then apply your new skills and understanding to the practice test.

After studying Chapter 17, I can successfully:

_____ Identify the components of nucleic acids RNA and DNA (17.1).

_____ Describe the nucleotides contained in DNA and RNA (17.1).

_____ Describe the primary structure of nucleic acids (17.2).

_____ Describe the structure of RNA and DNA; show the relationship between the bases in the double helix (17.3).

_____ Explain the process of DNA replication (17.3).

_____ Describe the structures and characteristics of the three types of RNA (17.4).

_____ Describe the synthesis of mRNA (transcription) (17.4).

_____ Describe the function of the codons in the genetic code (17.4).

_____ Describe the role of translation in protein synthesis (17.5).

_____ Describe some ways in which DNA is altered to cause mutations (17.6).

_____ Explain how retroviruses use reverse transcription to synthesize DNA (17.7).

Practice Test for Chapter 17

1. A nucleotide contains
- **A.** a nitrogen base.
- **B.** a nitrogen base and a sugar.
- **C.** a phosphate and a sugar.
- **D.** a nitrogen base and a deoxyribose.
- **E.** a nitrogen base, a sugar, and a phosphate.

2. The double helix in DNA is held together by
- **A.** hydrogen bonds.
- **B.** ester linkages.
- **C.** peptide bonds.
- **D.** salt bridges.
- **E.** disulfide bonds.

3. The process of producing DNA in the nucleus is called
- **A.** complementation.
- **B.** replication.
- **C.** translation.
- **D.** transcription.
- **E.** mutation.

4. Which occurs in RNA but *not* in DNA?
- **A.** thymine
- **B.** cytosine
- **C.** adenine
- **D.** phosphate
- **E.** uracil

5. Which molecule determines protein structure in protein synthesis?
- **A.** DNA
- **B.** mRNA
- **C.** tRNA
- **D.** rRNA
- **E.** ribosomes

6. Which type of molecule carries amino acids to the ribosomes?
- **A.** DNA
- **B.** mRNA
- **C.** tRNA
- **D.** rRNA
- **E.** protein

For questions 7 through 15, select answers from the following nucleic acids.

- **A.** DNA
- **B.** mRNA
- **C.** tRNA
- **D.** rRNA

7. _____ Along with protein, it is a major component of the ribosomes.

8. _____ A double helix consisting of two chains of nucleotides held together by hydrogen bonds between nitrogen bases.

9. _____ A nucleic acid that uses deoxyribose as the sugar.

10. _____ A nucleic acid produced in the nucleus that migrates to the ribosomes to direct the formation of a protein.

11. _____ It can place the proper amino acid into the peptide chain.

12. _____ It has nitrogen bases of adenine, cytosine, guanine, and thymine.

13. _____ It contains the codons for the amino acid order.

14. _____ It contains a triplet called an anticodon loop.

15. _____ This nucleic acid is replicated during cellular division.

For questions 16 through 20, select answers from the following.

A. —A—G—C—C—T—A—
      | | | | | |
    —T—C—G—G—A—T—

B. —A—U—U—G—C—U—C—

C. —A—G—T—U—G—U—
      | | | | | |
    —T—C—A—A—C—A—

D. —G—U—A—

E. —A—T—G—T—A—T—

16. _____ A section of mRNA

17. _____ An impossible section of DNA

18. _____ A codon

19. _____ A section from a DNA molecule

20. _____ A single strand that would not be possible for mRNA

Use the statements A–E to answer questions 21 through 25.

    A. tRNA assembles the amino acids at the ribosomes.
    B. DNA forms a complementary copy of itself called mRNA.
    C. Protein is formed and breaks away.
    D. tRNA picks up specific amino acids.
    E. mRNA goes to the ribosomes.

21. _____ First step

22. _____ Second step

23. _____ Third step

24. _____ Fourth step

25. _____ Fifth step

Answers to the Practice Test

1. E	**2.** A	**3.** B	**4.** E	**5.** A
6. C	**7.** D	**8.** A	**9.** A	**10.** B
11. C	**12.** A	**13.** B	**14.** C	**15.** A
16. B, D	**17.** C	**18.** D	**19.** A	**20.** E
21. B	**22.** E	**23.** D	**24.** A	**25.** C

Answers and Solutions to Selected Text Problems

17.1   Pyrimidines (cytosine, thymine, uracil) contain a single ring. Purines (adenine and guanine) have a two-ring system.

      a. Pyrimidine   **b.** Pyrimidine

17.3 DNA contains two purines, adenine (A) and guanine (G) and two pyrimidines, cytosine (C) and thymine (T). RNA contains the same bases, except thymine (T) is replaced by the pyrimidine uracil (U).

    a. DNA   **b.** Both DNA and RNA

17.5 Nucleotides contain a base, a sugar, and a phosphate group. The nucleotides found in DNA would all contain the sugar deoxyribose. The four nucleotides are: deoxyadenosine 5′-monophosphate (dAMP), deoxythymidine 5′-monophosphate (dTMP), deoxycytidine 5′-monophosphate (dCMP), and deoxyguanosine 5′-monophosphate (dGMP).

17.7   **a.** Adenosine is a nucleoside found in RNA.
      b. Deoxycytidine is a nucleoside found in DNA.
      c. Uridine is a nucleoside found in RNA.
      d. Cytidine 5′-monophosphate is a nucleotide found in RNA.

17.9

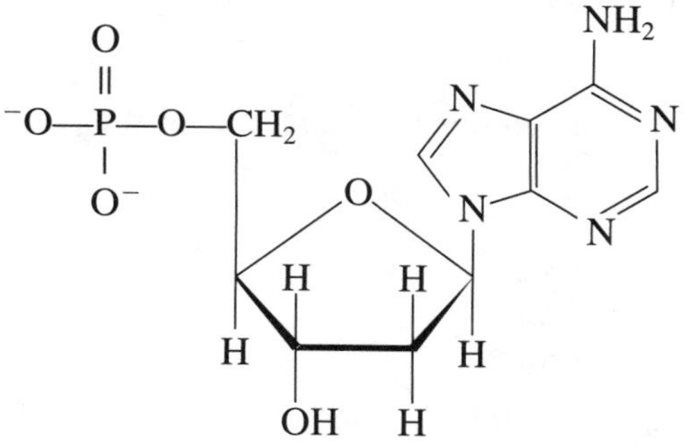

17.11 The nucleotides in nucleic acids are held together by phosphodiester bonds between the 3′-OH of a sugar (ribose or deoxyribose) and the 5′-phosphate group of another sugar.

17.13

Guanine (G)

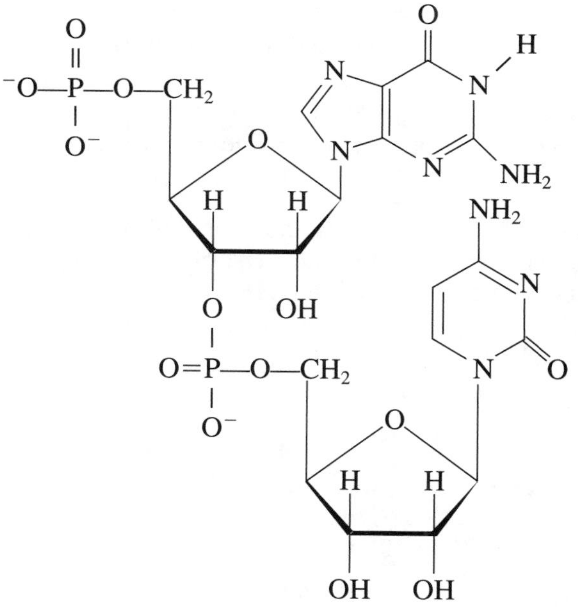

Cytidine (C)

17.15 The two DNA strands are held together by hydrogen bonds between the bases in each strand.

17.17   **a.** Since T pairs with A, if one strand of DNA has the sequence —A—A—A—A—A—A—, the second strand would be —T—T—T—T—T—T—.
      b. Since C pairs with G, if one strand of DNA has the sequence —G—G—G—G—G—G—, the second strand would be —C—C—C—C—C—C—.

    c. Since T pairs with A, and C pairs with G, if one strand of DNA has the sequence —A—G—T—C—C—A—G—G—T—, the second strand would be —T—C—A—G—G—T—C—C—A—.

    d. Since T pairs with A, and C pairs with G, if one strand of DNA has the sequence —C—T—G—T—A—T—A—C—G—T—T—A—, the second strand would be —G—A—C—A—T—A—T—G—C—A—A—T—.

17.19 Ribosomal RNA (rRNA) is found in the ribosomes, which are the sites for protein synthesis. Transfer RNA (tRNA) brings specific amino acids to the ribosomes for protein synthesis. Messenger RNA (mRNA) carries the information needed for protein synthesis from the DNA in the nucleus to the ribosomes.

17.21 In transcription, the sequence of nucleotides on a DNA template (one strand) is used to produce the base sequences of a messenger RNA. The DNA unwinds and one strand is copied as complementary bases are placed in the mRNA molecule. In RNA, U (uracil) is paired with A in DNA.

17.23 In mRNA, C, G, and A pair with G, C, and T in DNA. However, in mRNA U will pair with A in DNA. The strand of mRNA would have the following sequence: —GGC—UUC—CAA—GUG—.

17.25 A codon is the three-base sequence (triplet) in mRNA that codes for a specific amino acid in a protein.

17.27   **a.** The codon CUU in mRNA codes for the amino acid leucine.
    b. The codon UCA in mRNA codes for the amino acid serine.
    c. The codon GGU in mRNA codes for the amino acid glycine.
    d. The codon AGG in mRNA codes for the amino acid arginine.

17.29 When AUG is the first codon, it signals the start of protein synthesis and incorporates methionine as the first amino acid in the peptide. Eventually the initial methionine is removed as the protein forms its secondary and tertiary protein structure. In the middle of an mRNA sequence AUG codes for methionine.

17.31 A codon is a base triplet in the mRNA template. An anticodon is the complementary triplet on a tRNA for a specific amino acid.

17.33 The mRNA must be divided into triplets and the amino acid coded for by each triplet read from the table.
    a. The codon AAA in mRNA codes for lysine: —Lys—Lys—Lys—
    b. The codon UUU codes for phenylalanine and CCC for proline: —Phe—Pro—Phe—Pro—
    c. —Tyr—Gly—Arg—Cys—

17.35 After a tRNA attaches to the first binding site on the ribosome, its amino acid forms a peptide bond with the amino acid on the tRNA attached to the second binding site. The ribosome moves along the mRNA and a new tRNA with its amino acid occupies the open binding site.

17.37   **a.** By using the pairing: DNA bases   C   G   T   A

               mRNA bases   G   C   A   U

    the mRNA sequence can be determined: —CGA—AAA—GUU—UUU—.
    b. The tRNA triplet anticodons would be as follows: GCU, UUU, CAA, AAA.
    c. The mRNA is divided into triplets and the amino acid coded for by each triplet read from the table. Using codons in mRNA: —Arg—Lys—Val—Phe—.

17.39 In a substitution mutation an incorrect base replaces a base in DNA.

17.41 In a frameshift mutation, a base is lost or gained, which changes the codons, and therefore the amino acids, in the remaining polypeptide chain.

17.43 The normal triplet TTT in DNA transcribes to AAA in mRNA. AAA codes for lysine. The mutation TTC in DNA transcribes to AAG in mRNA, which also codes for lysine. Thus, there is no effect on protein synthesis.

17.45 **a.** —Thr—Ser—Arg—Val— is the amino acid sequence produced by normal DNA.

b. —Thr—Thr—Arg—Val— is the amino acid sequence produced by a mutation.

c. —Thr—Ser—Gly—Val— is the amino acid sequence produced by a mutation.

d. —Thr— STOP Protein synthesis would terminate early. If this mutation occurs early in the formation of the polypeptide, the resulting protein will probably be nonfunctional.

e. The new protein will contain the sequence —Asp—Ile—Thr—Gly—.

f. The new protein will contain the sequence —His—His—Gly—.

17.47 **a.** Both codons GCC and GCA code for alanine.

b. A cross-link in the tertiary structure of hemoglobin cannot be formed when the polar glutamic acid is replaced by nonpolar valine. The resulting hemoglobin is malformed and less capable of carrying oxygen.

17.49 A virus contains either DNA or RNA, but not both, inside a protein coating.

17.51 **a.** An RNA-containing virus must make viral DNA from the RNA, a process called reverse transcription.

b. A virus that uses reverse transcription is a retrovirus.

17.53 Nucleoside analogs like AZT or ddI mimic the structures of nucleosides that the HIV virus uses for DNA synthesis. These analogs are incorporated into the new viral DNA chain, but the lack of a hydroxyl group in position 3′ of the sugar stops the chain from growing any longer and prevents replication of the virus.

17.55 **a.**

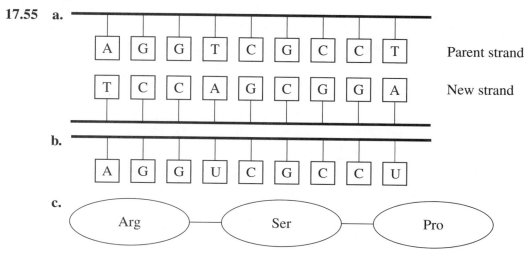

17.57 **a.** pyrimidine **b.** purine **c.** pyrimidine
d. pyrimidine **e.** purine

17.59 **a.** thymine and deoxyribose **b.** adenine and ribose
c. cytosine and ribose **d.** guanine and deoxyribose

17.61 They are both pyrimidines, but thymine has a methyl group.

17.63 They are both polymers of nucleotides, connected through phosphodiester bonds between alternating sugar and phosphate groups, with bases extending out from each sugar.

17.65 **a.** —C—T—G—A—A—T—C—C—G—

b. —A—C—G—T—T—T—G—A—T—C—G—T—

c. —T—A—G—C—T—A—G—C—T—A—G—C

17.67 **a.** tRNA **b.** rRNA **c.** mRNA

17.69 **a.** ACU, ACC, ACA, and ACG
b. UCU, UCC, UCA, UCG, AGU, and AGC
c. UGU and UGC

17.71 **a.** AAG codes for lysine.   **b.** AUU codes for isoleucine.
       c. CGG codes for arginine.

17.73 Using the genetic code the codons indicate the following: start(methionine)
—Tyr—Gly—Gly—Phe—Leu—stop

17.75 The anticodon consists of the three complementary bases to the codon.

     a. UCG   **b.** AUA   **c.** GGU

17.77 Three nucleotides are needed for each amino acid plus a start and stop triplet, which makes a minimum total of 33 nucleotides.

17.79 **a.** Because A bonds with T, T is also 28%. Since A and T = 56%, there is 44% for the other nucleotides or 22% G and 22% C.
       b. Because C bonds with G, G is also 20%. Since C+G make up 40%, the nucleotides A and T are 60%, or 30% A and 30% T.

17.81 A DNA virus attaches to a cell and injects viral DNA that uses the host cell to produce copies of DNA to make viral RNA. A retrovirus injects viral RNA from which complementary DNA is produced by reverse transcription.

18
Metabolic Pathways and Energy Production

Study Goals

- Explain the role of ATP in anabolic and catabolic reactions.
- Compare the structures and function of the coenzymes NAD^+, FAD, and coenzyme A.
- Give the sites, enzymes, and products for the digestion of carbohydrates, triacylglycerols, and proteins.
- Describe the key reactions in the degradation of glucose in glycolysis.
- Describe the pathways for pyruvate.
- Describe the reactions in the citric acid cycle that oxidize acetyl CoA.
- Explain how electrons from NADH and H^+ and FAD move along the electron transport system to form H_2O.
- Describe the role of oxidative phosphorylation in ATP synthesis.
- Calculate the ATP produced by the complete combustion of glucose.
- Describe the oxidation of fatty acids via β oxidation.
- Calculate the ATP produced by the complete oxidation of a fatty acid.
- Explain the role of transamination and oxidative deamination in the degradation of amino acids.

Think About It

1. Why do you need ATP in your cells?

2. What monosaccharides are produced when carbohydrates undergo digestion?

3. What is meant by *aerobic* and *anaerobic* conditions in the cells?

4. Why is the citric acid cycle considered a central pathway in metabolism?

5. What are the products from the digestion of triacylglycerols and proteins?

6. When do you utilize fats and proteins for energy?

Key Terms

1. Match the key terms with the correct statement shown below.

    a. ATP   **b.** anabolic reaction   **c.** glycolysis
    d. catabolic reaction   **e.** mitochondria

1. _____ A metabolic reaction that uses energy to build large molecules

2. _____ A metabolic reaction that produces energy for the cell by degrading large molecules

3. _____ A high-energy compound produced from energy-releasing processes that provides energy for energy-producing reactions

4. _____ The degradation reactions of glucose that yield two pyruvate molecules

5. _____ The organelles in the cells where energy-producing reactions take place

Answers   **1.** b   **2.** d   **3.** a   **4.** c   **5.** e

2. Match the key terms with the correct statement shown below.

    a. citric acid cycle   **b.** oxidative phosphorylation   **c.** coenzyme Q
    d. cytochromes   **e.** proton pump

1. _____ A mobile carrier that passes electrons from NADH and $FADH_2$ to cytochrome b in complex III

2. _____ Proteins containing iron as Fe^{3+} or Fe^{2+} that transfer electrons from QH_2 to oxygen

3. _____ Protons move from the matrix into the inner membrane space to create a proton gradient

4. _____ The synthesis of ATP from ADP and P_i using energy generated from electron transport

5. _____ Oxidation reactions that convert acetyl CoA to CO_2 producing reduced coenzymes for energy production via electron transport

Answers   **1.** c   **2.** d   **3.** e   **4.** b   **5.** a

18.1 Metabolism and ATP Energy

- Metabolism is all the chemical reactions that provide energy and substances for cell growth.
- Catabolic reactions degrade large molecules to produce energy.
- Anabolic reactions utilize energy in the cells to build large molecules for the cells.
- Energy is stored in ATP, a high-energy compound that is hydrolyzed when energy is required for the anabolic reactions that do work in the cells.
- The hydrolysis of ATP, which releases energy, is linked with many anabolic reactions in the cell.

◆ Learning Exercise 18.1A

Identify the stages of metabolism for each of the following processes.

    a. Stage 1   **b.** Stage 2   **c.** Stage 3

1. _____ Oxidation of two-carbon acetyl CoA that provides most of the energy for ATP synthesis.

2. _____ Polysaccharides undergo digestion to monosaccharides such as glucose.

3. _____ Digestion products such as glucose are degraded to two- or three-carbon compounds.

Answers   **1.** c   **2.** a   **3.** b

◆ Learning Exercise 18.1B

Complete the following statements for ATP.

The ATP molecule is composed of a nitrogen-base (1) _____ a (2) _____ sugar, and three (3) _____. ATP undergoes (4) _____, which cleaves a (5) _____ and releases (6) _____. For this reason, ATP is called a (7) _____ compound. The resulting phosphate group, called inorganic phosphate, is abbreviated as (8) _____. This equation can be written as (9) _____. The energy from ATP is linked to cellular reactions that are (10) _____

_____.

Answers   **1.** adenine   **2.** ribose   **3.** phosphate groups   **4.** hydrolysis
              5. phosphate   **6.** energy   **7.** high-energy   **8.** P_i
              9. $ATP + H_2O \rightarrow ADP + P_i + Energy$ (7.3 kcal/mole)   **10.** energy requiring

18.2 Digestion of Foods

- Digestion is a series of reactions that break down large food molecules of carbohydrates, lipids, and proteins into smaller molecules that can be absorbed and used by the cells.
- The end products of digestion of polysaccharides are the monosaccharides glucose, fructose, and galactose.
- Proteins begin digestion in the stomach, where HCl denatures proteins and activates peptidases that hydrolyze peptide bonds.
- In the small intestine trypsin and chymotrypsin complete the hydrolysis of peptides to amino acids.
- Dietary fats begin digestion in the small intestine where they are emulsified by bile salts.
- Pancreatic lipases hydrolyze the triacylglycerols to yield monoacylglycerols and free fatty acids.
- The triacylglycerols reformed in the intestinal lining combine with proteins to form chylomicrons for transport through the lymph system and bloodstream.
- In the cells, triacylglycerols hydrolyze to glycerol and fatty acids, which can be used for energy.

◆ Learning Exercise 18.2A

Complete the table to describe sites, enzymes, and products for the digestion of carbohydrates.

Food	Digestion site(s)	Enzyme	Products
1. Amylose			
2. Amylopectin			
3. Maltose			
4. Lactose			
5. Sucrose			

Answers

Food	Digestion site(s)	Enzyme	Products
1. Amylose	**a.** mouth **b.** small intestine	**a.** salivary amylase **b.** pancreatic amylase	**a.** smaller polysaccharides (dextrins), some maltose and glucose **b.** maltose, glucose
2. Amylopectin	**a.** mouth **b.** small intestine	**a.** salivary amylase **b.** pancreatic amylase	**a.** smaller polysaccharides (dextrins), some maltose and glucose **b.** maltose, glucose
3. Maltose	small intestine	maltase	glucose and glucose
4. Lactose	small intestine	lactase	glucose and galactose
5. Sucrose	small intestine	sucrase	glucose and fructose

◆ **Learning Exercise 18.2B**

Match each of the following terms with the correct description.

    a. stomach   **b.** small intestine

  1. _____ HCl activates enzymes that hydrolyzes peptide bonds in proteins.

  2. _____ Trypsin and chymotrypsin convert peptides to amino acids.

Answers   **1.** a   **2.** b

18.3 Important Coenzymes in Metabolic Pathways

- Coenzymes such as FAD and NAD^+ pick up hydrogen and electrons during oxidative processes.
- Coenzyme A is a coenzyme that carries acetyl (two-carbon) groups produced when glucose, fatty acids, and amino acids are degraded.

◆ **Learning Exercise 18.3**

Select the coenzyme that matches each of the following descriptions of coenzymes.

    a. NAD^+ **b.** NADH   **c.** FAD   **d.** $FADH_2$ **e.** coenzyme A

  1. _____ participates in reactions that convert a hydroxyl group to a C=O group

  2. _____ contains riboflavin (vitamin B_2)

  3. _____ reduced form of nicotinamide adenine dinucleotide

  4. _____ contains the vitamin niacin

  5. _____ oxidized form of flavin adenine dinucleotide

  6. _____ contains the vitamin pantothenic acid, ADP, and an aminoethanethiol

  7. _____ participates in oxidation reactions that produce a carbon–carbon double bond (C=C)

  8. _____ transfers acyl groups such as the two-carbon acetyl group

  9. _____ reduced form of flavin adenine dinucleotide

Answers   **1.** a   **2.** c, d   **3.** b   **4.** a, b   **5.** c
               6. e   **7.** c   **8.** e   **9.** d

18.4 Glycolysis: Oxidation of Glucose

- Glycolysis is the primary anaerobic pathway for the degradation of glucose to yield pyruvic acid.
- Glucose is converted to fructose-1,6-diphosphate that is split into two triose phosphate molecules.
- The oxidation of the three-carbon sugars yields the reduced coenzyme, 2 NADH, and 2 ATP.
- In the absence of oxygen, pyruvate is reduced to lactate and NAD^+ is regenerated for the continuation of glycolysis.
- Under aerobic conditions, pyruvate is oxidized in the mitochondria to acetyl CoA, which enters the citric acid cycle.

◆ Learning Exercise 18.4A

Match each of the following terms of glycolysis with the best description.

    a. 2 NADH   **b.** anaerobic   **c.** glucose   **d.** two pyruvate
    e. energy invested **f.** energy generated **g.** 2 ATP   **h.** 4 ATP

1. _____ the starting material for glycolysis

2. _____ steps 1–5 of glycolysis

3. _____ operates without oxygen

4. _____ net ATP energy produced

5. _____ number of reduced coenzymes produced

6. _____ steps 6–10 of glycolysis

7. _____ end products of glycolysis

8. _____ number of ATP required

Answers   **1.** c   **2.** e   **3.** b   **4.** g   **5.** a
           6. f   **7.** a,d,g   **8.** g

◆ Learning Exercise 18.4B

Fill in the blanks with the following terms.

    lactate NAD^+ acetyl CoA

    $NADH + H^+$ aerobic   anaerobic

When oxygen is available during glycolysis, the three-carbon pyruvate may be oxidized to form (1)

_____ + CO_2. The coenzyme (2) _____ is reduced to (3) _____. Under

(4) _____ conditions, pyruvate is reduced to (5) _____.

Answers   **1.** acetyl CoA **2.** NAD^+ **3.** $NADH + H^+$
           4. anaerobic   **5.** lactate

◆ Learning Exercise 18.4C

Essay: Explain how the formation of lactate from pyruvate during anaerobic conditions allows glycolysis to continue.

Answer   Under anaerobic conditions, the oxidation of pyruvate to acetyl CoA to regenerate NAD^+ cannot take place. Then, pyruvate is reduced to lactate using $NADH + H^+$ in the cytoplasm and regenerating NAD^+.

18.5 The Citric Acid Cycle

- Under aerobic conditions, pyruvate is oxidized in the mitochondria to acetyl CoA, which enters the citric acid cycle.
- In a sequence of reactions called the citric acid cycle, acetyl CoA combines with oxaloacetate to yield citrate.
- In one turn of the citric acid cycle, the oxidation of acetyl CoA yields two CO_2, GTP, three NADH, and $FADH_2$. The phosphorylation of ADP by GTP yields ATP.

◆ Learning Exercise 18.5

In each of the following steps of the citric acid cycle, indicate if oxidation occurs (yes/no) and any coenzyme or direct phosphorylation product produced ($NADH + H^+$, $FADH_2$, GTP).

Steps in citric acid cycle	Oxidation	Coenzyme
1. acetyl CoA + oxaloacetate → citrate	_____	_____
2. citrate → isocitrate	_____	_____
3. isocitrate → α-ketoglutarate acid	_____	_____
4. α-ketoglutarate → succinyl CoA	_____	_____
5. succinyl CoA → succinate	_____	_____
6. succinate → fumarate	_____	_____
7. fumarate → malate	_____	_____
8. malate → oxaloacetate	_____	_____

Answers
1. no
2. no
3. yes, $NADH + H^+$
4. yes, $NADH + H^+$
5. no, GTP
6. yes, $FADH_2$
7. no
8. yes, $NADH + H^+$

18.6 Electron Transport

- The reduced coenzymes from glycolysis and the citric acid cycle are oxidized to NAD^+ and FAD by transferring protons and electrons to the electron transport system.
- In the electron transport system, electrons are transferred to electron carriers including flavins, coenzyme Q, iron-sulfur proteins, and cytochromes with Fe^{3+}/Fe^{2+}.
- The final acceptor, O_2, combines with protons and electrons to yield H_2O.

◆ Learning Exercise 18.6A

Write the oxidized and reduced forms of each of the following electron carriers.

1. flavin mononucleotide   oxidized _____   reduced _____

2. coenzyme Q   oxidized _____   reduced _____

3. iron-protein clusters   oxidized _____   reduced _____

4. cytochrome *b*   oxidized _____   reduced _____

Answers
1. oxidized: FMN   reduced: $FMNH_2$
2. oxidized: Q   reduced: QH_2
3. oxidized: Fe^{3+} S cluster   reduced: Fe^{2+} S cluster
4. oxidized: cyt *b* (Fe^{3+})   reduced: cyt *b* (Fe^{2+})

◆ **Learning Exercise 18.6B**

Identify the protein complexes and mobile carriers in electron transport.

a. cytochrome c oxidase (IV) **b.** NADH dehydrogenase (I) **c.** cytochrome c
d. coenzyme Q-cytochrome c reductase (III) **e.** succinate dehydrogenase (II) **f.** Q

1. _____ FMN and Fe-S clusters

2. _____ $FADH_2 + Q \rightarrow FAD + QH_2$

3. _____ mobile carrier from complex I or complex II to complex III

4. _____ electrons are passed from cyt a and a_3 to $^1/_2 O_2$ and $2H^+$ to yield H_2O

5. _____ mobile carrier between complex III and IV

6. _____ cyt b and Fe-S clusters

Answers **1.** b **2.** e **3.** f **4.** a **5.** c **6.** d

◆ **Learning Exercise 18.6C**

1. Write an equation for the transfer of hydrogen from $FMNH_2$ to Q.

2. What is the function of coenzyme Q in electron transport?

3. What are the end products of electron transport?

Answers **1.** $FMNH_2 + Q \rightarrow FMN + QH_2$
2. Q accepts hydrogen atoms from $FMNH_2$ or $FADH_2$. From QH_2, the hydrogen atoms are separated into protons and electrons, with the electrons being passed on to the cytochromes, the electron acceptors in electron transport.
3. $CO_2 + H_2O$

18.7 Oxidative Phosphorylation and ATP

- The flow of electrons through the electron transport system pumps protons across the inner membrane, which produces a high-energy proton gradient that provides energy for the synthesis of ATP.
- The process of using the energy of electron transport to synthesize ATP is called oxidative phosphorylation.
- The oxidation of NADH yields three ATP molecules, and $FADH_2$ yields two ATP.
- The complete oxidation of glucose yields a total of 36 ATP from direct phosphorylation and the oxidation of the reduced coenzymes NADH and $FADH_2$.

◆ **Learning Exercise 18.7A**

Match the following terms with the correct description below.

    a. oxidative phosphorylation   **b.** ATP synthase
    c. proton pumps   **d.** proton gradient

1. _____ The complexes I, III, and IV through which H^+ ions move out of the matrix into the inner membrane space

2. _____ The protein tunnel where protons flow from the inner membrane space back to the matrix generates energy for ATP synthesis

3. _____ Energy from electron transport is used to form a proton gradient that drives ATP synthesis

4. _____ The accumulation of protons in the inner membrane space that lowers pH

Answers   **1.** c   **2.** b   **3.** a   **4.** d

◆ **Learning Exercise 18.7B**

Complete the following:

Substrate	Reaction	Products	ATP produced
1. Glucose	glycolysis (aerobic)		
2. Pyruvate	oxidation		
3. Acetyl CoA	citric acid cycle		
4. Glucose	glycolysis (anaerobic)		
5. Glucose	complete oxidation		

Answers

Substrate	Reaction	Products	ATP produced
1. Glucose	glycolysis (aerobic)	2 pyruvate	6 ATP
2. Pyruvate	oxidation	acetyl CoA + CO_2	3 ATP
3. Acetyl CoA	citric acid cycle	2 CO_2	12 ATP
4. Glucose	glycolysis (anaerobic)	2 lactate	2 ATP
5. Glucose	complete oxidation	6 CO_2 + 6 H_2O	36 ATP

18.8 Oxidation of Fatty Acids

- When needed for energy, fatty acids link to coenzyme A for transport to the mitochondria where they undergo β oxidation.
- In β oxidation, a fatty acyl chain is oxidized to yield a shortened fatty acid, acetyl CoA, and the reduced coenzymes NADH and $FADH_2$.
- The energy obtained from a particular fatty acid depends on the number of carbon atoms.
- Two ATP are required for activation. Then each acetyl CoA produces 12 ATP via citric acid cycle, and electron transport converts each NADH to 3 ATP and each FADH to 2 ATP.

◆ **Learning Exercise 18.8A**

1. Write an equation for the activation of myristic (C_{14}) acid: CH_3—$(CH_2)_{12}$—$\overset{\overset{\displaystyle O}{\displaystyle \|}}{C}$—$OH$

2. Write an equation for the first oxidation of myristyl CoA.

3. Write an equation for the hydration of the double bond.

4. Write the overall equation for the complete oxidation of myristyl CoA.

5. a. How many cycles of β oxidation will be required?

  b. How many acetyl CoA units will be produced?

Answers

1. CH_3—$(CH_2)_{12}$—$\overset{\overset{\displaystyle O}{\displaystyle \|}}{C}$—$OH$ + HS—CoA + ATP → CH_3—$(CH_2)_{12}$—$\overset{\overset{\displaystyle O}{\displaystyle \|}}{C}$—$S$—$CoA$ + AMP + $2P_i$

2. CH_3—$(CH_2)_{12}$—$\overset{\overset{\displaystyle O}{\displaystyle \|}}{C}$—$S$—$CoA$ + FAD → CH_3—$(CH_2)_{10}$—$CH{=}CH$—$\overset{\overset{\displaystyle O}{\displaystyle \|}}{C}$—$S$—$CoA$ + $FADH_2$

3. CH_3—$(CH_2)_{10}$—$CH{=}CH$—$\overset{\overset{\displaystyle O}{\displaystyle \|}}{C}$—$S$—$CoA$ + H_2O → CH_3—$(CH_2)_{10}$—$\overset{\overset{\displaystyle OH}{\displaystyle |}}{CH}$—$CH_2$—$\overset{\overset{\displaystyle O}{\displaystyle \|}}{C}$—$S$—$CoA$

4. myristyl (C_{14})—CoA + 6 CoA + 6 FAD + 6 NAD$^+$ + 6 H$_2$O →
  7 Acetyl CoA + 6 FADH$_2$ + 6 NADH + 6 H$^+$

5. a. 6 cycles   **b.** 7 acetyl CoA units are produced

◆ **Learning Exercise 18.8B**

Lauric acid is a 12-carbon fatty acid: $CH_3(CH_2)_{10}COOH$.

1. How much ATP is needed for activation?

2. How many cycles of β oxidation are required?

3. How many NADH and $FADH_2$ are produced during β oxidation?

4. How many acetyl CoA units are produced?

5. What is the total ATP produced from electron transport and the citric acid cycle?

Answers   **1.** 2 ATP   **2.** 5 cycles
3. 5 cycles produce 5 NADH and 5 $FADH_2$ **4.** 6 acetyl CoA
5. 5 NADH $\times$ 3 ATP = 15 ATP; 5 $FADH_2$ $\times$ 2 ATP = 10 ATP;
6 acetyl CoA $\times$ 12 ATP = 72 ATP; Total ATP = 15 ATP + 10 ATP + 72 ATP −
2 ATP (for activation) = 95 ATP

18.9 Degradation of Amino Acids

• Amino acids are normally used for protein synthesis.
• Amino acids are degraded by transferring an amino group from an amino acid to an α-keto acid to yield a different amino acid and α-keto acid.
• In oxidative deamination, the amino group in glutamate is removed as an ammonium ion, NH_4^+.
• α-Keto acids resulting from transamination can be used as intermediates in the citric acid cycle, in the synthesis of lipids or glucose, or oxidized for energy.

◆ **Learning Exercise 18.9A**

Match each of the following descriptions with transamination (T) or oxidative deamination (D).

1. _____ produces an ammonium ion, NH_4^+

2. _____ transfers an amino group to an α-keto acid

3. _____ usually involves the degradation of glutamate

4. _____ requires NAD^+ or $NADP^+$

5. _____ produces another amino acid and α-keto acid

6. _____ usually produces α-ketoglutarate

Answers   **1.** D   **2.** T   **3.** D   **4.** D   **5.** T   **6.** D

◆ **Learning Exercise 18.9B**

1. Write an equation for the transamination reaction of serine and oxaloacetate.

2. Write an equation for the oxidative deamination of glutamate.

Answers

1. $$\overset{\overset{\displaystyle NH_3^+}{|}}{HO-CH_2-CH-COO^-} + \overset{\overset{\displaystyle O}{\|}}{^-OOC-C-CH_2-COO^-} \rightarrow$$

$$\overset{\overset{\displaystyle O}{\|}}{HO-CH_2-C-COO^-} + \overset{\overset{\displaystyle NH_3^+}{|}}{^-OOC-CH-CH_2-COO^-}$$

2. $$\overset{\overset{\displaystyle NH_3^+}{|}}{^-OOC-CH-CH_2-CH_2-COO^-} + NAD^+ \text{ (or } NADP^+) + H_2O \rightarrow$$

$$\overset{\overset{\displaystyle O}{\|}}{^-OOC-C-CH_2-CH_2-COO^-} + NH_4^+ + NADH \text{ (or } NADPH) + H^+$$

Checklist for Chapter 18

You are ready to take the practice test for Chapter 18. Be sure that you have accomplished the following learning goals for this chapter. If you are not sure, review the section listed at the end of the goal. Then apply your new skills and understanding to the practice test.
After studying Chapter 18, I can successfully:

_____ Describe the role of ATP in catabolic and anabolic reactions (18.1).

_____ Describe the sites, enzymes, and products of digestion for carbohydrates (18.2).

_____ Describe the coenzymes NAD^+, FAD, and Coenzyme A (18.3).

_____ Describe the conversion of glucose to pyruvate in glycolysis (18.4).

_____ Give the conditions for the conversion of pyruvate to lactate and acetyl coenzyme A (18.4).

_____ Describe the oxidation of acetyl CoA in the citric acid cycle (18.5).

_____ Identify the electron carriers in the electron transport system (18.6).

_____ Describe the process of electron transport (18.6).

_____ Explain the chemiosmotic theory whereby ATP synthesis is linked to the energy of electron transport and a proton gradient (18.7).

_____ Account for the ATP produced by the complete oxidation of glucose (18.7).

_____ Describe the oxidation of fatty acids via β oxidation (18.8).

_____ Calculate the ATP produced by the complete oxidation of a fatty acid (18.8).

_____ Explain the role of transamination and oxidative deamination in degrading amino acids (18.9).

Practice Test for Chapter 18

1. The main function of the mitochondria is
   A. energy production.   **B.** protein synthesis.   **C.** glycolysis.
   D. genetic instructions.   **E.** waste disposal.

2. ATP is a(n)
   A. nucleotide unit in RNA and DNA.
   B. end product of glycogenolysis.
   C. end product of transamination.
   D. enzyme.
   E. energy storage molecule.

Use one or more of the following enzymes and end products for the digestion of each of items 3 through 6.
   A. maltase   **B.** glucose   **C.** fructose   **D.** sucrase
   E. galactose   **F.** lactase   **G.** pancreatic amylase

3. sucrose _____   **4.** lactose _____

5. small polysaccharides _____   **6.** maltose _____

7. The digestion of triacylglycerols takes place in the _____ by enzymes called _____.
   A. small intestine; peptidases   **B.** stomach; lipases   **C.** stomach; peptidases
   D. small intestine; lipases   **E.** All of these

8. The products of the digestion of triacylglycerols are
   A. fatty acids.   **B.** monoacylglycerols.   **C.** glycerol.
   D. diacylglycerols.   **E.** All of these

9. The digestion of proteins takes place in the _____ by enzymes called _____.
   A. small intestine; peptidases   **B.** stomach; lipases   **C.** stomach; peptidases
   D. small intestine; lipases   **E.** stomach and small intestine; proteases and peptidases

10. Glycolysis
   A. requires oxygen for the catabolism of glucose.
   B. represents the aerobic sequence for glucose anabolism and ATP production.
   C. represents the splitting off of glucose residues from glycogen.
   D. represents the anaerobic catabolism of glucose to pyruvate.
   E. produces acetyl units and ATP as end products.

Associate the following coenzymes A–E with the correct descriptions 11 through 15.

A. NAD^+ **B.** NADH **C.** FAD **D.** $FADH_2$ **E.** Coenzyme A

11. _____ converts a hydroxyl group to a C=O group

12. _____ reduced form of nicotinamide adenine dinucleotide

13. _____ participates in oxidation reactions that produce a carbon–carbon double bond (C=C)

14. _____ transfers acyl groups such as the two-carbon acetyl group

15. _____ reduced form of flavin adenine dinucleotide

Answer the following questions for glycolysis.

16. _____ Number of ATP invested to oxidize one glucose molecule

17. _____ Number of ATP (net) produced from one glucose molecule

18. _____ Number of NADH produced from one glucose molecule

19. Which is true of the citric acid cycle?

    A. Acetyl CoA is converted to CO_2 and H_2O.
    B. Oxaloacetate combines with acetyl units to form citric acid.
    C. The coenzymes are NAD^+ and FAD.
    D. ATP is produced by direct phosphorylation.
    E. All of the above

Match the compounds A–E with statements 20 through 24.

    A. malate   **B.** fumarate   **C.** succinate
    D. citrate   **E.** oxaloacetate

20. _____ formed when oxaloacetate combines with acetyl CoA

21. _____ H_2O adds to its double bond to form malate

22. _____ FAD removes hydrogen to form a double bond

23. _____ formed when the hydroxyl group in malate is oxidized

24. _____ the compound that is regenerated in the citric acid cycle

25. One turn of the citric acid cycle produces

    A. 3 NADH.   **B.** 3 NADH, 1 $FADH_2$.   **C.** 3 $FADH_2$, 1 NADH, 1 ATP.
    D. 3 NADH, 1 $FADH_2$, 1 ATP, 2 CO_2.   **E.** 1 NADH, 1 $FADH_2$, 1 ATP.

26. How many electron transfers in electron transport provide sufficient energy for ATP synthesis?

    A. none   **B.** 1   **C.** 2   **D.** 3   **E.** 4

27. Electron transport

    A. produces most of the body's ATP.
    B. carries oxygen to the cells.
    C. produces CO_2 + H_2O.
    D. is only involved in the citric acid cycle.
    E. operates during fermentation.

In questions 28 through 30, match the components of the electron transport system with the following activities.

    A. NAD^+ **B.** FMN   **C.** FAD   **D.** Q   **E.** cytochromes

28. _____ a mobile carrier that transfers electrons from FMN and $FADH_2$ to cytochromes

29. _____ the coenzyme that accepts hydrogen atoms from NADH

30. _____ the electron acceptors containing iron

Select the correct answer A–F for the number of ATP produced in the descriptions 31 through 35.

    A. 2 ATP   **B.** 3 ATP   **C.** 6 ATP   **D.** 12 ATP   **E.** 24 ATP   **F.** 36 ATP

31. One turn of the citric acid cycle (acetyl CoA $\rightarrow 2\,CO_2$)

32. Complete combustion of glucose (glucose $+\ 6\,O_2 \rightarrow 6\,H_2O\ +\ 6\,CO_2$)

33. Produced when NADH enters electron transport

34. Produced from glycolysis (glucose $+\ O_2 \rightarrow 2$ pyruvate $+\ 2\,H_2O$)

35. Produced from oxidation of 2 pyruvate (2 pyruvate $\rightarrow 2$ acetyl CoA $+\ 2\,CO_2$)

36. Chylomicrons formed in the intestinal lining
    A. are lipoproteins.
    B. are triacylglycerols coated with proteins.
    C. transport fats into the lymph system and bloodstream.
    D. carry triacylglycerols to the cells of the heart, muscle, and adipose tissues.
    E. All of these

Consider the β oxidation of palmitic (C_{16}) acid for questions 37 through 40.

37. The number of β oxidation cycles required for palmitic (C_{16}) acid is
    A. 16   **B.** 9   **C.** 8
    D. 7   **E.** 6

38. The number of acetyl CoA groups produced by the β oxidation of palmitic (C_{16}) acid is
    A. 16   **B.** 9   **C.** 8
    D. 7   **E.** 6

39. The number of NADH and $FADH_2$ produced by the β oxidation of palmitic (C_{16}) acid is
    A. 16   **B.** 9   **C.** 8
    D. 7   **E.** 6

40. The total ATP produced by the β oxidation of palmitic (C_{16}) acid is
    A. 96   **B.** 129   **C.** 131
    D. 134   **E.** 136

41. The process of transamination
    A. is part of the citric acid cycle.   **B.** converts α-amino acids to β-keto acids.
    C. produces new amino acids.   **D.** is not used in the metabolism of amino acids.
    E. is part of the β oxidation of fats.

42. The oxidative deamination of glutamate produces
    A. a new amino acid.   **B.** a new β-keto acid.
    C. ammonia, NH_3.   **D.** ammonium ion, NH_4^+.
    E. urea.

43. The purpose of the urea cycle in the liver is to

    A. synthesize urea.
    B. convert urea to ammonium ion, NH_4^+.
    C. convert ammonium ion NH_4^+ to urea.
    D. synthesize new amino acids.
    E. take part in the β oxidation of fats.

44. The carbon atoms from various amino acids can be used in several ways such as

    A. intermediates of the citric acid cycle.   **B.** formation of pyruvate.
    C. synthesis of glucose.   **D.** formation of ketone bodies.
    E. All of these

45. Essential amino acids

    A. are not synthesized by humans.   **B.** are required in the diet.
    C. are excreted if in excess.   **D.** include leucine, lysine, and valine.
    E. All of these

Answers to the Practice Test

1. A	**2.** E	**3.** D, B, C	**4.** F, B, E	**5.** G, B
6. A, B	**7.** D	**8.** E	**9.** E	**10.** D
11. A	**12.** B	**13.** C	**14.** E	**15.** D
16. 2	**17.** 2	**18.** 2	**19.** E	**20.** D
21. B	**22.** C	**23.** E	**24.** E	**25.** D
26. D	**27.** A	**28.** D	**29.** B	**30.** E
31. D	**32.** F	**33.** B	**34.** A	**35.** C
36. E	**37.** D	**38.** C	**39.** D	**40.** B
41. C	**42.** D	**43.** C	**44.** E	**45.** E

Answers and Solutions to Selected Text Problems

18.1   The digestion of polysaccharides takes place in stage 1.

18.3   A catabolic reaction breaks down larger molecules to smaller molecules accompanied by the release of energy.

18.5   The phosphoric anhydride bonds (P—O—P) in ATP release energy that is sufficient for energy-requiring processes in the cell.

18.7   Digestion breaks down the large molecules in food into smaller compounds that can be absorbed by the body. Hydrolysis is the main reaction involved in the digestion of carbohydrates.

18.9   **a.** The disaccharide lactose is digested in the small intestine to yield galactose and glucose.
          b. The disaccharide sucrose is digested in the small intestine to yield glucose and fructose.
          c. The disaccharide maltose is digested in the small intestine to yield two molecules of glucose.

18.11   The bile salts emulsify fat to give small fat globules for lipase hydrolysis.

18.13   The digestion of proteins begins in the stomach and is completed in the small intestine.

18.15   In biochemical systems, oxidation is usually accompanied by gain of oxygen or loss of hydrogen. Loss of oxygen or gain of hydrogen usually accompanies reduction.
          a. The reduced form of NAD^+ is abbreviated NADH.
          b. The oxidized form of $FADH_2$ is abbreviated FAD.

18.17   The coenzyme FAD accepts hydrogen when a dehydrogenase forms a carbon–carbon double bond.

18.19   Glucose is the starting reactant for glycolysis.

18.21 In the initial reactions of glycolysis, ATP molecules are required to add phosphate groups to glucose.

18.23 ATP is produced directly in glycolysis in two places. In reaction 7, phosphate from 1,3-bisphospho-glycerate is transferred to ADP and yield ATP. In reaction 10, phosphate from phosphoenolpyruvate is transferred directly to ADP.

18.25 a. In the phosphorylation of glucose to glucose-6-phosphate, 1 ATP is required.
   b. NADH is produced in the conversion of glyceraldehyde-3-phosphate to 1,3-bisphosphoglycerate.
   c. When glucose is converted to pyruvate, two ATP and two NADH are produced.

18.27 A cell converts pyruvate to acetyl CoA only under aerobic conditions when sufficient oxygen is available.

18.29 The overall reaction for the conversion of pyruvate to acetyl CoA is:

$$\underset{pyruvate}{CH_3-\overset{\overset{O}{\|}}{C}-COO^-} + NAD^+ + HS-CoA \rightarrow \underset{acetyl\ CoA}{CH_3-\overset{\overset{O}{\|}}{C}-S-CoA} + CO_2 + NADH + H^+$$

18.31 When pyruvate is reduced to lactate, the NAD^+ is used to oxidize glyceraldehyde-3-phosphate, which recycles NADH.

18.33 One turn of the citric acid cycle converts 1 acetyl CoA to $2\ CO_2$, 3 NADH, $FADH_2$, GTP (ATP), CoA, and $2H^+$.

18.35 The reactions in steps 3 and 4 involve oxidative decarboxylation, which reduces the length of the carbon chain by one carbon in each reaction.

18.37 NAD^+ is reduced by the oxidation reactions 3, 4, and 8 of the citric acid cycle.

18.39 In reaction 5, GDP undergoes a direct substrate phosphorylation to yield GTP, which converts ADP to ATP and regenerates GDP for the citric acid cycle.

18.41 a. The six-carbon compounds in the citric acid cycle are citrate and isocitrate.
   b. Decarboxylation reactions remove carbon atoms as CO_2, which reduces the number of carbon atoms in a chain (reactions 3 and 4).
   c. The one five-carbon compound is α-ketoglutarate.
   d. Several reactions are oxidation reactions;
   isocitrate $\rightarrow$ α-ketoglutarate; α-ketoglutarate $\rightarrow$ succinyl CoA;
   succinate $\rightarrow$ fumarate; malate $\rightarrow$ oxaloacetate.
   e. Secondary alcohols are oxidized in reactions 3 and 8.

18.43 The Fe^{3+} is the oxidized form of the iron in cytochrome *c*.

18.45 a. The loss of $2H^+$ and $2e^-$ is oxidation.   b. The gain of $2H^+ + 2e^-$ is reduction.

18.47 NADH and $FADH_2$ produced in glycolysis, oxidation of pyruvate, and the citric acid cycle provide the electrons for electron transport.

18.49 FAD is reduced to $FADH_2$, which provides $2H^+$ and $2e^-$ for coenzyme Q, then cytochrome *b*, and then cytochrome *c*.

18.51 The mobile carrier coenzyme Q (or CoQ) transfers electrons from complex I to III. It also transfers electrons from complex II to complex III.

18.53 When NADH transfers electrons to FMN in complex I, NAD^+ is produced.

18.55 a. $NADH + H^+ + \underline{FMN} \rightarrow NAD^+ + FMNH_2$
   b. $QH_2 + 2\ Fe^{3+}\ cyt\ b \rightarrow \underline{Q} + \underline{2\ Fe^{2+}\ cyt\ b} + 2\ H^+$

18.57 In oxidative phosphorylation, the energy from the oxidation reactions in electron transport is used to drive ATP synthesis.

18.59 Protons must pass through ATP synthase to return to the matrix. During the process, energy is released to drive the synthesis of ATP.

18.61 The reduced coenzymes NADH and $FADH_2$ from glycolysis and the citric acid cycle transfer electrons to electron transport, which generates energy to drive the synthesis of ATP.

18.63 **a.** 3 ATP are produced by the oxidation of NADH in electron transport.
b. 6 ATP are produced in glycolysis when glucose degrades to 2 pyruvate.
c. 6 ATP are produced when 2 pyruvate are oxidized to 2 acetyl CoA and 2 CO_2.
d. 12 ATP are produced in one turn of the citric acid cycle as acetyl CoA forms 2 CO_2.

18.65 Fatty acids are activated in the cytosol of the mitochondria.

18.67

a. and **b.** $$CH_3{-}(CH_2)_6{-}CH_2{-}\underset{\alpha}{CH_2}{-}\overset{\displaystyle O}{\overset{\displaystyle \|}{C}}{-}S{-}CoA$$
$$\phantom{CH_3{-}(CH_2)_6{-}CH_2{-}}\underset{\beta}{}$$

c. $$CH_3{-}(CH_2)_8{-}\overset{\displaystyle O}{\overset{\displaystyle \|}{C}}{-}S{-}CoA + NAD^+ + FAD + H_2O + SH{-}CoA \rightarrow$$

$$CH_3{-}(CH_2)_6{-}\overset{\displaystyle O}{\overset{\displaystyle \|}{C}}{-}S{-}CoA + CH_3{-}\overset{\displaystyle O}{\overset{\displaystyle \|}{C}}{-}S{-}CoA + NADH + H^+ + FADH_2$$

d. $CH_3{-}(CH_2)_8{-}COOH + 4\,CoA + 4\,FAD + 4\,NAD^+ + 4\,H_2O \rightarrow 5$ Acetyl CoA $+ 4\,FADH_2 + 4\,NADH + 4\,H^+$

18.69 The hydrolysis of ATP to AMP removes two inorganic phosphates from ATP, which provides the same amount of energy as the hydrolysis of 2 ATP to 2 ADP.

18.71 **a.** The β oxidation of a chain of 10 carbon atoms produces 5 acetyl CoA units.
b. A C_{10} fatty acid will go through 4 β oxidation cycles.
c. 60 ATP from 5 acetyl CoA (citric acid cycle) + 12 ATP from 4 NADH + 8 ATP from 4 $FADH_2$ − 2 ATP (activation) = 80 − 2 = 78 ATP

18.73 When glucose is not available for energy, fats undergo fatty acid oxidation. The accumulation of excess acetyl CoA results in the production of ketone bodies in the body. This may occur in starvation, fasting, and diabetes.

18.75 High levels of ketone bodies lead to ketosis, a condition characterized by acidosis (a drop in blood pH), and excessive urination and strong thirst.

18.77 The reactants are an amino acid and an α-keto acid, and the products are a new amino acid and a new α-keto acid.

18.79 In transamination, an amino group replaces a keto group in the corresponding α-keto acid.

a. $$H{-}\overset{\displaystyle O}{\overset{\displaystyle \|}{C}}{-}COO^-$$ **b.** $$CH_3{-}\overset{\displaystyle O}{\overset{\displaystyle \|}{C}}{-}COO^-$$

18.81 NH_4^+ is toxic if allowed to accumulate in the liver.

18.83 **a.** The three-carbon atom structure of alanine is converted to pyruvate.
b. The four-carbon structure of aspartate is converted to fumarate or oxaloacetate.
c. Valine is converted to succinyl CoA.
d. The five-carbon structure from glutamine can be converted to α-ketoglutarate.

18.85 Lauric acid, $CH_3—(CH_2)_{10}—COOH$, is a C_{12} fatty acid. $(C_{12}H_{24}O_2)$

a. and **b.**
$$CH_3—(CH_2)_8—\underset{\beta}{CH_2}—\underset{\alpha}{CH_2}—\overset{\overset{O}{\|}}{C}—S—CoA$$

c. Lauryl-CoA + 5 CoA + 5 FAD + 5 NAD^+ + 5 $H_2O \rightarrow$
   6 Acetyl CoA + 5 $FADH_2$ + 5 NADH + $5H^+$
d. Six acetyl CoA units are produced.
e. Five cycles of β oxidation are needed.
f. activation $\rightarrow -2$ ATP
   6 acetyl CoA $\times$ 12 $\rightarrow$ 72 ATP
   5 $FADH_2$ $\times$ 2 $\rightarrow$ 10 ATP
   5 NADH $\times$ 3 $\rightarrow$ 15 ATP
   Total   95 ATP

18.87 **a.** carbohydrate   **b.** fat   **c.** carbohydrate
   d. fat   **e.** protein   **f.** carbohydrate

18.89 $ATP + H_2O \rightarrow ADP + P_i + 7.3$ kcal/mole

18.91 Lactose undergoes digestion in the mucosal cells of the small intestine to yield galactose and glucose.

18.93 Glucose is the reactant and pyruvate is the product of glycolysis.

18.95 Pyruvate is converted to lactate when oxygen is not present in the cell (anaerobic) to regenerate NAD^+ for glycolysis.

18.97 The oxidation reactions of the citric acid cycle produce a source of reduced coenzymes for electron transport and ATP synthesis.

18.99 Electron transport regenerates the oxidized forms of the coenzymes NAD^+ and FAD for use again by the citric acid cycle.

18.101 In the chemiosmotic model, energy released by the flow of protons through the ATP synthase is utilized for the synthesis of ATP.

18.103 The oxidation of glucose to pyruvate by glycolysis produces 6 ATP. 2 ATP are formed by direct phosphorylation along with 2 NADH. Because the 2 NADH are produced in the cytosol, the electrons are transferred to form 2 $FADH_2$, which produces an additional 4 ATP. The oxidation of glucose to CO_2 and H_2O produces 36 ATP.

18.105 **a.** $1 \text{ day} \times \dfrac{24 \text{ h}}{1 \text{ day}} \times \dfrac{3600 \text{ s}}{1 \text{ h}} \times \dfrac{2 \times 10^6 \text{ ATP}}{1 \text{ s·cell}} \times 10^{13} \text{ cells} \times \dfrac{7.3 \text{ kcal}}{1 \text{ mole ATP}} \times$

   $\dfrac{1 \text{ mole ATP}}{6.02 \times 10^{23} \text{ ATP}} = 21 \text{ kcal}$

   b. $21 \text{ kcal} \times \dfrac{1 \text{ mole ATP}}{7.3 \text{ kcal}} \times \dfrac{507 \text{ g}}{1 \text{ mole ATP}} = 1500 \text{ g of ATP}$

18.107 **a.** 6 ATP/glucose   **b.** 3 ATP/pyruvate   **c.** 12 ATP/glucose   **d.** 12 ATP/acetyl CoA
   e. 44 ATP/C_6 acid   **f.** 3 ATP/NADH   **g.** 2 ATP/$FADH_2$

18.109 **a.** maltose   **b.** stearic acid   **c.** glucose
   d. caprylic acid   **e.** citrate

Answers to Combining Ideas from Chapters 16 to 18

CI.29   **a.** An alpha-galactosidase is a hydrolase.
       b. The substrate is the α-1, 4-glycosidic bond of galactose.
       c. High temperatures will denature the hydrolase so it no longer functions.

CI.31   **a.** citric acid cycle   **b.** electron transport   **c.** both   **d.** electron transport
       e. electron transport   **f.** both   **g.** citric acid cycle   **h.** both

CI.33   **a.** aminoethanethiol, pantothenic acid, phosphorylated adenosine diphosphate
       b. Coenzyme A carries acetyl group to the citric acid cycle for oxidation.
       c. The acetyl group links to the S atom in the aminoethanethiol part of CoA.
       d. 809 g/mole

CI.35   **a.**

$$
\begin{array}{l}
CH_2\!-\!O\!-\!\overset{\displaystyle O}{\overset{\|}{C}}\!-\!(CH_2)_{14}\!-\!CH_3 \\[6pt]
CH\!-\!O\!-\!\overset{\displaystyle O}{\overset{\|}{C}}\!-\!(CH_2)_{14}\!-\!CH_3 \;+\; 3H_2O \;\rightarrow \\[6pt]
CH_2\!-\!O\!-\!\overset{\displaystyle O}{\overset{\|}{C}}\!-\!(CH_2)_{14}\!-\!CH_3
\end{array}
\qquad
\begin{array}{l}
CH_2\!-\!OH \\[6pt]
CH\!-\!OH \;+\; 3\;HO\!-\!\overset{\displaystyle O}{\overset{\|}{C}}\!-\!(CH_2)_{14}\!-\!CH_3 \\[6pt]
CH_2\!-\!OH
\end{array}
$$

       b. glyceryl tripalmitate $C_{51}H_{98}O_6$ 807 g/mole
       c. Activation of palmitic acid -2 ATP

8 acetyl CoA $\times$ 12 ATP	96 ATP
7 FADH$_2$ $\times$ 2 ATP	14 ATP
7 NADH $\times$ 3 ATP	21 ATP
Total	129 ATP

       d. $1 \text{ pat butter} \times \dfrac{0.05 \text{ oz butter}}{1 \text{ pat butter}} \times \dfrac{454 \text{ g}}{16 \text{ oz}} \times \dfrac{80. \text{ g fat}}{100. \text{ g butter}}$

          $\times \dfrac{1 \text{ mole fat}}{807 \text{ g}} \times \dfrac{3 \text{ mole PA}}{1 \text{ mole fat}} \times \dfrac{7.3 \text{ kcal}}{1 \text{ mole PA}} = 40. \text{ kcal}$

       e. $45 \text{ min} \times \dfrac{1 \text{ h}}{60 \text{ min}} \times \dfrac{750 \text{ kcal}}{1 \text{ h}} \times \dfrac{1 \text{ pat butter}}{40 \text{ kcal}} = 14 \text{ pats of butter}$